Oceans in Decline

Sergio Rossi

Oceans in Decline

 Springer

Copernicus Books is a brand of Springer

Sergio Rossi
Institute of Environmental Science and Technology (ICTA)
Universitat Autònoma de Barcelona
Cerdanyola del Vallés, Barcelona, Spain

and

Dipartimento di Scienze e Tecnologie Biologiche ed Ambientali (DiSTeBA)
Università del Salento
Lecce, Italy

ISBN 978-3-030-02513-7   ISBN 978-3-030-02514-4   (eBook)
https://doi.org/10.1007/978-3-030-02514-4

Library of Congress Control Number: 2018967434

Translation from the Spanish language edition: El Planeta Azul: un universo en extinción by Sergio Rossi, © 2010. Published by Debate-Random House Mondadori/Penguin. All Rights Reserved.

This Copernicus imprint is published by the registered company Springer Nature Switzerland AG
The registered company address is: Gewerbestrasse 11, 6330 Cham, Switzerland

If you wanted a horror book, read this... the monster is us.

To my parents

Preface

Are the oceans dying? This is a question that many people are asking themselves more and more insistently. The answer is that in no case are they dying—but they are being transformed. Deeply. Human-induced changes across the globe affect marine more than terrestrial ecosystems. And at sea, there is a problem: because it is not our environment, it is not easy for us to see what is happening. The disappearance of large predators (whales, sharks, tuna fish, turtles, seals, swordfish, etc.), as well as the drastic reduction in many living structures on the seabed (coral reefs, algae meadows and higher plants, deep corals, etc.), has led to change in entire ecosystems to achieve a new balance on the basis of abundant small organisms and accelerated life. And there is more: persistent pollution, both chemical and biological, and the not fully understood but certain effects of climate change may be adding to the plight of our oceans as we know them. We cannot claim to know the full impact that such changes may have on the entire system, on the functioning of our planet and on our own survival.

Today, no one can ignore the fact that human beings have essentially impacted every habitat in the biosphere. More needs to be learned about concepts such as ecological economics, the persistence of species in the system, sustainable exploitation and recoverability of what we, as a species, now dominate: the planet in general, and oceans in particular. This implies knowing more about population sizes, genetic flows and the positive or negative relationships between species. It also means learning to adapt our way of life and understanding our role in nature. We are undoubtedly most backward in our knowledge concerning the marine environment, in this list of concepts, as it the most inaccessible and unknown. The problem is partly one of focus, of perspective, of the way we understand what surrounds us.

The various transformations that have taken place (and are taking place) in marine ecosystems deserve reflection, not only as a result of specific experimentation but also from observation of what was once in balance and that has now shifted, across the globe.

In this book, I try to give a vision that is as global as possible of the past, present and future of the ecosystems that cover the most surface area (70% of our planet) and volume (99% of the biosphere): those of the oceans. It is not intended to touch on every point, as I do not want to detail all the instances that are taking our ocean to a very different place from the one that our ancestors knew just a few thousand years ago. However, I do want to give four openings so that the reader understands the magnitude of the process of transformation that has taken, and is taking, place on our planet.

I begin with a brief review of the history of our seas, showing that transformations and cataclysms are the order of the day on Earth. The sea has undergone many changes since the start: the movements of continents, the rise and fall in sea level, the appearance of cyanobacteria capable of creating oxygen (poisoning any being that could not consume it), its acidification and the sudden freezing of its surface. All these disturbances have been faced by species, some disappearing, others migrating or adapting to the changing conditions. But man's intervention has changed things. The pressure that we have exerted and the profound transformation of our ecosystems began much earlier than we think. Hundreds of years ago (in some cases, even thousands), humans understood that the sea was an apparently inexhaustible source of food and began to exploit it systematically and without control. In the sea, unlike on land, large carnivores were usually coveted prey, so a hunt for large cetaceans, sharks, seals, tuna fish, cod and turtles began, and their populations were soon reduced to levels that relegated them down the food chain: mankind became the new regulator of the system, the new great predator. This first part of the book analyses in depth what this ancient and fundamental historical transformation of the sea has meant.

In the second part of this book, the focus is on industrial and other means of fishing that currently take place, and the effect that they have on the transformation of the oceans. In recent decades (even centuries), much has been written about which disturbances have affected the marine balance most. Today we can say, without a doubt, that fishing is the main disruptor of the balance between species and is the culprit behind the changes in the flows of matter and energy. It has been argued that overfishing has virtually eliminated predators and resulted in a resurgence of the Mesozoic system, dominated by jellyfish, cephalopods, echinoderms and crustaceans. Of all types of resource extraction, trawling from the early twentieth century onwards has definitely been the most damaging, and it continues to be so. Besides trawling, there are the large pelagic fisheries capable of encircling and capturing entire shoals of tuna; they are unable to escape from this technology, which does not give them a chance. Even animals as resilient as anchovies and sardines are incapable of the speed of reproduction at which we force the species to multiply, due to the eagerness of our unrestrained market. The sea is at the limit of its capacity to bear fruit, and specialists have long warned that many species are either on the verge of local extinction or are already, commercially and ecologically, extinct.

The third part of the book analyses a series of different problems: those arising from pollution. These may be chemical, such as mercury or petroleum products entering food chains, damaging specific areas, especially coastal areas, or something as unknown as the enormous amount of plastic that reaches the sea and is in the system for centuries, being ingested by animals such as fish, turtles and whales that die of suffocation or poisoning before it decomposes. Pollution may also be deeply involved in other problems, yet less obviously. An example is the increase in red tides all over the planet or in algal blooms that can poison the sea and, in many cases, smother the bottom by generating areas with low oxygen content where life is transformed, dominated by anaerobic bacteria (the only life capable of surviving in these 'dead zones' of low concentrations or zero oxygen). The transformation of the shore itself by intense maritime traffic is analysed to explain the proliferation of alien species, a form of biological pollution that has recently been accelerated by the flurry of moving goods and creating port structures capable of accommodating such species. These movements bring with them tenants (invasive species), travelling, for instance, from Hong Kong to Marseilles, which, if they find the right conditions, will invade territories not yet inhabited by them due to geographical barriers.

The last source of transformation that is analysed results from climate change. Attention is focused on several problems, most of which are unknown to many. The focus is not so much on whether there is climate change (which, for the author, is irrefutable) or its origins, but on the effort that thousands of scientists are making to understand its effects on marine biota. In the first part, we will examine how we know that there is change. Based on this evidence, scientists have predicted a series of accelerated transformations, mainly due to an increase in temperature. Here this book focuses on analysing systems, such as the poles, their temperature and climate regulating effect, not forgetting the effects on the conceptual water column and various species, whether by temperature, acidification or change in marine currents or dissolved oxygen. Climate change is seen as yet another environmental alteration, and there is great uncertainty about its future implications. However, it is certain that the transformation started in earlier times, due to issues such as intense fishing and pollution.

Finally, the book outlines some of the solutions that are already in place or that may help to preserve our seas. The first focus is on aquaculture, now widely used as a source of protein to replace the pure, hard 'game' for various species that is fishing, observing its pros and cons. Throughout the following chapters, solutions such as the recovery of habitats (so difficult at sea) or the extraction of energy in various ways are dealt with as a future prospect in which the sea will undoubtedly be the central axis. Without doubt, this future requires a different management of the oceans. This is why we are analysing the model that we should use in future, in my opinion to make forward progress instead of going backwards on issues such as fishing, coastal redevelopment or the need to extend protected areas so that the system can recover to bear fruit again. In the end, the educational factor is emphasized: we fail to protect the sea because we do not know it and because it is still difficult to access and penetrate its functioning. I do not aim to be a catastrophist, and the book is written so that we can learn and react in the strong belief that we cannot avoid this issue, no matter how much we may feel that the sea is the eternal unknown.

Barcelona, Spain   Sergio Rossi

Contents

Part I

Seas of Yesteryear

1

A Brief History of Our Seas

Our planet is the living image of change, of profound transformation through the geological ages. And the oceans have undoubtedly been major players in this complex scenario. Life began in the ocean, and also in the atmosphere as we know it today. Today's distribution of heat and our climate itself are the result of the dynamics of our seas. Initially, water emerged not only from Earth's interior but also from the comets that incessantly impacted the surface of our fragile planet, which was trying to survive in a violent environment. These same comets and other fireballs carried organic matter, the molecules that were to form the basis of life. Many fireballs that hit early Earth disintegrated or were exposed to high temperatures. Others, however, did so without destroying the delicate molecules that they carried inside them, creating some of the broth necessary for life to form. The other necessary ingredients came from our own planet, the result of chemical reactions stimulated by an atmosphere very poor in oxygen yet rich in methane and hydrogen. Experts have not yet agreed how the first step was taken. Some suggest the classic model of the 'quiet pond', in which organic molecules interact to create pseudo-membranes (called micelles) and self-replicating ribonucleic acid (RNA). Other experts, on the other hand, suggest a volcanic caldera, a place subjected to high temperatures and not isolated from the ferocious volcanic environment that dominated the very young Earth at that time. Yet others suggest an icy space—an icy blanket that might have covered part of the planet in the past. In that environment, in the low temperatures the molecules would have interacted slowly to create the first structures to which we owe the life that we observe today.

© Springer Nature Switzerland AG 2019
S. Rossi, *Oceans in Decline*,
https://doi.org/10.1007/978-3-030-02514-4_1

Not everyone is convinced that life originated on Earth. Some specialists speak of an exogenous origin: life from other planets or asteroids. However, the question remains the same, and at the moment there is no clear answer to how life originated. We know that RNA capable of replicating itself probably emerged about 4 billion years ago, and that the first single cells date back about 3,900 million years (Ma). Unfortunately, the remains are much altered due to the passage of time. But, from that moment on, life became an unstoppable phenomenon, destined to survive great change, meteorite impact and continental movements causing glaciation, desertification and spectacular volcanic eruption.

What we understand as life then emerged: a self-sustaining chemical system capable of replicating and evolving, in the Darwinian sense of the word. Methanogenic bacteria, the archetypes, were some of the first organisms in our oceans, thanks to their ability to take advantage of the atmosphere and the oxygen-deprived oceans that had reigned from the start. Organisms that have changed little or not at all since then surround us now—simple, imperturbable and capable of resisting practically everything. In that long period, the pre-Cambrian Period, the longest chapter in the history of life, steps were taken that would decide the future of the world's living matter. The first organisms to form stable populations were undoubtedly those adapted to high temperatures but, as the Earth gradually cooled, other organisms displaced these so-called thermophilic organisms.

An absence of oxygen was the norm until the first bacteria capable of photosynthesis emerged. Many specialists suggest a primeval soup of organisms that either nourished themselves with dissolved organic matter or were able to use hot springs for their metabolic functions, always surrounded by an environment lacking in oxygen.

Oxidization was a difficult process, as the organisms and chemicals were surrounded by electron-charged molecules. The appearance of the chlorophyll molecule added complexity to the system. The process of photosynthesis that results from this series of complex molecules is one of the most conservative biochemical processes on the planet: the organisms that use it have not changed their mode of operation in billions of years. We can imagine the strength of a process in which organic molecules and oxygen are produced from carbon dioxide and water. Oxygen, a mere waste product, began to invade first the oceans and, when they became saturated, Earth's atmosphere itself. Oxygen burns, rusting everything in its path. Therefore, although in a slow and stately way, all life had to adapt, to seek environments without oxygen or die from this new poison that was invading the planet. It has always surprised me how such a 'simple' change could transform the destiny of a planet. This revolution began about 3,500 Ma ago and culminated in a catastrophic setback for the anoxic world. The architects were the original cyanobacteria, which proliferated without ceasing, creating in some cases the stromatolites that we can still observe almost undisturbed along parts of Australia's coastline. It was undoubtedly a radical change, one of the deepest traumas that our living environment has suffered.

Another change was to put a stop to the dominion of prokaryotes, both bacteria and archaea: the arrival of eukaryotes. These 'higher' organisms (a term that seems absurd in the context of the type of life on our planet then) were probably the result of symbiosis between two or more prokaryotes. These were more complex and had something that the more primitive ones did not: sexual reproduction. It is something that seems trivial, yet it is not trivial at all. Among other things, sexual reproduction contributed something fundamental: an acceleration of diversity. Until then, cells had been dividing identical genetic material, altered only by casual mutations that could arise from gamma ray impact or another mutagenic actor. To this was added the exchange of information between cells, which promoted changes, adaptation … and thus evolution. Cyanobacteria and stromatolites began to be cornered by more advanced, more diverse plant cells capable of adapting to different temperatures, salinity, nutrient concentrations and carbon dioxide. Eukaryotic algae began to conquer the planet, along with other eukaryotes such as ciliate, flagellates and foraminifera.

Meanwhile, the oceans were constantly being reshaped by plate tectonics. Continents drifted, as unstoppable as the planet itself and its environmental conditions. The effect of the sun, at first weak (4,000 Ma ago, only 70% of the radiation that reaches us now reached the planet), grew in force by shining on Earth's surface in a more stable way. The ozone layer was consolidated, allowing minimal penetration of ultraviolet rays, which are harmful to life, especially on the surface. The Earth itself would slow its turning, going from 18-hour to our current 24-hour days, and alter its inclination to rotate around the king of all our stars: the sun.

Multicellular organisms began to emerge, with new metabolic and food requirements. The first animals took hundreds of millions of years to appear. The oldest fossil records date from about 550 Ma, in the Ediacaran Period. These were great changes to an environment in which life was gradually opening up an increasingly complex, increasingly interactive division between the various parties. In the Cambrian Period, 500 Ma ago, the rules of the game, as we know it today, were largely established. After an explosion of unparalleled diversity, life on the planet took on various forms: some survived, others perished, but the food chains (herbivores, predators, parasites;

commensalism or symbiosis) were consolidated at this time, when the first arthropods, the first protochordates and the first molluscs emerged. The oceans, then called Japeto and Panthalassa, bore no resemblance to the current configuration. The continental masses were much smaller than at present, concentrated in Laurentia, Baltica and the immense Gondwana, with some islands forming archipelagos. Continental drift (and the movements of the planet in its orbit) caused the appearance and disappearance of oceans, the regulation, acceleration or stagnation of currents, the existence of immense polar ice caps (such as that of the Carboniferous Period, 300 Ma ago) and the formation of hypersaline shallow seas, which were sometimes devoid of life. Transgressions (the sea flooding the land) or regressions (the formation of land from flooded territory) due to changes in sea level of more than 300 metres (m) in height were other factors that changed the life of the oceans—and, of course, of the mainland—throughout the geological eras.

And life always adapted to change, even if that change was abrupt, as in the Permian or Cretaceous extinctions. One such change was associated with the appearance of diatoms, microscopic algae whose oldest fossils are in fact quite recent, about 180 Ma ago. Diatoms are extremely productive. It is now estimated that 40% of the primary (photosynthetic) production of the oceans is due to these algae, accounting for up to 20% of the world's primary production. They are important producers of oxygen and organic matter, and their proliferation possibly stimulated the growth of small and large herbivores. The increase in the numbers of zooplankton, consumed by both diatoms and other microscopic algae, had also provided ideal conditions for marine food chains to diversify and for their actors to reach a remarkable size. The basis of production also took the form of a proliferation in multicellular algae and angiosperms, which, after their conquest of the mainland, returned to the sea in search of new territories and ecological niches.

Life came ashore about 430 Ma ago. First, it took the form of bacteria and algae in symbiosis with fungi (the most primitive lichens date from 280 Ma), moss-like plants, horsetails and ferns. Next, it took the form of arthropods (especially spider-like animals) and other heterotrophic organisms (i.e. needing to feed on other organisms, as unable to perform photosynthesis) that took advantage of the Eden that opened up to them on the land. In the Carboniferous Period about 300 Ma ago they reached their full height, thanks to a concentration of oxygen significantly higher than at present. The first vertebrates came ashore several million years later, causing the entire food dynamic of the planet another upheaval. The first fish to modify their fins possibly fled from predators without leaving their aquatic environment. Transformed into powerful limbs, these appendages allowed them to reach shallow waters that other animals could not reach. It is also possible that, in a world where terrestrial vegetation probably formed dense, invasive forests in shallow, swampy areas, these animals with modified fins were agile contortionists capable of dodging the branches and roots of both terrestrial and aquatic vegetation. Some began to develop something like prehensile limbs at the end of their fins, like fingers. They were therefore able to hold onto vegetation and rocks, keeping their position in strong currents and even guiding them, to some extent, in turbid waters.

When they came to possess primitive lungs, these tetrapods conquered the land. The most plausible first candidate so far is *Hypernerpeton*, an inhabitant of the seas, from about 365 Ma. This invasion of the land by life would change everything. Organisms, and especially plants, transformed the mineralogical composition of the soil (referring to the interaction of minerals and the living parts of plants) and its concentration of organic products in rivers or groundwater, ending up in the sea. Once again, life had to adapt to the changes brought about by increases in sediment, detritus, phosphorus and other nutrients, which would displace one yet give rise to another totally different form of life in an eternal transformation of a tireless system.

It is therefore important to maintain a dynamic view of the life of the oceans over the millions of years. Life, since it emerged from the early seas, has adapted to change of all kinds, ready to accept the challenge of surviving even major catastrophes that have put it in check, challenging time and again its balance and stability. I remember a conversation with a sadly deceased professor at the University of Barcelona, Javier Ruiz, who told me that 'as long as there is only one bacterium on the planet and there are environmental and energetic conditions adapted to the survival of life, the legacy of DNA will remain undisturbed'. Although it is not possible for humans to destroy life, the last ten thousand years have prompted a series of transformations that are once again challenging the balance that has endured until now.

But we are wrong if we think that these changes are due to our interference over the past century. Long before then we had begun to be one of the few organisms in Gaia's history capable of transforming the life and balance of an entire planet.

The Mother of All Extinctions

It is difficult to imagine an extinction in which 95% of marine organisms disappeared. But about 250 Ma ago that is what happened, according to the fossil record from the Permian Period. Some researchers conclude that it represented a real challenge to the more evolved life of both vertebrates and invertebrates, being more radical than the extinction in the Cretaceous Period 65 Ma ago. Entire reef structures disappeared with all their organisms, with some groups such as bryozoans, rugged corals or sea lilies forever displaced from their hegemony, relegated to smaller groups that, in some cases, still survive today. In less than a million years, nine-tenths of plant and animal life disappeared from the face of the Earth. In Italy, traces have been found of dead forests where fungi feasted on rotting primeval plants. After the extinction, dinosaurs emerged, among other organisms, to roam for more than 180 Ma as the first predecessors of the mammals that were to replace them.

But, while the extinction of these primitive saurians seems linked (although there is not full consensus) to the famous Chicxulub meteorite, the extinction in the Permian Period has several possible causes that are debated by specialists. On the one hand, it is known that there was incessant volcanic activity for no less than 100,000 years. This may have caused a cooling of the planet and subsequent desertification. In Siberia, one can still see primitive volcanic formations (traps) similar to those that would have been all over our planet at that time. Related to these changes in climate (in fact, potential glaciation), a stagnation of the oceans is referred to by some experts. This would have caused a deep anoxia (lack of oxygen) in the lower layers, rising to the surface to eliminate animals and plants at shallower depths. But the legacy of a meteorite is present here, too. In Australia, there is a crater of about 120 km in diameter caused by the impact of a 5-km bolide that would have devastated the Earth at that time. Apart from noxious gases put into circulation and the intense acid rain, the sun would have been hidden for months by a thick cloud of dust and incessant fires that could have devastated the planet. Without light, there were no photosynthetic organisms to feed the next level up in the food chain, resulting in extensive loss of life. In reality, the causes are not mutually exclusive. Many specialists talk about simultaneous factors. What is clear is that the history of our oceans (and of the whole of planet Earth) experienced a serious imbalance in the Permian Period, a time when the course of life was past turning back to other possibilities and other horizons.

Great Marine Beasts

One of the most impressive changes in the seafood chain has been the extinction of the great marine saurus. Like the land saurus, its decline some 65 Ma ago was due to the impact of the Chicxulub asteroid, leaving the way clear for predators such as sharks, fish and cephalopods for some time. Possibly, the collapse of primary production indirectly left the saurus without food, leading to its extinction. When evolutionary divergence began, scattering species and requiring adaptation to the different environmental conditions of the planet, some of these reptiles took advantage of shallow waters to return to the sea. Originally, some 250 Ma ago, the first dinosaurs to conquer the sea probably did so for two reasons: the food that the sea provided and an absence of predators. Moreover, new territories gave them a chance to flee from those who hunted them. The first were animals somewhat similar to a *Diplodocus* with fins. Beasts such as *Kronosaurus*, an elongated snout dinosaur with a skull 2 m long, more than 10 m long and weighing 10 tonnes, must have been a nightmare for fish, turtles, cephalopods and other aquatic dinosaurs. If these beasts grew to an unimaginable size, it was for one simple reason: there was food available to sustain their voracious appetite. Banana-sized teeth like those of *Nothosaurus* (one of the first saurian settlers more than 230 Ma ago) only develop if there is prey to devour.

The disappearance of saurians left an unfilled ecological niche. Bony fish and elasmobranchs (sharks) temporarily filled the gap, giving a further twist to the complex history of our oceans. Then, about 50 Ma ago, 15 Ma after their disappearance, the aquatic dinosaurs were finally replaced by organisms that had followed them from the land: mammals. The first prehistoric whales spent long periods on land and could not reproduce directly at sea. But the productivity of the seas was skyrocketing, and the opportunity was too good to pass up. That's why, 24 Ma ago, odontocetes (toothed whales) and mysticetes (baleen whales) reached similar proportions to our present-day whales, culminating in the largest vertebrates that have ever existed.

Survivors Born

At noon at the end of June, back in 2000, I went to lunch with my then head of research, Josep-María Gili. We sat down to eat a fixed-price menu in a restaurant near the old Institute of Marine Sciences in Barcelona's Plaza del Mar and he told me a fascinating story from the last Antarctic trip that we had shared just a few months earlier. What if the organisms that live at the bottom of the waters surrounding the 'white continent' were relics from tens of millions of years ago?

In Antarctica, living on the seafloor under hundreds of metres of water, are organisms that have been little modified by evolution over the last 100 Ma years. Prolonged continental drift has condemned this huge land to extreme isolation. When it began to drift south about 70 Ma ago, Antarctica's climate began to change, little by little. Some 40 Ma later it was totally isolated, due to the formation of a circumpolar current, such as that which today acts as an almost insurmountable physical barrier to most organisms in temperate or tropical latitudes. The formation of an immense mass of ice on land above sea level (in places up to several kilometres thick), due to the permanent extreme cold, created an environment of frost hostility paralleled only by certain areas of the North Pole. In that relentless external environment, along with the stability of the cold Antarctic water temperatures, the abundance of plankton established on the dark and impenetrable bottom an animal garden comparable in diversity and exuberance to coral reefs or the communities on the seabed of the Mediterranean. This community is made up of benthic suspensives, filter feeders that on live or dead suspended particles (debris, eggs, small crustaceans, etc.) without moving from the place, like animal trees. But they do it in a peculiar habitat: these animals (sponges, gorgonians, etc.) live where there is a soft bottom. What's so special about that? Gorgonians, corals and sponges can tolerate a moderate amount of sediment and other particles, although an excess simply chokes them. There are no rivers in Antarctica, and the amount of debris transported by glaciers from land is not comparable to that from rivers such as the Nile, Mississippi or Amazon.

In the past, animals such as these benthic suspension feeders at the bottom of the white continent's seas were found all over the planet, but little by little they have been displaced by animals such as bivalves or polychaete worms that (efficiently) filter those particles and adapt to soft bottoms full of sand and mud from the land. Many of the organisms that were dominant during the Cambrian Period or Palaeozoic Era were gradually cornered into deeper places where they could still dominate the space and go through their life cycles. The temperature in Antarctica, much lower than elsewhere on the planet, prevented the proliferation of large, fast and efficient predators such as sharks, large fish and decapod crabs (Fig. 1.1).

Another factor to take into account is that the ancestors of the organisms that we now observe on the Antarctic seabed survived one of the most traumatic cataclysms that Gaia has ever experienced: the impact of the meteorite some 65 Ma ago. But how? While the extinction of the Cretaceous Period had obvious consequences for the fauna of shallower parts of the planet, animals living at greater depth did not suffer as much. It has been inferred that the meteorite impact directly and indirectly caused a collapse in ocean primary production. But there, in the deepest parts of the ocean and in the Antarctic seas, sponges, gorgonians and many other organisms resisted

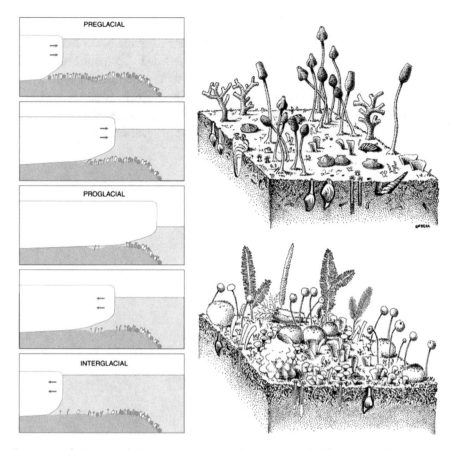

Fig. 1.1 Left image: glacier movement on the Antarctic shelf. Sessile organisms and associated fauna are dislodged and pushed to the shelf's edge; and the shelf once the glacier has retreated. Right image: structural and taxonomical similarities of ancient (Cretaceous) and actual benthic sessile fauna in the Antarctic benthos. *Source* Gili et al. (2006)

because they had adapted to food shortages: for months there is no light in Antarctica, therefore they have to manage without a continuous supply of microscopic algae, which need light to divide. In this, the most extreme area of the southern hemisphere, their diet is partly made up of fine particles on which other organisms can barely survive, and it has enabled them to resist the major fluctuations in food production that occur in the water column. In other words, not even the impact of the great meteorite could make them disappear. However, there remains the question of glaciation. Indeed, the advance and retreat of the glaciers and of the immense icecap, hundreds of metres deep, cleared them again and again from the continental shelf.

However, due to the ability of these organisms to take refuge in deep zones on the edge of continental shelves or in underwater canyons, where the ice was unable to scour the bottom, when the glacier receded they could reclaim the space that had been lost. That is why Antarctic organisms, adapted to a climate that is hostile in terms of temperature but rich in food (at least for a certain period in the annual cycle) and stable in many respects, have been able to withstand the adverse weather conditions of an ever-changing planet.

2

The Sea That Once Was

When I started my career in biology, I understood that mankind is one of the few species on the planet capable of changing the global environment. At the end of the last great glaciation about 11,500–10,000 years ago (the beginning of the Holocene), humans began to take up their position on what would be the great transformation of ecosystems wherever they set foot. Sea level began to rise again, as a result of the massive melting of the glaciers, in some places reaching tens of metres higher than previously, until quasi-stabilization about two or three thousand years ago. New submarine platforms were created, new shallow areas where humans began to settle to take advantage of the resources of the border between sea and land: the shore. The shore has the advantage of offering the fruits of both land and sea and, although they began to exploit the intertidal zone and the marshes more from necessity than curiosity, people soon learned that the sea was a very rich and varied resource.

Freed from the restraints of the impressive ice cover (more than 40 million km^2 at the end of the Pleistocene Period), the inhabitants of the Mediterranean basin and other places such as Central Europe found that the stabilization of the climate represented an opportunity to expand their dominion thanks to agriculture, grazing and, gradually, the centralization of activity in cities. Much earlier, human migrations had already begun to account for large mammals in various parts of the world. Some 40,000 years ago, the last great migrations from the African continent were reaching where they had never been able to before, exploiting resources to maintain settlements first through hunting and fishing and then through agriculture and grazing. Climate change accompanied the agricultural expansion, with a gradual rise in temperatures and the extension of wooded, moist areas, in

© Springer Nature Switzerland AG 2019
S. Rossi, *Oceans in Decline*,
https://doi.org/10.1007/978-3-030-02514-4_2

contrast to the great desert areas typical of the end of the last glaciation. This warm climate accompanied mankind about nine thousand years ago, reaching its peak about five thousand years ago, when it gradually cooled down again gently, stabilizing to some extent about four thousand years ago. Even then, the coast was one of the areas under most pressure from agricultural and urban populations. So, to what extent is what we see today natural, in the midst of our Anthropocene epoch?

Historical ecology is a very young discipline. The real impact of mankind, not only in the last fifty to a hundred years but long before, is a slippery and difficult subject to interpret. Perhaps because of this it is attractive in the eyes of a scientist. I have been very interested in it over the last few years, especially after a talk by a scientist such as Jeremy Jackson who made me ask the simple question, whether anything that we see at the moment can actually be considered 'natural'.

Let's take an example: some 7,500 years ago, the inhabitants around the Wadden Sea, between the coasts of Netherlands and Denmark, saw groups of grey whales and fin whales approaching to feed in shallow waters rich in small crustaceans. Probably due to the inaccessibility of the sea in those initial times, their feeling must have been one of fear and curiosity. In any case, these large animals were not the object of active hunting due to the impossibility of reaching them without appropriate boats and gear. But other animals were more within reach of their spears, bows and traps. During the Roman era, grey seals, large fish such as sturgeons, salmon and rays, large birds and other animals began to be hunted constantly, and hunting gradually spread to the whole of Europe. The Middle Ages are considered to be the turning point in the systematic exploitation of the sea. It is at this time that fishing gear such as traps or nets were optimized, but above all it was the moment when cities started in earnest, increasing their populations and therefore their pressure on the environment. In medieval markets it was common to find porpoise meat and other small cetaceans that had swum in the coastal areas of the Wadden Sea, and the exploitation of cod and herring had already entered into an industrial stage of exploitation. The first to suffer the consequences were river and estuarine fish, some of whose populations disappeared or were reduced between the twelfth and fifteenth centuries. Not only fishing but the management of the riverbeds and their pollution began to affect the organisms that lived there. 'Holland towards 850 AD had almost 100% of uncultivated land: forests, steppes, marshes,' claims Heinke Lotze of the Alfred Wegener Institute in Germany, 'but by 1350 only 50% was still "virgin" in some sense.' But one of the direct or indirect effects of coastal exploitation was the loss of so-called 'complex' ecosystems. Thus, the oyster banks that were already consistently exploited in 1000 AD became over-exploited in the seventeenth century, reaching near collapse in the nineteenth century, a pale reflection of what they were in the past. Another example is the tubeworms that formed authentic reefs, which have since disappeared, thus all their associated fauna, due to collapse by sedimentation and trawler fishing; or plants such as *Zostera marina*, which went from occupying 150 km^2 around 1870 to less than a square kilometre today, due to active commercial exploitation (more than 750 tonnes were extracted annually between 1869 and 1930). Mussel banks fared no better, and in the Netherlands alone were reduced from more than 4,000 ha to under 200 ha today. With the destruction of these habitats, of course, the associated species that were actively exploited through small-scale local fishing disappeared. Little by little, all the organisms disappeared that had been controlling the levels of microscopic algae, sedimentation and recycling of matter in coastal areas. The most recent changes were those to red and brown algae, which are longer lived but less resistant to organic and inorganic pollutants than the weeds and slime that now festoon the shores of the Wadden Sea.

The vision we have of this and other seas at present is completely different from what it would have been a few hundred years ago. Some 350 years ago, the inhabitants of this northern European coast saw the last of the whales, never to return because they had become the main target for hunters. The descendants of those who were surprised to see those giants so close to the coast would never see such a sight off their shores.

Because of this profound transformation, some 300–400 years ago the first Dutch and English settlers who arrived in Chesapeake Bay (eastern United States) could not believe what they saw: an enormous bay of more than 165,000 km^2 (more than a third of the surface area of Spain), full of cetaceans, birds and fish, populating the paradise that they were entering to transform irredeemably. The Indians had already transformed the system, but the rapid expansion of the colony and a vision of the system as seemingly inexhaustible were to affect that area of the planet in just a few hundred years. The new settlers had something that the Indians did not: boats that could travel away from the shore to increase the catch of fish or large cetaceans. For the men and women who arrived, wildlife was simply a commodity to be exploited, backed by the Christian doctrine of the subjugation of nature to mankind's power and needs.

Settlers along the Potomac, James and Susquehanna Rivers soon learned to exploit salmon, sturgeon, trout, cod and other fish, which were abundant and shared the habitat with whales and other cetaceans. There are reports of sturgeons over 5 m long, weighing almost five tonnes: in a single day, a group of five to ten fishermen could fish over 600 specimens of varying sizes. There are also reports of fishing or similar extractions of other organisms, but the one that undoubtedly had the greatest impact on the overall functioning of the bay was the destruction of the oyster beds. 'There is historical evidence of huge banks of oysters, hundreds of metres wide and kilometres long, that were found in shallow areas along the 300 km offshore that make up the Chesapeake Bay area', explains Jeremy Jackson of the American Scripps Institution of Oceanography: 'The boats even ran aground on these dangerous banks.'

Oysters were systematically exploited from 1800 until they were removed by mechanical means (excavators) in 1870, cleaning the channels of these bivalve molluscs. Little by little, the regression of the oysters had unexpected effects. These animals are active filters, acting as the kidneys of the system. Before 1850, they were able to filter all the water in the bay in a week. Recent estimates of the density and distribution of these creatures put the time needed to do the same task at between 46 and 50 weeks. Proliferations, or 'blooms', of single-celled algae have always been frequent in this area, because these microscopic plants are highly productive and have a high concentration of nutrients. Oysters (and other organisms) had been responsible for transforming many of these living particles into living or dead biomass by filtering and retaining them. Agricultural and urban development waste provided more nutrients that made these blooms more frequent, persistent and concentrated in terms of algae per cubic metre. In 1930, the first symptoms of anoxia began to be detected in the deeper areas of Chesapeake Bay. After the spring/summer bloom, some of the algae remained unconsumed by the now-extinct oyster banks and other organisms that had suffered a similar fate, and hung in the background. Here, the concentration of organic matter increased further, and bacteria found it an ideal substrate for uncontrolled growth. In order to consume organic matter, bacteria have to breathe, so one of the immediate effects was the elimination of the oxygen necessary for other organisms. Today, dead zones of between 15 and 50 m deep have been created in which organisms such as fish, molluscs or crustaceans cannot live. Those that are able, such as blue crabs, move to shallower areas, while others, such as sponges, gorgonians or sessile bivalves, perish due to lack of oxygen. Chesapeake Bay is now just a relic of what it was in the not-so-distant past.

The same can be said of seas such as the Mediterranean and the Caribbean, in fact in each and every place on the planet, because our influence on the sea has been decisive for much longer than it seems. But what is wrong with such changes, apart from not seeing large animals and the earlier pristine beauty?

What's the problem? As we will see, the damage to a number of ecosystems has led to a marked reduction in their functioning.

Statistics are being revealed that show that we live surrounded by an 'unnatural' ocean, as Professor Jeremy Jackson insisted at a dinner party after his talk in Menorca that I was lucky enough to attend. 'Knowledge of the ocean is based on a completely distorted picture of what once was the actual functioning of ecosystems', Jackson added. Many habitats such as coral reefs, underwater plant meadows or estuaries do not function as they used to, and this is having an impact not only on the amount of protein that we can extract from the sea in the form of fish, cephalopods or crustaceans but also on the planetary dynamics themselves. Changes in our oceans generally result in a loss of diversity, structural complexity, functional skills and economic services, because complex systems have the capacity to absorb the carbon that is generated in the atmosphere on the one hand, and to store life that can then be exploited in various ways on the other. The Mediterranean shore, one of the coastlines most altered by humans in the last ten thousand years, has more than 22,000 km^2 covered in asphalt, concrete and cement, which have changed the way the system works. It has been a major blow to a system already suffering from overfishing and the destruction of complex habitats: 'In just ten years, from 1990 to 2000, 10% of this transformation has taken place in the form of ports, urbanizations, roads … a series of artificial habitats that have a decisive influence on the global functioning of coastal ecosystems, and never in a positive way', explains Laura Airoldi of the Italian University of Bologna.

The most vulnerable animals and plants are precisely those that live the longest and, in turn, are the most capable of sequestering nutrients and carbon from the system. What we see today is a series of systems dominated by small fish and invertebrates, where cycles are very short and microbes are the real masters of the landscape. We will have to wait decades or centuries to see the return of megafauna (whales, sharks, tuna, groupers, turtles and manatees), while small invertebrates increase everywhere, as they occupy the niche that the larger ones have not been able to dominate for centuries, due to over-exploitation. 'In the end, what we see is a series of "two-dimensional" habitats where a third dimension of large herbivores and carnivores, large coral reefs and extensive macroalgae and spermatophyte meadows give our oceans a completely different meaning and functionality', Jackson comments. But are these factors so important to the functioning of the living planet? What role can they play in the cycles of matter and energy?

The Mysterious History of the Black Sea

One of the last climatic 'catastrophes' that mankind experienced during the Holocene was the formation of the Black Sea. Several archaeological and geological investigations have shown the existence of a civilization that fled from a flood of biblical proportions some 7,500 years ago. In fact, if you look at a core of local sediment from a depth of more than 150 m, you can see a lighter layer with freshwater bivalve mollusc shells. In this area, several kilometres off the current Turkish coast, there was a lake that occupied only about two-thirds of the current surface area. And on that shore lived people, taking advantage of both the resources of the lake itself and the fertile lands that bordered it. Beyond that, the land was arid, so they must have seen the lake as a gigantic oasis where they could start their first crops, graze their primitive livestock and build settlements. Looking at the same sediments, we see that about 7,500–5,000 years ago these freshwater bivalve molluscs were replaced by common mussels and the sediments became darker, thus richer in organic matter. According to American geologists Walter Pitman and William Ryan, the Black Sea ceased to be a lake when the last glaciation came to an end some 11,000 years ago. Continental ice had begun to melt, and the water ended up in the Mediterranean and raised its level. The Marmara Sea, which was very deep, saw saltwater ingress in gigantic quantities, equivalent to some 200 Niagara Falls, then the Bosphorus Strait gave way to this flood from the Mediterranean, connecting it to the Black Sea.

The inhabitants of those primitive settlements must have seen with great concern a daily increase of about 15 cm of water. In just a few months, more than 150 m depth of water came in until the water bodies of the Marmara and Mediterranean were in equilibrium. The salt water went to the bottom and, due to the absence of currents without a strong temperature gradient to promote circulation, everything below 200 m deep died, leaving only bacteria that were capable of living without oxygen.

A sea can also change in a short time, unpredictably, due to weather. Far longer ago, the Mediterranean had experienced other abrupt changes. The connections and disconnections with other bodies of water due to fluctuations in sea levels created a unique environment, depopulated and re-conquered several times due to the Mediterranean's isolation. The only natural point of contact was (and is) the Strait of Gibraltar. Perhaps the most dramatic of these crises was the Messinian crisis. About 6 Ma ago there were drastic falls in sea level, and this led to a disconnection between the Mediterranean and the Atlantic Ocean, which then could not feed this closed sea with water. The Mediterranean nearly dried up completely. An accumulation of salt on the bottom (calculated as thousands of cubic metres) was accompanied by the formation of submarine canyons carved by rivers such as the Nile and the Rhone, now under hundreds of metres of water. Excessive concentrations of organic matter can be seen in geological cores from this time, before the desiccation: so-called sapropels. All around, the environment was desert-like and hostile, and areas of Central Europe went from a cool, humid climate to a rather arid climate. After about a million years (i.e. 5 Ma ago), the waters returned in force in the form of enormous waterfalls from the Atlantic through the same strait as long ago. The fauna and flora that we find today on the Mediterranean seabed and in its waters are derived directly from this new invasion, which had had the time to develop a high percentage of endemic species, some of which were recently discovered in precisely those submarine canyons that were the first to be flooded. The case of the Black Sea is much more recent, and it is more than likely that it was in the time of humans. Who knows, perhaps this natural catastrophe was the origin of the legend not only in the Bible but in countless remote Western and Eastern versions of the Universal Flood?

The Red Forest that Disappeared

The disappearance of the red coral forests, corals shaped like a small vermilion tree that are endemic to the Mediterranean and part of the eastern Atlantic, is perhaps one of the least known and saddest phenomena to have taken place in our seas. Precious corals, as we shall see below, have been traded since the earliest times. In particular, the exploitation of red coral (*Corallium rubrum*) dates back to the Palaeolithic period. Remains of objects made of red coral have been found from more than 25,000 years ago, and in the Neolithic age it was already being traded by the Minoan and Mycenaean civilizations. The first traders must have found whole branches of this coral after strong easterly storms because, even in the twentieth century until a few decades ago, it could be found points on the beaches of Cape Creus after gales. It became an essential material in Mediterranean culture and later was exported to places such as Nepal, China and Japan, Yemen, Persia and Jordan, where it was used for decorative objects, as amulets, in potions and as a status symbol. We have found engravings from that time showing people extracting large branches of red coral by free diving, so it must have been at shallow depths. The Romans used it widely to cure various diseases and to concoct love potions, taking it from nearby shores, starting with those of Lazio and Tuscany. This did not decline with the arrival of Christianity, which used it, especially in the beginning, as a protection against Satan. Extraction spread, gradually becoming industrialized and no longer from shallow depths by free diving, because the red coral that could be reached that way had probably been exhausted in accessible places. Instead, devices similar to a 'St Andrew's cross' were used, a device in the form of crossed sticks with hanging nets that dragged the bottom to pull up coral branches and bring them aboard. Deeper and deeper extraction was taking place, and more and more men and boats were being used to obtain the precious 'red gold'. Begur, in Catalonia, was perhaps one of the first industrial centres of production and trade between the fifteenth and seventeenth centuries, and there were other Mediterranean cities, such as Genoa, Pisa and Marseilles, that challenged its monopoly of a trade that extended to the New World. The exploitation, more mining than fishing, consisted of finding a coral bank and extracting coral it until it was exhausted, then moving to the next. The *coralines* (boats dedicated to coral extraction) reached their maximum numbers in the nineteenth century, when the craft and market were concentrated in Torre del Greco, near Naples. In 1862 there were about 350 ships, and in just two years these increased to more than 1,200 for the systematic plunder of the known banks (Fig. 2.1).

Red coral began to disappear from certain areas, from where it would not recover, and in others it was then exploited until the forest of red 'trees' of

Fig. 2.1 Red coral being harvested in the shallow waters off the Sicily coast. As can be seen, in the sixteenth century the precious coral was gathered by free diving, nowadays impossible. *Source* Tescione (1973)

about 20–30 cm in height (branches up to 50 cm and trunks of over 3 cm in diameter are known) was reduced to a field of mere blades of grass just 4–8 cm in height and a few mm at the base. The last major bank to be exploited was Alborán in the 1980s, where *coralines*, especially from Sicily, plundered the unspoilt resource without control. The arrival of scuba diving in the 1950s and 1960s was the final blow, because divers could penetrate the coral reefs where the bars and beams of the St Andrew's cross-style gear and nets could not: caves, cracks, ceilings of large rocks and walls. In the Medes Islands (Girona), in the 1960s and 1970s, the Cow Tunnel was an underwater cathedral, its walls almost entirely lined with crimson forest. Within a few weeks, divers had destroyed the red trees, removing everything. Today there is hardly any coral—here and there, in the tunnel, twigs of a few centimetres in length try to survive and grow.

A few millennia ago the hard substrate, the rock of our Mediterranean coasts, must have been widely populated by this red forest, a living stone paradise formed by the slow-growing organism, red coral. As an eco-engineering species, its three-dimensional structure added a great deal of complexity to the system. It served to retain particles and transform currents, had a great impact on the recycling of matter and was a refuge for countless organisms. But the vermilion forest has disappeared, and we shall probably never see the like of this splendour again.

The Closest Thing to the Past

In order to understand how the sea has been transformed and to be able to manage it properly, it essential to have something as a baseline; that is, a pristine system in which nature might be observed without direct human interference. This type of ecosystem is practically impossible to find, although some places exist beyond the fishing and tourist zones and commercial routes. One such is the reef off Kingman Island, as studied by Stuart Sandin (of California's Scripps Research Institute) and others. This reef, more than 1,800 km from Hawaii in the middle of the Pacific, has not yet been disturbed by humans because of its remote location.

What we see there has nothing to do with the coral reefs to which we are accustomed: white-tipped sharks, large snappers and other large carnivores prowl around hungrily, incessantly watching the movements of the smaller fish. The world seems the other way around here: the biomass of predators far exceeds that of herbivores and coral reefers, thus challenging the food chains to which we are accustomed (few predators and many prey). If we weighed the total mass of these large predators it would be more than four tonnes per hectare, while the small- and medium-sized fish do not total even one tonne per hectare. What happens in this system? Why is it upside down?

The answer, in fact, is simple: it is our other systems that are really upside down. The scientists who conducted this study compare the reef to an island not too far away yet where the anthropogenic impact is great: Kiritimati Island, where the human population exceeds 5,000, unlike the human desert of Kingman Island. Fishing and tourism are fully developed on this small island of Kiritimati, alongside agriculture and urban management. Diving in its waters, we find that the biomass of large predators per hectare is less than 0.2 tonnes, while the rest, the mobile prey, totals more than a tonne per hectare. But that is not all. The most important thing is that the seabed coverage of slow-growing corals and *coralines* algae off Kingman Island is just 70%, indicating a mature, complex and healthy system, while in Kiritimati macroalgae and fast-growing fleshy algae cover almost 80%. The latter is a much poorer, more disturbed system in which overfishing allows only small, fast-growing animals to grow—those capable of adapting to over-exploitation, and which feed on an immature bottom in perpetual regeneration due to the direct hand of humans. Off Kingman Island, meanwhile, many slow-growing, long-lived predators inspect the nooks and crannies of the reef for a population of prey that has adapted to a life without quarter, continually renewing itself under pressure from carnivores, just as it must have in seas everywhere on the planet many centuries ago.

3

Hunted to Extinction

In our scheme of how the ocean works, we need to include both large predators and large herbivores, and in much greater numbers than currently. Their role in ecosystems has never been proven, therefore their true influence is difficult to quantify. What we do know is that long ago large animals were commonplace, so what we see now cannot be described as 'pristine' at all. As discussed in the previous chapter, large creatures are not just big eaters but key players in the regulation and maintenance of the marine system. For example, a tiger shark weighs about the same as a hundred groupers or ten thousand ladyfish. Therefore, the effect of just one on the system can be much greater than that exerted by the range currently inhabiting our seas, which have been taken over by small fish and organisms. Throughout history, we have been eliminating large animals to eat them, for their products (hides, fat, feathers, bones, etc.) or simply because they were dangerous or competed with us for space and prey. But what is the effect on the entire ecosystem if we take them out of the equation?

In the first expeditions to the remote islands of the northern island arc from the Kamchatka Strait to Alaska, fur hunters found dozens of sea otters in the small bays and beaches that they visited. Once base camp was set up, they would walk along the beaches in search of these creatures for their skins, as well as pinnipeds and other animals. In a diary written in the early 1700s during one of these expeditions, a sailor related how easy it was to catch these unsuspecting animals, which had never seen man and therefore did not identify him as an enemy: 'In December we did not go beyond 4 km from the base camp to hunt our prey; in January we had to travel about 6–8 km, and in February there was no choice but to move more than 20 km to make

© Springer Nature Switzerland AG 2019
S. Rossi, *Oceans in Decline*,
https://doi.org/10.1007/978-3-030-02514-4_3

hunting profitable.' Within three months they had eradicated the population of otters (and other animals) in a place that is difficult for humans to reach. However, there is evidence of previous populations, from Japan through the arc of the Aleutian Islands. Today, we find them in the remote Kuril Islands, the Commander Islands, some parts of the Aleutians and southern Alaska. There is also a stronghold of these animals in Baja California, and they have been successfully reintroduced to the coasts of Oregon, Washington and British Columbia, but in numbers well below the tens or hundreds of thousands before contact with humans.

How important is it that sea otters disappeared? Otters feed mainly on large marine invertebrates, such as large sea urchins, crabs and cephalopods, and fish. Each of them ingests daily the equivalent of 20–23% of their body mass. 'Considering a weight of 23 kg per otter, a conservative estimate leads us to an annual intake of about 35,000 kg of prey per square kilometre by these animals with a moderate density, which we find on any island where they are currently found', explains James Estes of the Biological Resource Division of the University of California. Sea urchins and other invertebrates feed on the algae that populate the bottoms of these cold seas, forming thick forests that capture carbon dioxide, oxygenate the water and give structure and complexity to the environment. The 35 tonnes of sea urchins, abalone and other herbivores eaten by this animal per square kilometre per year exert strong control over the food chain.

Sea urchins are the most voracious predators of algae, especially their smaller and more vulnerable young forms. Having a large number of otters reduces the population, leaving only the smaller specimens to seek refuge in cracks and crevices. It has been proven that where there are no otters there are whole carpets of sea urchins, constantly advancing, feeding on the algae and leaving behind a bare, rocky substrate. However, centuries ago there were large herbivores such as sea cows that also fed on algae. That they would have exerted some pressure on the system is unquestionable, but these animals feed on the growth tops of the algae, while sea urchins forage for absolutely everything, right down to the smallest algae that have just arrived on the rock's surface, preventing new plants from becoming established and thus renewing the population. The type of pressure is different, and the regulation of each and every link in the chain is necessary if the system is to mature and remain as complex as the environment itself (nutrients, temperature, food, etc.). Therefore, this classic example of interaction shows not only the importance of a single animal to the system, but above all the complexity of the interactions between organisms and how artificial is our current system.

In many places, otters stopped being hunted in the early twentieth century. After a century of recovery, in the 1990s a sharp decline was detected in certain areas. This was incomprehensible, since no disease had been detected and they apparently had enough food. However, anomalous behaviour had been detected among the orcas in these areas. They approached the shore but, instead of attacking sea lions, they rushed to hunt the otters that they found swimming calmly on the surface: 'Killer whales have been able, in six years, to reduce the sea otter population by 78%', reported Estes. There is evidence that the two species had lived together peacefully for decades, ignoring each other. So, what happened? The killer whales had simply run out of prey, due to the failure of the sea lion populations to recover and, above all, due to the overfishing of large fish—their prey of choice and essential to their survival. If you have nothing to eat, you start to look elsewhere, even where you never imagined there would be food. The recovery of the otter population relies not only on protecting the animal itself. It will be necessary to protect the entire ecosystem, and even adjacent ecosystems such as pelagic and benthic systems far from the shore, so that the killer whales can find their preferred food.

The same applies to large herbivores. Towards the 1980s, one of the organisms that controlled the macroalgae of the Caribbean reefs was the sea urchin, *Diadema antillarum*. This urchin suffered mortality of up to 99% in certain areas, clearing the way for uncontrolled growth of the algae that covered the coral, suffocating them. However, experts like Professor Jeremy Jackson wonder if the problem did not originate in more distant times: 'Coral reefs in the Caribbean and elsewhere in the world are very different from those that once structured tropical coastal ecosystems', Jackson explains in several scientific articles. To illustrate this change, he gives the example of dugongs and sea turtles. Both are large herbivores capable of ingesting a huge amount of sea grass in the *Thallassia* meadows that stretch across bays and reefs throughout the Caribbean. The animals' consumption helped the renewal of the system and prevented excessive plant biomass, which otherwise ends up growing and detaching, accumulating a large amount of organic matter that favours the proliferation of bacteria and other organisms.

The establishment of new settlers in areas such as Jamaica completely changed the landscape in a few decades. The human population grew both because of the settlers who lived in the area and also the slave trade, as these archipelagos were an intermediate point before moving slaves to the continent. When the English conquered Jamaica around 1655, the main supply of protein, especially for slaves, was from the turtles grazing on the seagrass beds. Historical records estimate that between 1688 and 1730 more than 13,000 turtles a year were caught in the Cayman Islands alone (north west of

Jamaica). In about forty years, some 550,000 turtles were hunted in this archipelago of less than 300 km^2 of land. This is a conservative estimate, as the size of catches is based on what was legally taken, not taking into account either poaching or uncontrolled fishing, whether for subsistence or illegal trade. Testimonies from that time speak of a real 'turtle soup' on bays and reefs. In the Cayman Islands alone there were perhaps 2.5 million turtles, suggesting 33–39 million throughout the Caribbean, in a conservative extrapolation from the catch reports of the day. However, taking into account models of phanerogamous (spermatophyte) grassland extension, data on these herbivores and their natural history, before human intervention the number could have been as high as 660 million. This does not include the population of manatees, which may have reached similar heights in the past yet was exploited for their meat and other products.

Today, the turtles number only a few tens of thousands, and in the case of manatees a few thousand. As can be easily understood, the change to the working of the system that arises by extracting these large herbivores is complete. The dynamics are completely different from what once existed. The accumulation of organic matter and the drastic increase in bacterial biomass and other micro-organisms may be affecting the basis of the entire system; that is, the plants themselves and the corals that inhabit this area. The final insult to this already afflicted system is the increase in agriculture, tourism and of course fishing pressure, in an area where the population has grown exponentially during the past century. Around 1870, some half a million people lived in Jamaica, a pressure already considerable for the area, but the 2012 data show a stable population of nearly three million, to which must be added the fluctuating population of tourists.

'In some parts of Jamaica, the invasion of fleshy algae has reached over 90% of the surface area, replacing corals and other growing algae', notes Terence Hughes of the Department of Marine Biology at James Cook University in Australia. The transformation of the system has been ongoing for several centuries, but human pressure has been increasing fast, leaving no room for recovery. 'Diadema's disappearance and the increase in hurricanes was the *coup de grâce* for a system that is not recovering because of the great distortion it suffered after the fall of the different links that maintained it', adds Hughes. Without large herbivores and other controllers such as sea urchins or fish, algal blooms (going from 4% in 1970 to over 92% by 2000, in some areas) occupy everything from the surface to more than 40 m deep.

Therefore, we should speak not so much of the extinction of species (manatees and turtles are still present in the Caribbean and other areas), but of something much more worrying: functional extinction. This type of extinction occurs when one or more species do not exercise, by their number and biomass, the same role in controlling the system as they once did.

In all ecosystems, there is a tendency for some species to reach considerable size if the production of the system permits; that is, if there is enough food in the lower links of the food chain. Humans have always exploited these species first and then moved on to smaller species once the population was depleted. In all our seas, history has followed a very similar course: the disappearance of large carnivores and herbivores (both on land and at sea) has been a constant feature since ancient times. Large groupers inhabited the coasts of Cyprus at very shallow depths 8,000 years ago, and in Andalusia it has been proven that, 12,000 years ago, people were already fishing for large benthic animals near the shore, using rudimentary gear. Just as an increase in the body mass of some animals and plants is a constant evolutionary process, it is a constant in the evolution of a civilization to harness these great sources of protein, carbohydrate and fat for its own benefit. That is why we must stop thinking about the oceans as we have done so far. Hundreds of years ago, the seas ceased to be 'the last frontier'. As we will continue to see in the following chapters, our profound ignorance and irrational exploitation of the sea have given rise to a system that is completely altered, and little or nothing remains of what it was like. The most serious aspect is that we are not in a position to know how the various changes suffered by the sea are changing the planetary dynamics themselves, much less how they will affect other anthropogenic changes such as climate change, a serious blow to an already much-transformed system.

The Tragic Story of the Sea Cow

One of the most well-known and unambiguous tragedies of uncontrolled predation is the story of the sea cow. The settlers who arrived in the most remote islands of the Aleutians must have been surprised by such an animal, a species of manatee that in some cases exceeded 9 m in length and 2.5 m in diameter, weighing between 4 and 10 tonnes. *Hydrodamalis gigas* grazed peacefully in herds in bays and sheltered beaches, constantly browsing the seaweed that grew on the bottom at shallow depths. The sea cow was not new to human beings: due to them having no fear of man, whom they did not see as an enemy, the natives of the Pacific coast and fishermen off Japan, Korea and the Aleutians Islands had already made serious inroads into the population several thousand years earlier for their valuable flesh.

In 1740, the arrival of fur hunters in remote destinations in the island arc that connects Asia with America precipitated the extinction of what has probably been one of the Holocene's largest upper-plant marine herbivores. The new settlers began to hunt sea cows, not without difficulty due to their large size, as a source of protein and fat. Their skin was extremely strong and the taste of their flesh similar to ox. The German naturalist George Steller, who embarked on an expedition with Captain Vitus Bering, provides the only reliable zoo testimony from that time. Curiously, like other naturalists of the day he described them in detail and with admiration, then provided comprehensive details of how they should be hunted, cut, preserved and even cooked. In 1768, on the island of Bering, the last sea cows were killed. It had taken less than three decades to exterminate the animal. There had been an unsuccessful attempt to regulate the hunting of this and other animals, such as otters and sea lions, in the area, but the recent discovery and annexation of the islands had made the territory an almost lawless place, with frenetic exploitation of resources.

It is known that in a single small bay there might have been between one and two hundred specimens of sea cow. An adult manatee (of about 500 kg) of those still surviving in the Caribbean can eat about 50 kg of plants each day. Making an approximation by weight (the diet being slightly different, because it is a boreal animal), a 7-tonne sea cow could have ingested 700 kg of plants daily. If we multiply by about a hundred individuals, a conservative estimate is that in a single day these animals could have been extracting about 70 tonnes of algae. They were therefore a powerful controller of the system, a regulator of algae growth and a considerable agent of renewal of the sea bottoms that they inhabited.

We will never fully understand the role of these large animals, because their disappearance has left us with a completely different picture from that which we might have seen. What I do see is that the sum of extinctions is capable of changing a system not only locally, but globally. That is why we must be increasingly cautious in assessing the true meaning of any kind of extinction.

The Extinction of the Great Auk

Like other extinct, flightless birds like the dodo or moa, the great auk (*Alca impennis*) has a sad history. This bird, typically about 50–70 cm in height (it reached a maximum height of a metre and sometimes weighed more than 5 kg), was actively hunted from the Neolithic period. There is fossil evidence that it was spread across the East and West Atlantic, from Florida to

Newfoundland and from the Portuguese coast to the Baltic Sea. It was also observed in the waters of the Western Mediterranean until about four or five thousand years ago. It has been speculated that during the twelfth and thirteenth centuries great auks migrated northwards in a time of climatic warming: this would have led them to Greenland. In the islands of the British archipelago the bird served as currency for the inhabitants along the coast in purchasing grain and other products that had to be obtained from afar, but the people who exploited it the most in recent history were sailors fishing in remote areas, or those involved in naval conflicts or simply passing along the trade routes.

The bird, always trusting, laid a single egg (much appreciated for its taste) from which a chick emerged quickly, in about six weeks, then went to sea with its parents. In the water it was an unbeatable bird, able to dive for several minutes to catch fish or squid, but it had to go ashore to nest, making it highly vulnerable. Its large size and abundant numbers certainly put great pressure on the schools of fish in the areas where it lived, especially on herring and other pelagic fish. The most widespread form of hunting was simply to beat them to death, although there are reports of large numbers being caught alive and transported to other places where they were kept for a long time until they were killed for food. In some places they were accidentally taken in large numbers in fishing nets, but generally their extinction was due to systematic exploitation by sailors and coastal inhabitants. When they were no longer abundant, the total lack of regulation by the authorities was a serious disadvantage, in many cases because they lived in places where there was no clear jurisdiction over resource exploitation: the great auk was everyone's, and everyone had the right to hunt them down.

Little by little they disappeared from places where they had been common. When Linnaeus described and named them in 1758, they were already extremely rare along the coasts of Europe. In the Baltic the last was sighted in 1790, and by 1800 they had disappeared from North America, where they were known to be once abundant north of New York. Slowly but inexorably, the colonies became isolated in remote places in the northernmost parts of the Atlantic. Between 1808 and 1813 ships in the Napoleonic conflict used the few remaining specimens in the north for supplies, leaving only a few in Iceland. As great auks became so rare, their value to collectors rose many greatly. Around 1830/1831, a series of expeditions associated with museums and private collectors wreaked havoc among the last few specimens that were taking refuge on islands around Iceland. In 1830, an earthquake sank the island of Geirfunglaster, the site of one of the largest colonies still in existence in the area.

At this point, the inhabitants of the surrounding islands suddenly began to see great auks returning to places from where they had disappeared decades earlier. They were hunted again. The sailors who frequented those seas considered them to have become extinct around 1840, yet in 1844 the last expedition was mounted to find live great auks. The price had risen astronomically, reaching 100 Danish kroner at that time for the capture of a live specimen. The last pair were found by this same expedition on the island of Eldey. The five men of the expedition sighted the great auks, went down to the beach where they were standing, and killed them. They were aware that they were probably the last great auks, or boreal penguins, in existence.

The irony is that shortly after this last sighting, when their extinction had been recorded, thanks to the work of Darwin and Wallace debate on the evolution of species started. At that time, various scientific forums asked the question, 'Are great auks an example of maladjustment? Is this a clear case of survival of the strongest and disappearance of the weakest?' This represents one of the earliest examples of a misunderstanding of evolutionary theory due to human imbecility.

The Sad Seal

Another example of much more recent near extinction is that of the Mediterranean monk seal (*Monachus monachus*). The first time I clearly saw a picture of this creature's face, it seemed to me the expression of a wise and tired animal—almost sad (in Catalan, it is known as *vell marí*, or old sailor). This seal is considered to be one of the most endangered mammals on the planet, as there are only 600 individuals between the coasts of the East Atlantic and the Mediterranean. There is evidence that they have been hunted for more than 6,500 years: on the coast of Malaga there are remains that indicate their use not only for food but also for making utensils. During the Roman Empire they were systematically hunted throughout the conquered lands. The chronicles of ancient Rome speak of monk seal populations along beaches throughout the Mediterranean. In 625 BC there were coins circulating in the Eastern Mediterranean that bore an image of the monk seal, and already the seals had begun to be exploited commercially in many places for their fat, meat and skin. After the fall of the Roman Empire, populations that had been systematically hunted, especially in accessible places, recovered, only to fall again in the Middle Ages when commercial hunting was once again on the rise. Until the fifteenth and sixteenth centuries, these pinnipeds still gave birth on beaches and were widespread along the French, Spanish and Italian

Fig. 3.1 Monk seal. This warm temperate-water seal is an example of vanished fauna that have no real role in the ecosystem nowadays. Other big predators followed the same steps during recent centuries. *Credits* ADOBE STOCK

coasts, where they were under great pressure. Little by little, they began to seek refuge on cliffs difficult for people to reach. A typical instance of this gradual flight to protected sites is the Spanish coast, where the monk seal was found along beaches until it was displaced to more remote areas, such as Cabo de Gata or the Costa Brava, in the mid-twentieth century. Then, tourism completely prevented them from taking refuge in these places, as they no longer had the peace and quiet that they needed to complete their life cycle and feed. Only 40 years ago they were present in Sardinia or Corsica, but gradually they have become completely extinct in the Western Mediterranean and are rarely seen by chance in a few remote locations in the Balearic or Sardinian archipelagos. Monk seals, therefore, ceased centuries ago to be a functional species for the system. They are animals that no longer exert the same pressure as they did a few thousand years ago, and have been replaced by the far more predatory human beings themselves. It remains to be seen whether the few hundred individuals are capable of recovering the populations, since dispersion and inbreeding can favour a rapid disappearance due to a lack of genetic flow; that is to say, a lack of information exchange that permits the genetic means, through sexual reproduction, to combat disease, infection or massive mortality due to external pathogens. I, in particular, consider the animal to be virtually extinct. All that remains is to await the total disappearance of the species, such as happened to the Caribbean seal a few years ago (officially) or the Yang-Tse dolphin (Fig. 3.1).

4

Whatever Happened to Cod?

Giovanni Caboto, a native of Genoa who adopted the Anglo-Saxon name John Cabot, described in his ship's logbook a sea off the coast of Maine in today's United States that was so full of cod (*Gadus morhua*) that the vessels almost tripped over them. His journey, which took place at the end of the fifteenth century and was intended to rival that of Columbus in search of the remote lands of Asia, describes the exuberant hunting and fishing along the coasts that he visited. Bartholomew Gosnold, in 1602, also in search of alternative routes and new places, came across present-day Cape Cod (known as Cape San Jaime by Esteban Gómez), a place where cod 'could be removed by submerging a basket with a weight'. Little by little the merchants of Bristol, London and other parts of Northern Europe followed in the footsteps of John Smith and the other explorers who had mapped the area, especially Chesapeake Bay, finding countless natural harbours, backwaters and perfect locations for fully fledged colonial settlement. Many were beginning to realize what lay ahead, a sea of seemingly inexhaustible resources, including cod. Long before, the Vikings and Basques had already discovered these apparently endless shoals of cod, yet the resources that the first 'official' visitors saw were still so rich that they would not have believed that, due to overfishing, practically all the cod fishing grounds would by now have collapsed.

Whatever happened? When did the near total functional extinction of the species occur? How did it get to this point? I am reminded of a meeting in Bergen (Norway), when I was talking to a cod farming specialist at a facility near the university. He told me that at that time (2002) the business was investigating completing the cod reproductive cycle in captivity, with all the problems of infection, feeding, optimum temperatures, and so on. He

© Springer Nature Switzerland AG 2019
S. Rossi, *Oceans in Decline*,
https://doi.org/10.1007/978-3-030-02514-4_4

explained that cod was already a coveted creature and that the only chance of recovering the catch, in his opinion, was through aquaculture. Natural stocks could be considered almost extinct, from a commercial point of view.

Cod is a benthic fish, with a large mouth capable of eating almost anything. With their mouths open, swimming a few metres from the bottom and even along it, cod gather up and eat other fish, cephalopods and crustaceans. It is no fun being one of these creatures when a shoal of cod is approaching through the shallow areas along the coast. Cod flesh is white and almost shiny, with 80% protein even at fresh weight, and this is somewhat unusual in a fish of this type. Some have come to regard it as a kind of 'sea pig', its body taking full advantage of almost everything that it consumes, and it has fewer small bones than other fish. The cod has a musculature adapted to sudden turns, and its agile yet unstructured movements resist taking the hook for a long time, be it a rod or a longline hook. Fishermen say that is not much of a fighter, and even large specimens are bought to the surface easily. There was a time when there were cod over 2 m long. Records from 1895 still describe cod of that size, and lobsters of over 1.5 m, off the coast of North America. But that's history, now.

Cod prefer shallow water, ranging from 30 to 100 m deep, although they can be found at more than 600 m. During the spawning season they move to shallow areas. A female a metre long can lay up to three million eggs, but more than nine million have been counted. They prefer low-temperature areas with plenty of food, which is why they are so abundant on the east coast of North America. Here, areas such as Georges Bank, where the platform extends to an area larger than Newfoundland itself (more than 20,000 km^2), with depths of a few tens to a hundred metres, cod has always been abundant. Hundreds of vessels have been operating there since the fifteenth and sixteenth centuries, and Viking and Basque sailors used to fish the areas closest to the coast. By the early 1700s cod had become a food staple in Europe and the emerging United States and Canada. They were fished, dried, salted and even transported alive to ports, to be left in large boxes for weeks to be consumed later. 'It is estimated that by 1500 there were up to 7 million tonnes of cod in this part of the world', says George Rose of the Fisheries Conservation Unit at the Fisheries Institute in Newfoundland, Canada. 'Based on the fishing records and documentation of the time, as well as the biology of the species itself, we estimate that the stocks suffered no more than 5% annual fishing mortality per year, which allowed for an almost immediate recovery of the stock.'

It worked well until the seventeenth century, but the human population was gradually increasing, demand was increasing and the technological power of fishing vessels was increasing. During the last phase of the so-called Little

Ice Age (a period in which overall temperatures dropped significantly), there was a period of decline in the fisheries in this and another area, culminating in a fall in catches between 1800 and 1880. However, experts do not consider this to have been decisive for cod stocks on the North American east coast. In 1885, the Canadian Minister of Agriculture said, 'Despite the fall we have suffered in recent years, our fishing grounds will be fertile and inexhaustible for the next few centuries.' The fishing was still going well; sometimes fishermen couldn't hook six-footers anymore, but never mind! They still found large fish in abundance. Large banks such as Grand Banks or Terranova were seen as inexhaustible sources of protein.

At the beginning of the twentieth century, the trawling technique arrived. Steamships (another significant step in the fishing industry) had already used this technique in North European seas a few decades earlier, and now it was time to implement it in the US and Canadian fishing industry. The first reaction by fishermen in the area was rejection, but when some shipowners saw the amount of fish that the technique caught everything changed drastically, and the fishing industry in that part of the world turned to systematic trawling off the coast of North America (and of course elsewhere). By 1920, trawling was consolidated in the two northernmost countries of the American continent: hundreds of vessels trawled again and again, dragging their nets in search of the coveted cod stocks. In 1929, 129,000 tonnes of cod were fished in large shoals alone, a peak in production that would never happen again.

The scientists of the time gave the first warning signs about the coasts of Great Britain as early as the early 1900s. In 1900, an extensive article by Walter Garstang in the *Journal of the Marine Biological Association of the United Kingdom* warned of the collapse of trawl fisheries and the impoverishment of the seas in that area. No one listened to either him or the American scientists, who were concerned that what had already happened in England and elsewhere (especially in the North Sea) was now happening near home.

Large-scale fishing is a fairly recent phenomenon almost everywhere in the world. The industrial revolution provided both a market and improved communication to permit increasingly globalized exploitation: more people, more consumption, more needs. However, despite the fact that already at the beginning of the twentieth century there was cause for alarm, history across the world repeats itself: once fishing grounds become too small or do not provide enough fish (as the fishing fleet has developed), new fishing grounds are sought.

Soon after dragging, thanks to high-powered electricity from diesel engines, industrial refrigeration came into play. Large freezers could hold huge quantities of fish. That meant two things: one, the ability to go further afield to fish and for longer; and two, the ability to market the fish in the depths of Kansas, where some people didn't even know what the sea was. Fish could be salted, vacuum packed or frozen, and there were no longer obstacles to trade. Sometimes very fine mesh nets were used, capturing absolutely everything in their path, and the aforementioned diesel engines began to grow to exert incredible power that would have been unthinkable only three decades earlier. In 1934, only 28,000 tonnes of cod were caught in the same area as the Grand Banks. Other fish also began to disappear or become scarcer but, as the power of the machines and the distance from the shore increased, there was always fish to catch. Something was happening to the stocks but, although alarm bells were beginning to sound, the shoals were still seen as inexhaustible.

Most of the fishing grounds in Northern Europe at that time were already exhausted. Due to fishing pressure, individual specimens in 1970 were, at equivalent size, less fertile in 1970 than in 2003. This is not a contradiction; the pressure of fishing was effectively selecting the smaller specimens, which were the ones left in the sea when larger ones were caught and consumed. 'The genetic change may be due to a response to the non-random, weight-selective and measure-selective extraction of fish', says Midrio Yoneda of the Fisheries Research Services Marine Laboratory in Britain. By selecting the larger ones through fishing, those that reproduce earlier gradually take over. In addition, they find more food when their competitors, the larger fish, are outnumbered. Changes in maturity have been recorded in cod (and other fish) since reliable statistics on biological characteristics of the species have become available in Europe; that is, since about 1890. But in many places fishermen continued to use 60 and 90 mm mesh nets for fishing, giving even the small- and medium-sized fish no chance.

By 1960, the Grand Banks and surrounding area off the east coast of North America became one of the favourite trawling areas for Soviet, French, Spanish and Portuguese shipowners. In 1965, the Soviet Union alone used 106 large trawlers as factory vessels and 425 small trawlers connected to 30 mother ships to freeze and store the fish. The world population continued to grow, and our hunger for cod increased unabated. With up to 300 vessels concentrated in an area of less than 5,000 km^2, the population structure itself and the seabed began to change. In 1965, the Soviet fleet alone was responsible for extracting 872,000 tonnes of fish off the North American coast, while other countries caught more than 600,000 tonnes. There was close coordination among the fishermen: when a large shoal of fish was found, the vessel responsible called on the others to increase the efficiency of the work and let no prey escape. The dragging continued, time and time again, destroying the life of the seabed, profoundly transforming the animal populations that lived fixed to the bottom of the sea without mobility, yet also those that could move about. Little by little, they were decreasing in size and weight. In 1974, foreign vessels fishing in this area caught about ten times as much as New England vessels and three times as much as Canadian vessels. This had to stop, not for reasons of conservation or ecology but because both the United States and Canada considered that the cod were theirs. For them, large fishing fleets so close to their shores represented a violation of their territory.

In early 1977, Canada and the United States declared a 200-mile economic exclusion zone. They drove everyone from the waters around their shores. But this measure, which could have alleviated the plundering, did nothing but replace it. The 825 US 'national' vessels operating in 1977 increased to 1,400 in 1982. There was no recovery, and no opportunity was given to cod and, in general, to the area of the Grand Banks, which was now very different from that fished by the fleets of the nineteenth century. The market began to adapt to all types of fish, whatever their size and, although there were biological strikes, fishing restrictions and new regulations, the decline in the area became widespread. More and more heavy equipment, more powerful engines and more precise probes were being used to pick up the fish signal. By the late 1980s, cod catches in the area had fallen to just 22,000 tonnes of fish. If we compare it with 7 million tonnes in 1500, an approximate but conservative estimate, the exploited stocks had fallen to 0.3% of what they were at the beginning of the sixteenth century. This was compounded by the fact that the cod caught had been reduced to smaller average size, which had also drastically reduced its reproductive capacity. In the mid-1980s the system was being fished to three to five times its carrying capacity, partly due to the forecasts of fisheries experts that the stocks would be rebuilt within a few years. They had based their calculations on past stocks, so they estimated a catch of about 350,000 tonnes cod in a short while. That figure was never reached, and the banks were never recovered.

Once the cod had disappeared, there was an explosion of invertebrates and small benthic fish, as now they were obviously without one of their most important predators, without one of their major stock regulators. And in 1992 there was a moratorium on the whole area, which is still dragging on today, because what was once a paradise for cod, raising many animal species by virtue of the enormous abundance of microscopic algae (the basis of the food chain), had become a real desert for fishing. More than 40,000 people lost their jobs and there was a forced change of job for those in the towns and cities that been supported by this and other species, in both small- and large-scale concerns. The moratorium continues more than seventeen years later, because stocks have not recovered. Most of the fishing grounds have reached the virtual economic extinction of the species, the point where more is invested in finding the resource than the benefits it can bring: functional extinction, economic extinction.

Canadian cod, in many places, is protected as a species due to the risk of extinction. Locally, it has disappeared completely, or its populations are so fragmented and impoverished that they are unlikely to recover in a human timescale. Over-exploitation of all adult-size specimens, the destruction of habitats, even a certain change in climate and water temperatures, as well as the disappearance of the prey on which cod were mainly fed, have dealt the lethal blow to one of the most important species in the history of human food. What seemed unimaginable happened, and possibly it will take decades or hundreds of years to return to a similar situation… if ever.

Basques and Cod

Although the Basques were not the first to cure and commercialize cod (it was probably the Vikings), their history awards them a place of honour in the story of this fish, so important for human history. Long before Columbus arrived in America, the Basques were familiar with the area and were actively fishing in Newfoundland, the 'new land' that provided them with huge quantities of this seafood. Previously, the Vikings had been there from Norway and Iceland, looking for this valuable fish off the Canadian coast and Greenland. But while the Scandinavians did it in a somewhat disorderly way and took advantage on many occasions to commit acts of piracy, the Basques made the business profitable in a discreet way; so discreet, no one knew where the fish came from. It was a well-kept secret. Columbus received a letter before he left in which he was clearly told that there were already people working where he intended to go, but he ignored it and the Basque shipowners of the time were concerned about him discovering the place where fishing seemed to have no end. When men like John Cabot or Amerigo Vespucci proclaimed their recent and fascinating discoveries to the four winds upon their arrival, the men of the Basques watched in horror as the secret slipped from their hands.

Great must have been the surprise when Jaques Cartier saw with his own eyes, just decades after the (official) discovery of America, hundreds of Basque ships fishing in the Gulf of the Saint Lawrence River in Canada. These men of the sea, born to this element, had crossed the inhospitable Atlantic years earlier to hunt whales and fish for cod, ending up in one of the most hostile weather zones on the planet. They carried with them something that the Vikings did not have, which allowed them to keep the fish until they reached market: salt. Cod has a very low percentage of fat (0.3–0.5%). This preserves it, allowing it to be stored for long periods. About 1000 AD, the Basques had an extensive market along the Atlantic coast and were famous as absolute masters of the art of crossing the ocean, dealing with storms and finding and catching fish.

Curiously enough, the Catholic Church was to promote this trade in fish enormously. Abstinence from eating meat at least once a week (in the strictest seasons until the middle of the year) forced the faithful to look for other sources of food, so cod became one of the most commercialized seafood products in Europe. Until the eighteenth century, the Basques continued to fish in the coastal and oceanic waters of eastern North America, until the Treaty of Utrecht between France, England and Spain in 1713 considerably reduced the fishing opportunities in this area. The ban on fishing had a considerable impact on the Basque economy, largely based on fishing, where small-scale operations, master shipowners, fishermen and traders lived off the profits of fishing grounds that seemed to have no end. As Francis Bacon said, 'cod fisheries are more profitable than all the mines in Peru for Spain'. Although these benefits were not equitable (fishermen and seafarers were not exactly rich), the reality is that cod and other seafood were an important pillar of Basque society, as well as feeding half of Europe.

The Hated Competitors

In March 2009 I read in the press that there was to be another 'routine' seal slaughter (about 280,000) off the coast of Canada. Fishermen have always claimed that seals seriously interfere with the fishing of cod and other fish, preventing the recovery of stocks, so they hunt them legally from time to time to control their numbers, and for other reasons. Already in 1988, when I was living in Montreal, discussions on the role of seals in the lives of fishermen were a topic quite beyond me (not having as much information about it then as I do now). Is it true? Are these pinnipeds really to blame for the lack of recovery in cod and other species' stocks? The vast majority of scientific studies carried out in various parts of the North Atlantic disagree with this view, widespread in the fishing world. It is obvious that seals are predators at the top of the food chain, controlled from the sea only by large sharks (temperate waters) and killer whales (colder waters), and from land mainly by polar bears in the northernmost latitudes. However, cod do not appear to be the preferred prey in the current diet of seals in these northern latitudes. In areas where cod and seals overlap, the annual loss of cod stocks has declined by up to 17% per year, while those of seals have reached 12% growth in the same seasonal period. In 25 years, the increase in seals has been spectacular. 'Taking into account the energy requirements of seals, their direct feeding, the abundance of prey and observing indirect indices such as biochemical markers, we have been able to observe that the diet of grey seals between 1993 and 2000 does not exceed 5% in cod', says Kurtis Trzcinski of the Belford Institute of Oceanography in Canada.

For Trzcinski and other authors involved in the issue, the collapse and lack of recovery are, in any case, not due to the actions of seals. Seals travel long distances to feed and opportunistically prey on the what they encounter; if small fish, squid or faster-growing organisms abound, they will take advantage of them. 'Even if all the seals were removed from the system,' continues Kurtis Trzcinski, 'the cod stock would probably not recover.' In addition, large cod have a diet quite similar to that of seals, so their disappearance would undoubtedly have prompted an increase in pinnipeds. The only time that they could be interfering with stock recovery is at the time of cod reproduction, but no one has provided conclusive data in this regard. Alida Bundy of the Belford Institute of Oceanography in Canada does not seem to see seals as the problem in stock recovery: 'In the models applied to understand the current functioning of the ecosystems in which cod are present, other factors such as overfishing to the point of exhaustion or the destruction of the habitat itself or the lack of food due to the disappearance of its prey are much more important.'

At the beginning of the twentieth century, in the port of Gloucester, a bastion of cod fishing in Massachusetts since time immemorial, a bounty of up to five dollars at the time was paid for a seal's nose to prove its elimination. The belief that seals are to blame for the lack of fish is widespread in all cultures. In the Mediterranean, monk seals raid the trammel nets of Greek fishermen, but the losses do not exceed 3–5% of the total catch and the government is willing to pay compensation (late, as always, but it does compensate them). However, coexistence is difficult and, until a few decades ago, fishermen themselves did not hesitate to kill them in order to eliminate a possible competitor in their fishing territory. In any case, it is easy to blame those who see them taking the fish, as it is a behaviour that obliterates any competition.

The Upcoming Changes

Like all animals and plants on the planet, cod respond to climate change. They have always done so, but at present the position may be, in many places, more precarious due to overfishing. Cod inhabit waters that range from −1 °C to a maximum of 20 °C. However, it has been shown that the places where it is most comfortable are those where temperatures vary between 0 and 12 °C. For this reason, following the North Atlantic warming maps, Kenneth Drinkwater of the Bjerknes Marine Research Institute in Norway has made a series of predictions from the current stocks about the distribution of cod, in which he takes into account factors such as salinity and potential food change (although the latter is much more difficult to predict): 'The waters that are going to heat up the most,' Drinkwater writes, 'are those of the Arctic and Sub-Arctic, but in certain areas, such as the Celtic or Irish seas, the water can reach temperatures in 2100 that displace cod stocks, which could make them disappear from these areas.' The cod would move northwards, occupying large areas of the Barents Sea and shallow areas of the Arctic shelf. We would find them more frequently in Greenland and in the northernmost parts of Russia and Canada, where they are now scarce. Spawning sites would move north, because the present waters where the species most frequently spawns will increase by between 2 and 4 °C over the next few decades. 'It is not a new phenomenon', adds Drinkwater. Between 1920 and 1930, the warming of Greenland's waters caused the banks to shift northwards up to 1,200 km, as recorded in fishing statistics. The ice limit would act as the edge of the displacement of cod, and in transformed habitats production would rise, giving them food and shelter. Places where it is too warm for the species will entail a metabolic cost (mainly related to increased breathing and acceleration of physiological processes) that will make life uncomfortable in many places where it survives. Spring migrations may be brought forward or delayed due to rising temperatures, but wherever they go they are likely to find more food because of increased plankton production, as well as ideal conditions for accelerated gonad production. Therefore, where the water temperature exceeds 12 °C, cod will tend to disappear in search of new survival platforms. This is a factor for those who now search these waters for this precious creature to take into account, as they may find themselves fishing for other species more in line with the change in temperature and food.

5

The End of the Cetaceans' Reign

One of the first global maritime industries was undoubtedly whaling. It is not known for sure when systematic whaling began, but there is written evidence that in the ninth century AD there were already what we might call small businesses that chased whales from onshore. In Korea there are caves describing the capture of whales around 6000 BC, and the Romans rounded up and killed whale specimens for trade. However, as the chronicles show, it was more sporadic than a flourishing business. Hunting them down was certainly a lucrative, but also a dangerous and strenuous enterprise. In the early Middle Ages whaling boats followed the large cetaceans long distances by rowing, first harpooning them then waiting for the huge animal to come up to the surface to harpoon securely its prominent back. It was normal to wait until it was exhausted before attempting to turn it over and open so it would bleed to death and could be dragged to a larger boat or to the shore to be cut up. It was not an easy task, because whales have immense force. They are combative, and the waters where the first hunters ventured were usually icy. The documents speak of intense hunting in the North Sea and the English Channel around the eleventh and twelfth centuries AD, where sailors, especially Basques, spent long periods of time chasing cetaceans. By the fourteenth century, there were already signs of a steep decline in the populations of several species in this part of Europe, particularly along the coasts of Normandy, Flanders and England. Some of them, such as the Atlantic grey whale, ceased to exist around the eighteenth century.

As in other fisheries around the world, the Basques were pioneers and were systematic. In the sixteenth century more than thirty galleons, with some two thousand crew specialized in capturing cetaceans, were based in Labrador and

© Springer Nature Switzerland AG 2019
S. Rossi, *Oceans in Decline*,
https://doi.org/10.1007/978-3-030-02514-4_5

set out north to hunt in remote Greenland. The chronicles speak of places where thousands of harbour porpoises, belugas and other small odontocetes (with teeth) were concentrated. Those same chronicles speak of areas where the great baleen cetaceans (without teeth) were grouped in hundreds of specimens, so hunting, despite all the difficulties, was quite simple due to the great numbers. In the northern hemisphere the English and Dutch displaced the Basques, as the latter could not easily access the areas to hunt whales or fish for cod without entering their territory. Around the island of Svalbard in northern Norway, the Dutch had over 240 whaling ships in 1684. The stocks at that time were already beginning to show signs of decline.

That was a reason for searching for unexplored seas, areas where the systematic hunting of large cetaceans had not yet arrived. The history of whaling is one of shipbuilding companies always in search of new fishing grounds, such as those in Patagonia, which began to be exploited methodically in the early 1700s. Whalers (and other fishermen) were pioneers, often arriving in places where even explorers had not arrived, such as the inhospitable waters of the Antarctic Peninsula. They kept secret their discoveries (as we have seen with cod) to be able to exploit the fishing grounds for longer, since many of the products derived from whales had become essential to an increasingly populated and sophisticated world. The oil from blubber had become essential for lamps and lighting systems, and the bones for various types of structures, even corsets.

Everything was taking advantage of the cetaceans, who found a new enemy in 1840 in the form of explosives. The first harpoons with a small charge on their tips were more efficient and lethal, which considerably reduced hunting time. At that time whalers began to hunt cetaceans systematically in the Pacific, first on the coast of California and in the Gulf of California, where they were grouped in large numbers after long migrations. In 1872 the migration was practically nonexistent, as the whalers of the area had exhausted the stocks. But new fishing grounds had to be found, as it was essential to reach other places where the 650 American vessels, with their more than 13,500 sailors, could fish. The business had to continue, so the Aleutians and the North Pacific, one of the last whaling destinations, were plundered. By the end of the nineteenth century, more than half of the places where whales were known to swim were empty: the populations were either extinct or so small that they were not worth the effort.

Then whaling began to go for anything, any cetacean. Whale oil was not much appreciated, because it was more expensive than mineral oil, but products such as whale meat, fat or bones continued to be in great demand. Upon the introduction of the diesel engine and other technologies, whales of every kind had their last moments, even in remote Antarctica. The Japanese came into play. They have always had a whaling tradition, but their catches had been mainly focused on the Japanese Sea between Korea and the Japanese archipelago. Whaling intensified, especially after the Second World War, whales being a cheap source of protein and fat for Japan (and other countries). But a decline came a few decades later. Japan went from a peak catch of 226,000 tonnes in 1962 to 15,000 tonnes in 1985 (just before it was banned worldwide). The decline in stocks indicated the imperative to stop whaling. The Japanese research institutes, closely linked to the country's whaling industry, were gagged, unable to provide clear statistics or even guesswork about the collapse of cetacean populations of all kinds: 'While in 1960, up to 23,000 dolphins were hunted in only two specific areas of the Sea of Japan,' says expert Toshio Kasuya of Japan's Nagayama Institute, 'in the same area they did not reach 1,000 in 1983.' Other cetaceans, such as sperm whales, went from an annual catch of about two thousand in the 1960s to ceasing to exist between Japan and Korea as a commercial species by 1970.

Then, in just two decades, instead of the moderate and more or less stable increases that had taken place between 1910 and 1950, the size of the catches soared. In 1976, after the ensuing drastic decline in their numbers, most whale species in the North Pacific were protected. But it was not until the early 1980s that a moratorium prohibited the hunting and trade of most cetacean species. Even now, in the Pacific, populations of some species such as the minke whale (*Balaenoptera acutorostrata*) continue to be under pressure from the whaling industry under a scientific banner. 'We have seen that both whales caught and by-catch by pelagic nets and those hunted for scientific purposes come from areas that are theoretically entirely reserved for population recovery', says Vimoskaselhi Lukoschek of the School of Biological Sciences at the University of Auckland, New Zealand. 'The Japanese stock is considered to be on the verge of collapse, with little chance of recovery'; the 'genetic fingerprint' indicates that these whales are present in Japanese dishes throughout the territory.

Japanese, Russians, Norwegians and Icelanders, each in their own way, have insisted on the need to revise the whaling ban treaties, arguing that there are now instruments to make whaling more sustainable. But, despite the fact that some populations are showing clear signs of recovery, protection measures are still in place for these animals whose populations are far below those in the historical records of the past.

From historic information (navigation logs of sailors from previous centuries and the docking of ships), it has been concluded that in the sixteenth century there may have been up to 36 million specimens of fin whale scattered around the planet, and up to 24 million humpback whales. However, these data have been proven to be unreliable. Using a much more sophisticated method, that of population genetics, a figure ten times higher has been calculated; that is to say, there were about 500 million whales, taking into account just these two species. The numbers, very approximate and probably a little inflated, once again allow us to glimpse a completely different reality from the current.

The question, once again, is, so what? What difference does it make? Whales are majestic, beautiful animals, a symbol for many people of the essence of the ocean, but does their disappearance mean anything? As in the previous examples of seals, sea cows, great auks, seals and turtles, what if they disappear?

I will try to illustrate with a concrete example what the loss of these organisms means. In just 370 km of coastline in the Aleutian Islands, 62,858 whales were hunted in a declared manner between 1949 and 1969, representing some 1.8 million tonnes of meat, bones and fat. At that time there was also a decline in phocids, sea lions and sea otters, although they were protected. No one fully understood what was happening and, although it was also known that fish in many areas were much less abundant, it did not appear to be sufficient cause for the pinniped and musselid populations' failure to recover. In some areas they disappeared completely. Alan Springer and his collaborators from the Institute of Marine Sciences at the University of Alaska saw it clearly: 'Whales, especially whale calves, are a very important part of the killer whale diet. Some species seek out protected areas in warmer waters even in the tropics where food is much scarcer but where they can give birth to their young and protect them from the killer whales.' 'Killer whales', or orcas, indeed used to be known by whalers as 'whale killers' but, if there are no whales, these creatures look for other prey. The almost 4,000 people who inhabit these sparsely populated 370 km of coastline began to change their diet. They replaced whale meat with that of seals, sea lions and otters, a much less calorific meal and much less appetizing, yet more affordable and always more profitable than fish. 'The problem is that killer whales have less and less food and attack any prey', adds Springer. The biomass of whales is 60 times greater than that of pinnipeds. We can look at the numbers any way we like, but the reality is that by removing so many whales we have removed another key piece of the system. Whales ingest enormous amounts of food, defecate immense amounts of faeces, supply much food to other organisms and, being so long-lived, retain much carbon in their structures. Our seas, increasingly dominated by short-lived, accelerated-cycle beings, are transformed, unable to retain carbon in the same way as long-lived, complex, accumulating animals or plants in the form of their tissues. We are faced with a system that cannot recycle matter in the same way as before. It is changing, and we do not know for certain where it is going and what role this type of organism plays in regulating ecosystems and the climate in general.

Disclosing Irregularities

I am sometimes surprised by people's capacity for deception and lack of remorse in continuing to exercise their 'right' as a predator, even when it has been shown that one or more species are heading for collapse due to their unscrupulousness and limited vision when exploiting a reserve. When I read that Scott Baker of the University of Victoria (Australia) had carried out an exhaustive study on the recovery of humpback whales back in 1991, I was disconcerted by what he had found. I realized that data concealment was going to be the key to why these large cetaceans were no longer crossing the New Zealand Strait, stretching from the Antarctic continent to Oceania. Could an environmental phenomenon be involved? Had there been a change in the diet of the whales that had led them astray? The local inhabitants insisted that hundreds of whales used to pass through every year and that, despite protection, the populations did not seem to be recovering.

Almost twenty years later, Phillip Clapham of the Seattle Marine Mammal Institute revealed the secret. 'The Russians (and other countries) were illegally and unregistered hunting thousands of whales in that area of the world', he reports. From 1947 to 1991 the then Soviets had certain hunting quotas. In fact, on board each ship there had to be an official marine biologist, a professional who would report all the catch data of the whalers operating in the area. The data were transferred to the International Whaling Commission (IWC), which was responsible for sorting and using the data to track stocks and assess whale stocks around the planet. However, there was double counting. In one book, they wrote down the real catches: all kinds of whales, without looking at size or species, but scrupulously recording the real biological data. These data were passed on to a department of the Ministry of Fisheries and to the KGB itself. In the other book, the 'official' version, fictitious catches were noted, those that fell within the provisions of the quotas and the regulations. The biologists signed the KGB's own documentation, in which they undertook not to say anything about this double counting. The information, passed on from various Russian institutions during 2008, has cost the job of more than one biologist and official. Some of them photographed parts of the original files (in total, more than 60 thousand pages) with their mobile phones and passed them to their foreign colleagues from their office computers. From 1959 to 1971, for example, the Russians hunted more than 48,500 whales in Antarctic waters, while the official IWC disclosure was no more than 3,000.

Humpback whale populations were already in decline, but the Soviet fleet had delivered the final blow in this part of the world, where control is difficult. 'There are now between 3,000 and 5,000 humpback whales in that area, 20–25% of the original population, being optimistic,' says Scott Baker: 'With all the restrictions that exist today (and taking into account a certain percentage of poaching), these populations may not recover in this area of the planet until well into 2050.' Hiding the information has consequences for IWC ecologists and fishing technicians working with models of recovery of large cetaceans: 'The numbers we have done so far did not take into account this type of mass poaching', says Vernon Smith of the US National Oceanic and Atmospheric Administration (NOAA). It is clear to the scientists involved that self-regulation is no use: there must be an effective supranational body to monitor the various countries involved in the capture of cetaceans.

Scientific Hunting

An article in a prestigious polar scientific journal suggests that in sub-Antarctic waters minke whales (one of the smallest baleen cetaceans) have lost 9% of their body fat over the last eighteen years due to the increasing shortage of krill. 'The increase in the number of predators of this crustacean and the decrease in its biomass are possible causes of the decrease in stocks', writes Kenji Konishi of the Tokyo Cetacean Research Institute, in collaboration with the Institute of Medical Science of the University of Oslo in his *Polar Biology* article: 'They could have lost as much as 0.02 cm per year.' The data are interesting, but the controversy lies in what scientists call sample size and the type of sample used to arrive at these conclusions: 'To avoid statistical errors, 2,890 mature male whales and 1,814 pregnant females have been hunted', Konishi and his colleagues report in the article. More than 4,500 minke whales have been hunted between 1988 and 2005, increasing the numbers from about 250 to more than 400 per year.

Some scientists are deeply concerned about this article. *Polar Biology* is the most prestigious polar journal of marine biology in the world. By accepting this article, two serious mistakes are being made: the first is to tender for whaling, which is illegal, through work that is published in a rigorous scientific journal; the second is to suggest that in science 'anything goes' to achieve a specific objective—that the means justifies the ends. It is as if it is now being said that, in order to study what seals or penguins eat, we need to sacrifice five or six thousand specimens in search of a scientific article. What is the ethics? Any scientific work must first go through referees, two or more anonymous scientists who give their opinion, then correct and even reject an article according to the rigour of the data and the approach. Having published the work in an internationally recognized scientific journal, the Japanese and Norwegians have a tool to defend their covert hunting policy.

The minke whale (*Balenoptera acutorostrata*) is one of the most hunted cetaceans on the planet. According to the Species Regulatory Body (CITES), this whale, not being a threatened species, is at the limit of its exploitation, and the evolution of its populations must be closely monitored. In 2007, the Japanese government intended to catch 935 specimens, but the actions of conservation groups caused this number to fall to 551: 'We had to stop catching 45% of what we had planned', said a representative of the Japanese Fisheries Agency: 'We don't have enough time for the investigation because of the assaults and impediments caused by the conservationists.' The whales processed by the scientists are transferred to a company that finances the Institute itself and sells the meat. 'Over the past few years there has been a remarkable increase in catches', says Junichi Sato of Greenpeace.

Conservationists are most concerned about this upward trend, especially in Antarctic waters. I would query whether the Antarctic Treaty does not specify that fishing in its waters is prohibited for these and other species? As we can see, we respect nothing. The problem, in my view, is that targets can easily be shifted to other species. 'This year, 50 humpback whales are planned to be hunted in the waters of the white continent,' says Geoffrey Palmer of the New Zealand Whaling Committee, 'and some countries have already established areas where we have banned the hunting of all whales.' Australia, for example, has declared a 200-mile zone around its coastline to be free from hunting for these animals, considering it an inviolable sanctuary. The controversy continues, but several countries have criticized the fact that, with a huge fishing sector and a large number of cetaceans being hunted, there has been a meagre and questionable number of scientific works (less than 60 in more than twenty years).

I go beyond that. I think it is incredible that a scientist should have this work in his hands and not clearly denounce this farce, whereby nothing significant has been promoted to increase our understanding of the dynamics of the species, at the cost of the whales that have ended up being eaten in oriental restaurants.

Whales and Climate Change

Like all organisms on this planet, climate change affects the immediate future of cetaceans. These large, warm-blooded vertebrates, already under pressure from other factors such as hunting, pollution (especially plastics), lack of food resources or disease, have to withstand oceanic transformations on a small, medium and large scale. Many species of whales migrate extensively, so changes in temperature and productivity of the waters will affect them but, according to specialists, they may be the ones that are most likely to adapt to new routes. 'Migratory species are an enigma, because it is not known if their ability to move will allow them to quickly change the latitude at which they seek food or a place where they can give birth in peace', says Mark Simmonds of the Whale and Dolphin Conservation Society of WWF-International. In fact, there are cetaceans like the Gulf of California porpoise, the vaquita, that literally live trapped in an environment from which they cannot escape. If water temperature changes rise and significant oceanographic changes occur as predicted by the various specialists, the species will have no escape and will either adapt to the new food and thermal regime or become extinct. The same will happen to freshwater dolphins, such as those of the Ganges or the Amazon, which are restricted, by their evolution, to specific places from which they cannot escape.

There is no doubt that the most important effect will come from changes in the food chain. 'In both the North Sea and the North Pacific, the trophic structure of the system is already changing,' adds Simmonds, 'and it is only a matter of time before other seas are affected by a similar trend.' But these changes, replacing species phytoplankton and zooplankton of colder habits with species accustomed to higher temperatures, are not bringing more energy to the system. In other words, they are not a substitute in terms of either abundance or calories, and they are not a more succulent snack. On the contrary, it is being proved that, in general, they are poorer in energy.

The habits and diet of those who depend on the dynamics of ice for their livelihoods will also be severely disrupted. In this case, because the formation and melting of ice year after year forces the system into specific patterns of light and nutrients, all organisms (including cetaceans) that depend on its dynamics will be severely affected. Ice reduction may be critical to many specialist organisms, which will be restricted in their distribution and replaced by competing species. The most obvious cases are those of the narwhal (*Monodon monoceros*), beluga whales (*Delphinapterus leucas*) and the Greenland whale (*Balaena mysticetus*), the first being the most vulnerable due to its intimate relationship with the icy layers of the oceans.

We must not forget the other hemisphere, perhaps even more vulnerable because of its intense over-exploitation in past decades. Here, no less than half of the biomass of marine mammals from all over the world is concentrated, all dependent in one way or another on one type of small crustacean: krill. In the Antarctic seas, changes in the dynamics of ice and its regression in certain areas are affecting this small crustacean on which fish, penguins, seals and, of course, many species of cetaceans (especially baleen whales) depend. The recovery of large cetacean populations from intense hunting may be affected by a lack of adequate food. Applying the precautionary principle to the management of these cetaceans is therefore essential, since the changes already suffered by their populations are compounded by one that is even more global and difficult to predict: climate change.

Part II

Effects of Overfishing

6

Major Fisheries and the End of Innocence

Man has never stopped hunting at sea. Although we have cultivated the land and raised livestock for millennia, in the sea we have not taken up mariculture to replace the fishing that is carried out all over the world, although it is increasingly important, as will be seen in future chapters. This reflection has always confirmed to me how divorced we are from the marine environment and how far we are from managing it sustainably. We employ many means to extract various types of fish. Most have been refined to hitherto unthought of heights in recent decades. The last fishing peak was in the 1960s and 1970s, when the maximum quotas were recorded since scientific monitoring of fish stocks taken from the sea began.

Fishing exploitation is defined as the organized extraction of renewable marine resources or, in other words, all actions aimed at fishing a given area. Most fisheries, based on fishing for certain species, are either at the shore or on the continental shelf. The current estimate of the protein that goes to our table from these fisheries is 16–18%, varying greatly by region of the world. The top ten countries that produce (or catch) the most fish are China, Peru, Japan, USA, Chile, Indonesia, Russia, India, Thailand and Norway. It is important to note that these ten account for more than half of the world catch and that China's catch is by far the largest. Only 10% of the catch is from inland waters (rivers, lakes, reservoirs), and the rest is caught at sea.

We have already seen that it was in the second half of the twentieth century that fishing became an industrialized and globalized commercial venture. Between 1945 and 1960, several events took place to transform the ocean scene, such as the systematic introduction of diesel engines in large vessels, which made the fleets grow significantly in several countries of the world, or

© Springer Nature Switzerland AG 2019
S. Rossi, *Oceans in Decline*,
https://doi.org/10.1007/978-3-030-02514-4_6

the technological development that allowed far more autonomy to those vessels, which began to organize themselves for deep-sea fishing. Even then, pelagic fish were over-exploited.

In a short time, we went from about catches of about 18–30 million tonnes of catch on the planet. Between 1960 and 1970 there was a decisive geographical expansion of fishing grounds, and fleets were consolidated in waters far from their country of origin, the catches growing from 30 to 60 million tonnes worldwide. Between 1970 and 1980, each country's own jurisdictional waters (the famous 200 miles) were delineated and changes began to be detected at the ecosystem level. Exploitation reached 68 million tonnes, and the 'maximum sustainable catch' model of the past was openly criticized. It was between 1980 and 1990 that the concept of sustainability and environmental impact was introduced, as well as the term 'by-catch': organisms (including many fish) that are returned to the sea, almost all dead, because they are not the main target of fishing. During this period, the operation increased to 85 million tonnes, a figure that stabilized to 90 million tonnes. Let us not forget that more than 40 million people worldwide depend directly on these marine resources, and it is estimated that the numbers living off them indirectly may easily exceed 200 million.

The science of the exploitation and viability of fishing involves the disciplines of biology, ecology and oceanography, on the one hand, and anthropology, economics and sociology, on the other, to make estimates and models of how much fish can be extracted and under what conditions. But even though there are a number of studies and approaches, the main mistake that has been made over and again is to apply a model of continuous catch without considering when the exploitation of the stocks began. This is mainly due to the fact that the analysis and valuation of fisheries resources worldwide face serious problems and difficulties. This is largely due to the difficulty of obtaining reliable data and information, both in terms of overall catch and the species caught. The ultimate aim of exploitation is to market the fish harvested locally, regionally or globally and, as Veijo Kaitala of the Integrative Ecology Unit at the University of Helsinki explains, 'any fishery has a high degree of uncertainty as to the stocks and carrying capacity of the system'.

The fishing effort, which also needs to be considered in order to get a firm idea of trends, is often another obstacle to be overcome. We certainly lack a holistic view of the resources, the ecosystem and its impact. However, fisheries science has come a long way in the past two decades, and the best part is that there seems to be a better understanding between fishermen and scientists or analysts. This has become apparent in recent years not least because fishermen and related companies themselves have become aware of the real risks of collapse to economies around the world. The estimation of resources is done in two ways: onboard inspections (where the fish are measured, weighed, etc.) and data collection at the auction centres. The two combine to give increasingly accurate figures for the world fish market, coordinated by the Food and Agriculture Organization of the United Nations (FAO). These figures warn of collapses of species in various areas of the world (Fig. 6.1).

In my opinion, the model applied so far, in which an isolated species is identified for management, is completely obsolete. It does not go anywhere by isolating the individual from its environment, because the interactions are multiple and the health of the ecosystem itself is essential to the maintenance of a fishery. I am surprised, most surprised, that only in the past two decades has this become regarded as obvious and that it has not been taken into account by professionals responsible for the management of the renewable resources of the sea. According to Louis Botsford of the Department of Wildlife, Fish, and Conservation Biology at the University of California and Juan Carlos Castilla of the Catholic University of Chile, 'Unfortunately, when you approach a fishing collapse, you must take into account new generations or spatial variations in production or interactions with other species in the ecosystem'. In this regard, Jake Rice of the University of Ottawa's Fisheries and Oceans, Canada, insists that the ecosystem approach is the only one to provide a reliable response to resource exploitation in environments as complex as the coral or algae beds around the world. 'Specialists

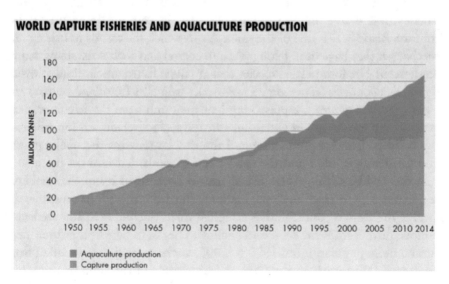

Fig. 6.1 Fisheries and aquaculture production from 1950–2014. The almost exponential increase of aquaculture can easily be seen. *Source* FAO (2016)

are expected to be able to integrate the whole in order to be able to respond to how stocks of particular species are exploited; this is very complex but essential work if this is to work, and it seems that the fisheries sector has begun to understand it', says Rice.

I understand that changes in fisheries policy are sensitive, as they are always under much social pressure. Subsidy policies do not help, since instead of redirecting the problem and looking for alternatives they often try to sustain it until there is nothing left to fish for. In the United States alone, there are more than 150 federal plans that include definitions of overfishing and stipulate actions to prevent collapse, but in most cases we are not aware that the stock is over-exploited until it collapses, often irreversibly in our short-term perception of timescales.

We are now fully aware that in any fish stock the primary source of mortality is the fishing itself. It can be said that the fishing grounds of virtually the entire planet are either at the limit of exploitation or have clearly gone beyond this point (some of them irreversibly, centuries ago): Half are fully exploited, one-quarter are exhausted or over-exploited and only 20% are moderately exploited or under-exploited—none are considered to be virgin. The trend is to look for fish deeper, with trawling being the most developed technique for resource extraction, along with driftnets and industrial long-lines. The effects are now being seen, as a large part of the fishing grounds that have exceeded these exploitation limits cannot be recovered.

Perhaps the Mediterranean, despite its modest contribution to world fisheries (less than 2 million tonnes per year, in combination with catches from the Black Sea) is one of the examples in which the trends in fishing and the changes that have been taking place in coastal and shelf ecosystems can be clearly seen. The four pelagic species that are most fished are sardines (*Sardina pilcardus*), anchovies (*Engraulis encrasicolus*), sardinellas (*Sardinella aurita*) and sprats (*Sprattus sprattus*), representing half the catch from the Mediterranean coast. These species are essential links in the seafood chain, an intermediate step between primary production and the tiny crustaceans that make up the crustacean zooplankton and the fish or cephalopods that feed on them.

In the 1940s, sardines were fished mainly because, as a more coastal species, it was easy to reach their populations. But from the 1960s onwards, the fleets of the various countries that exploited them (including Spain) were able to move further from the coast and anchovies began to take over as their flesh became more popular. From 1990 to 2005, anchovy catches have fallen from 7,000 to 4,000 tonnes in the Gulf of León, while in the Catalan Sea, especially in the banks facing the Ebro Delta, the fall in this same period has been from 13,000 to just over 500 tonnes. Of course, although trends are downward, there is a significant fluctuation that depends not only on fishing but on environmental factors. And this is the point that several scientists at the Institute of Marine Sciences (ICM-CSIC) in Barcelona have begun to understand in depth. 'The nutrients coming from the Ebro River have a decisive influence on the production of the platform that the river itself has formed over millennia,' explains Isabel Palomera, 'and these nutrients in turn have an impact on the entire food chain: the more nutrients and certain physical conditions, the greater the production of phytoplankton (micro-scopic algae) and, in turn, zooplankton, on which both juveniles and adults feed on these pelagic fish.' But more telling is the importance of winds, temperature and the consequent upwellings to explain this legacy of nutrients. 'Anchovies alone consume up to 22% of the production of the system below them, while sardines consume up to 25%', explains María Pilar Olivar of the same institute. Therefore, the impact on the various stages of something as 'abstract' as nutrients and winds can be fundamental to fisheries management, and the flexibility of the model has to be more concerned with the potential burden of the system under these new approaches than with a mere fixed-catch quota without any ecosystem basis.

Anchovies and sardines have similar diets, so overfishing of one species by market demand may mean an increase in the other. And they are not the only species whose production is affected by the influence of rivers or winds. Many others, both pelagic and benthic, are directly affected by a series of complex environmental variables, as Josep Lloret of the University of Girona has shown. Beyond these two species, data indicate that the whole of the Catalan Sea (and many other seas) is becoming impoverished. Changes in the quantity and quality of the fish that arrive at fish markets are detected when a reliable series of data is analysed, such as the one that Marta Coll mentions. 'There are direct and indirect impacts on food chains, with the proliferation of species of low commercial value and lower trophic levels, such as certain benthic scavenger or detritivorous invertebrates (species that eat detritus of mainly plant origin)', she says: 'There is a move towards less ecosystem diversity and an increase in "accelerated", or short-cycle and opportunistic species.'

All the specialists' opinions converge on one point: ecosystem management must be achieved, far beyond mere management of the individual species, taking into account each and every one of the components of the communities. Many working groups are coming to the same conclusion, such as the Japanese groups studying the exploitation of Japanese sardines (*Sardinops melanostictus*) or Pacific mackerel (*Scomber japonicus*). They have found that in many areas the stocks are far below that necessary for the species to survive fishing pressure, from an economic point of view, so 'we have to decentralize decision-making, make it flexible, adaptable to local trends', says Yukimasa Ishida of Japan's Tohoku National Fisheries Research Institute: 'We should not be afraid to debate, to discuss at any given time with reliable scientific data in hand what is best to exploit fish stocks.' Transparency and valuing scientific effort are fundamental tools, but so is accepting the fact that there must be closures, readjustments and large areas where fishing simply cannot take place. Although this may seem almost obvious, it has only begun to be taken into consideration a few years ago, and it has certainly not yet reached beyond the administration level.

What is clear at this stage is that fishing is influencing both the overall functioning of marine ecosystems and the very evolution of species. Despite the fact that in the mid-1980s some specialists, rescuing old texts from the beginning of the twentieth century, commented that fishing could act as a source of evolutionary changes in species, only a decade and a half ago did this start to be considered as the uncontroversial reality. Fishing is selective; in one way or another it has already resulted in the extraction of larger specimens, leaving others that are not desirable due to small size or because they are not commercially interesting. An example of such an evolution is the acceleration of the reproductive cycle and, above all, how it has forced smaller specimens to reproduce earlier. In evolutionary theory, any increase in mortality, even if uniform over time, favours the maturation of younger, smaller, earlier specimens. The ones that survive are the small ones, and above all those that are able to survive leave behind a reproductive legacy in their offspring. The models indicate that this 'forced evolution' may change the performance of the fishing grounds,' says Mikko Heino of the University of Bergen in Norway, 'but we don't know exactly how.' Species, or individuals within the species, that are able to escape fishing gear will also be favoured, because of their speed, flexibility, size or simply because they are in remote and inaccessible areas. In fact, the case of those individuals who are less affected by extraction is perfectly in line with the Darwinian concept of biological adaptation. What is becoming increasingly clear is that fisheries-induced developments can have implications for biodiversity and a non-reversible impact on population dynamics, demography and biomass, and thus on the economic performance of stocks.

Today, individuals survive in an increasingly unnatural environment, so in the long term their populations may be less able to adapt to the slow ecosystem changes. In other words, we are changing the functioning of these ecosystems radically by the pressure that we exert on them, and reversing the process has a high degree of uncertainty. For example, fish harvesting involves taking 8–10% of the world's primary production from the sea; that is, the matter and energy produced by single- or multi-celled plant organisms on the planet. This makes us, as a species, capable of changing the flows of matter and energy on a planetary level. Few scientists have really stopped to think about what this global impact means; that is, to what extent we are moving a resource in such a way as to change the trends in biogeochemical cycles that affect the carbon footprint itself.

I cannot forget that marine ecosystems produce what are called 'ecosystem services' of direct or indirect use to society, so their management must be more delicate, more studied and better optimized than mere extraction without the capacity to plan a medium- or long-term future. The new fishing models, which have been much developed over the last decade, need to implement multi-species approaches, to consider the effects of the environment on population fluctuations and have greater flexibility and control in adopting measures. I believe that a precautionary principle is fundamental in any discipline: if we don't know what can happen, we had better think about it before undertaking mineral extraction, constructing a dam, building a road network, and so on. I wonder why this principle is not applied in fishing. I think the answer is that we are basically not applying it to anything that we manage.

Squid Take up Positions

The cephalopods match fish in terms of their presence in our oceans: the amount of biomass of squid, cuttlefish or octopus on the planet is equivalent to that of anchovies, tuna or groupers. Undoubtedly, this group of invertebrates has had spectacular evolutionary success, with massive reproduction and highly accelerated growth. Overfishing of cephalopods' fellow sea creatures, fish, has greatly influenced the explosive rise in the productivity of these creatures, which are capable of reproducing in many cases after just a year of intense development. An adult needs to ingest an amount equivalent to 30% of its weight per day to survive, but a juvenile can reach maturation at 70% of adult body weight. In addition, cephalopods behave as both predators and prey, being an intermediate link in the food chain that now has fewer and fewer predators to capture them. Because of their size or agility, fish, cetaceans, seals, and so on, are in many cases extinct, from a functional point of view, in the ecosystem, so this increase is somewhat predictable thanks to cephalopods' enormous adaptability to new environmental conditions.

In the English Channel, the 25% drop in fish catches from 1983 to 2003 was made up for by a 300% increase in cephalopod catches (from around

8,000–23,000 tonnes). In this change is, of course, a change in the habits and marketing of the product, but it cannot be concealed that a fast-growing animal such as the squid, which can reach maturity before many species of fish, has replaced one type of catch with another. In some places, the collapse of cephalopod species such as the Japanese squid (*Todarodes pacificus*) has already occurred, but in others, such as the Mediterranean Sea, a discrete fishing pattern has been maintained, not exceeding 2% of the world's catch of 50,000–60,000 tonnes. The consumption of these animals is a cultural issue, deeply rooted in Southeast Asia (the main consumers) but less so in places such as England or Scandinavia, where people prefer fish.

Despite the opportunism and adaptive potential of various cephalopod species, there are questions about their adaptation to changes in the ecosystem due to anthropogenic pressure or climate change. 'Over the next 100 years we will see 100 generations of many of these species,' writes Gretta Pecl of the Tasmanian Aquaculture and Fisheries Institute of Australia, 'and we will not have to wait 100 years to see the effects of fishing and climate change on the transformation of populations and species themselves.' Temperature increases could lead to a reduction in the size of individuals and a lower survival rate as their metabolism accelerates, requiring more oxygen and more food, according to several experts. Others indicate that, because there are fewer competitors for food (fish), their cycles would not be much altered, although their distribution would be. 'Cephalopods can be an indication of environmental changes due to their short life cycle,' says Ángel Guerra of ECOBIOMAR of the CSIC, 'seeing both rapid declines and readjustments to the new system faster than in other organisms.' As for other animals, the environment and fishing set the pace.

However, direct extraction is also a cause of change in the population of certain cephalopods. Despite their rapid growth, behaviour at the time of reproduction can mean an unbalanced capture of males and females. This is the case of the southern squid (*Sepioteuthis australis*), an animal whose populations are generally inclined in favour of the male sex (there are about seven to nine males per female). 'The fishing is more selective both in terms of size and especially in terms of the movements of the male', says Ty Hibberd of the Institute of Antarctic and Southern Ocean Studies in Tasmania, Australia. This fact not only changes the population structure, but also, as mentioned earlier in the general section, it changes the evolutionary pattern of the species. The changing environment, species, rates of renewal, the role in the ecosystem and interactions with other species are all factors that must be taken into account in order for these fast-growing and highly adaptable organisms to survive the increased extraction that they suffer, to replace in part the protein that had been derived from the many fish species that have already become exhausted.

El Niño and the Peruvian Anchovy

The South American Pacific coast is one of the most productive areas on the planet. Of the approximately 82 million tonnes of fish extracted from the sea in 2011, more than 8 million tonnes are anchovy from these seas (mainly off the coast of Peru). That said, it is easy to understand why a cyclical disturbance such as El Niño, which affects this and many other species, is such an important phenomenon here and around the globe. The change in temperature and, above all, the decline in the concentration of nutrients in the upwelling areas produce a cascade effect that radically transforms ecosystems. It is not only that many species disappear (or fall to minimum production levels), but also that, in a way, the seas become 'tropical', with many species being introduced from the north during the El Niño phenomenon (i.e. from Central America and Colombia on the Pacific side). People who live largely from either commercial and small-scale fishing are accustomed to these effects.

El Niño occurs when the coastal winds that carry the waters from the coast to the sea weaken, allowing a tongue of warm water from Indonesia to be carried as a Kelvin wave. These waves are produced by a change in water temperature that induces an imbalance (something like a gradient), causing a massive displacement of water from one point to another until it encounters an obstacle (like a giant current from the west of the Pacific colliding with the coasts of Peru and Chile). Under normal conditions, it is dry winds that cause the upwelling of deep water: they displace surface water from the coast to the sea and, in the sea, when you displace something it has to be replaced—in this case by deeper water. But the propagation and arrival of this Kelvin wave, this immense mass of warm water (which can be up to 8 °C above the average temperature) does not allow these deep waters to rise, therefore their nutrients never reach the surface. The system is no longer fed, and production is altered.

Both El Niño and La Niña (the opposite effect, in which the waters become much colder) greatly affect the fishing and agricultural economy of the area, but it has been shown that, when the phenomenon is virulent, in practice its shockwave affects the entire planet. In the case of Chile, Peru and Ecuador, the impact can be as much as 11–12% of the gross domestic product, with recovery being very slow. When fishing for anchovy (*Engraulis ringens*), for example, the catch can be reduced to less than a third due to ecosystem changes that leave the fish without food. Associated with the phenomenon, the 'normal' dynamics of reproduction and recruitment fail due to lack of food. During this period, creatures in these warmer and poorer waters have to feed more on zooplankton than on phytoplankton in their primary growth stages, and a large number are unable to capture the 'diluted' prey due to lack of nutrients.

Anchovy is a typical case of 'explosive' fishing. Peru has put enormous effort into its capture, processing and export, which is not surprising as it once accounted for 25–30% of export earnings. In the 1950s the country was already preparing anchovy, mainly into fishmeal and oil. Like other large pelagic fisheries, the maximum yield was obtained in the 1960s, with more than 150 centres specializing in processing products, from no less than 18% of the world's fish catch and making 50% of the world's fishmeal, at almost 12 million tonnes at its peak in 1970. When I was collaborating in the CENSOR project on the El Niño phenomenon, these figures confirmed that the system was able to withstand clear over-exploitation thanks to the richness of nutrients and the biology of the pelagic species. In discussion with Jurgen Laudien of the Alfred Wegener Institute (Germany), we understood the implications of the change from this climatic phenomenon for the people of this area, who live from the exploitation of anchovy and other species. The whole fishing sector was at full capacity, but what did not work out in the end was the system, which collapsed suddenly. It went from more than 10 million tonnes in the early 1970s to just 1.3 million tonnes in 1973, an order of magnitude less. The FAO then warned that the 9.5 million tonne maximum limit must be respected if anchovy stocks were not to be exhausted. But, as Jurgen told me, the Peruvian fleet moved south in search of new fishing grounds. Successive blows to the industry were caused by successive occurrences of El Niño, especially in 1982/83, when thousands of people lost their jobs, dozens of factories closed, and production again fell.

However, the stocks of anchovy, a fish that forms very dense shoals, recover, and the authorities learn. The Peruvian IMARPE (Peruvian Fishing Institution) set a maximum catch level of 7 to 8 million tonnes and production stabilized in 1999. Today, the products of Peruvian anchovy provide 12% of the profits from exports to Peru (about €800 million), and many years ago the government started a programme to use this fish for human consumption, especially in poor areas such as the mountains of the interior of the country. This is because the catch of this fish, the creature that is most captured in our oceans, is converted almost entirely into flour and oil— among other things, for feed in marine and lake fish farms.

Cantabrian Anchovy

Having seen at first hand the collapses in the stocks of other clupeids such as the Peruvian anchovy itself, as mentioned above, or the case of the Alborán Sea, where anchovy stocks disappeared a few years ago due to overfishing, why was the Cantabrian anchovy fished almost to exhaustion? In June 2006 this species was considered to be commercially extinct. Only about 1,200 tonnes were fished, after about 20,000 tonnes had caught annually in the last few decades, with peaks of 80,000 tonnes in the 1960s. This is a pattern that was followed, as we have seen, by most fisheries in the twentieth century. The political decision to reopen fishing after the end of 2005 proved to be a mistake, and of the 5,000 tonnes that Spanish vessels were allowed to fish, only about 720 tonnes were caught after much effort. This is a typical case of a decision taken more to satisfy everyone than to consider the data from many specialists. Fishing technicians, scientists and even the fishermen themselves had advised against recommencing fishing, but in spite of everything they went back to sea and found that the recovery had not taken place.

Since 2001, there had been reduced regeneration until the collapse of 2005. 'More than half of the variability in recruitment is due to the outcrop of water in that area', says Ángel Borja of ATZI in the Basque Country: 'if the amount of nutrients supplied by the deeper waters is low, recruitment will be lower due to a lack of sufficient production of microscopic algae, the basis of the entire food chain.' Environmental changes are once again fundamental to fisheries management. In this case, the dilution of the food and surely its quality have influenced the population at its most vulnerable stage: larvae and juveniles. Small fish are very vulnerable to environmental changes. The winds from the coast in spring favour this upwelling in the spawning season. In the case of anchovy, as we have seen, the temperature, the direction of the currents and the turbulence (movement of the water and the particles within it) are decisive for the development of new generations. In spite of being creatures with short cycles and therefore adaptable to changes in climate or new environmental conditions, anchovies have a limit to exploitation, like any other organism. And we must reiterate that many species depend on this fish, from above and below in the food chain.

Once again, we have an environmental picture: interactions in the ecosystem and the effect of fishing. Until 1978 there was no specific plan for the extraction of this anchovy, the Spanish fleet being the first to exploit it. In 1945–1950, some ten to twenty thousand tonnes were caught, but drastic improvements in the fleet resulted in a fourfold increase in the catch. The

French joined the fishing effort to a lesser extent, although the financial yield was much higher as it was sold almost as an luxury item. While in 2004 anchovies were sold at around €3 to €4 per kilo in Spain, the French sold them for €25 per kilo: they did not need so many to make a similar profit. Spain still has a 90% catch quota, while France contents itself with 10%. What happens is that while 100 tonnes in our market means about €350,000, the French are reported to make about €2.5 million. In any case, the benefits reported for this organism are very high, apart from the fact that it is an iconic fish in the culture of northern Spain.

In 2009, it was found that the biomass of Cantabrian anchovy had reached 33,000 tonnes, which would allow the limit on fishing to be waived. But, while the Spanish government urgently needed to recommence fishing in January of the following year, the French authorities continued to ask for one more breeding year; that is, to commence in July 2010. Like any industry, fishing is not free from political haste, social pressure and media impact. But now that we have started to learn, now that we have begun to understand how to improve and get the best out of the system without collapsing it, isn't it worth starting to apply our new fishing models so that people can continue to live off this resource for many more decades? It would be good to think that certain resources are actually scarce yet could be highly profitable if they were recognized as what they really are: items that are not basic to survival but are a luxury, or a whim. Let us not forget that we continue to need to hunt in the sea.

7

The Sea's Harvest

It's an early start for a couple of amateur fishermen who decide to try their luck with tuna off the coast of England. It is the beginning of July 1928, and both of them are euphoric because they know that they may be lucky: tuna have been seen in that area since late May and now seem to be abundant. Who doesn't like to eat a good plate of grilled tuna? It is also Atlantic bluefin tuna (*Thunnus thynnus*), one of the finest on the market. When several boats are engaged in fishing for them in the open sea, the news spreads like wildfire and many professionals (and also recreational boats) concentrate on fishing for tuna and opening up a market that, until then, had not been presented to them. In Norway, northern France and Denmark, they are also engaged in this new type of fishing. They are not experts, but they learn quickly and use increasingly sophisticated methods to catch a fish that is undoubtedly elusive because of its agility and power.

Between 1923 and 1931, bluefin tuna appeared in the waters of the North Sea in large numbers. The Norwegians made the most of it, fishing more than a thousand tonnes a year between 1930 and 1932 and catching, once they had understood how, more than ten thousand tonnes in 1950. Sport fishing also became widespread, and a fish that was once hooked only by chance (i.e. as by-catch or discard) in herring fishing had now become a first-class catch. Everything was going smoothly until 1960, when it disappeared. The data indicate that this tuna was making a so-called trophic migration (i.e. in search of food) from the central Atlantic to its northernmost parts to feed in the waters of the North Sea. They chased mackerel and herring, devouring them as they passed like a well-organized pack, ready to give these shoals of smaller fish a good run. In terms of herring and other small pelagic fish alone, it is estimated

© Springer Nature Switzerland AG 2019
S. Rossi, *Oceans in Decline*,
https://doi.org/10.1007/978-3-030-02514-4_7

that around 1950 the tuna consumed about 150,000–200,000 tonnes, assuming an annual consumption of 30% of stocks. As is normal for a fish whose place is at the top of the food chain, tuna are only bothered by predators such as killer whales, sharks and—for the smallest specimens—dolphins, if they can reach the 70–75 km per hour needed to catch them. But the systematic fishing that was just beginning to be established in these waters eliminated tuna in a few decades. Today, it is very rare to see a *Thunnus thynnus* off the Norwegian coast, and it has nothing to do with temperature, which is actually rising in this part of the world.

Tuna, animals with a highly accelerated metabolic system, need temperatures above 17 °C and below 30 °C to be comfortable, although they can withstand low temperatures just above 3 °C. Despite the fact that they usually swim at the surface, these fish can go down to a thousand metres deep, perhaps in search of prey such as shoals of pelagic fish or squid. It is a voracious fish that eats other fish, jellyfish, squid and crustaceans and, in general, everything. A tuna of large size and several decades old can reach up to 6 m and weigh a tonne, travelling thousands of miles tirelessly to reach its spawning grounds in the Gulf of Mexico and in the Mediterranean. The males fertilize the females in an ancestral frenzy that the fishermen know well, following their destiny towards the places where they have been hunted since time immemorial. Spawning lasts between one and two days, and hundreds of millions of fertilized eggs hatch from there (an adult female can lay millions of eggs). Maturity, as in many of these organisms, comes after several seasons, at about four to five years and more than a metre in length, when weighing about 30–35 kg, although certain species may take more than eight years to become fully mature. Their physiological characteristics and their need to move continuously at high speeds make their muscle tissue dense, strong and very compact, and their flesh, almost devoid of bones apart for the spine and near their fins, so valuable.

It's a fish as valuable as silver. It has always been appreciated for its flesh, but it has never before been seen as of so much value as it is today. It is sold all over the world in one form or another (fresh, frozen, canned, etc.), and it is in Japan and China that development of various ways of treating this coveted animal has become most intense in recent decades. In the case of Atlantic bluefin tuna, they can reach figures in excess of €6,000 per kilo. In 2013, an exorbitant €1,350,000 was paid for a 220-kg specimen on the Tokyo fish market. It is a very profitable business and, like all businesses that are highly profitable in the short term, blind and certainly corrupt. In 2007, bluefin tuna stock was estimated at about 78,700 tonnes, compared to the more than 300,000 tonnes caught in 1958. It has gone into rapid collapse, and it is known that, data in hand, the decline in these catches between 1957 and 2007 has been 74.2%. Since then, it has continued to decline: 'We have gone from tuna of about 190 cm in length on average in the 1970s to tuna of about 150 cm', says Dr Siskey of the Chesapeake Biological Laboratory in the United States.

All contribute to the decline, although some countries have more restrictions and controls than others. Turkey, for example, has 52% of the total number of vessels in the Mediterranean fleet (more than 170 in 2008), but its catches are minimal compared to those of Spain (887 tonnes in 2006 compared to Spain's quota of 5,378 tonnes that same year and a much smaller fleet). Something doesn't add up. It is known that legal activity is excessive, yet the illegal activity may (at a conservative estimate) be double that of the allocated quotas. 'Already in 2002 the quota was 2.5 times higher than recommended by the scientists,' explains Sergi Tudela of the World Wildlife Fund Mediterranean Department, 'but the trend remains the same: the goal for 2010 is to reach 25,000 tons, when all experts say it would be to lead the species to collapse if the catch exceeds 15,000 tons'.

If you keep digging, you realize that now bluefin tuna no longer stand a chance. First, their migratory routes are monitored from the air, and aircraft are chartered exclusively for that purpose. Large-tonnage vessels move towards the tuna shoals, which they will find after about 20–40 days of navigation (depending on the fleet and the area) then extend immense nets that, literally, make whole shoals of scombrids (and whatever accompanies them) disappear, and put them in floating cages. The cages are transported, because nowadays it is much more profitable to wait six to twelve months, artificially feeding them to reach more decent sizes at an accelerated rate. We have grown from around 14,600 tonnes in 2002 to over 26,300 tonnes in 2007. In the Mediterranean alone, half of the tuna caught are destined for these farms, which are scattered throughout the Mediterranean, including in countries where the fishing culture of this animal had been practically nonexistent. The market appreciates the growth in captivity of such scombrids, rearing specimens that are caught at 30–40% below the legal size. Spain joins in the practice, as the opportunity is too good to lose. Here, 90% of the 13,800 'cultivated' tonnes in 2007 were produced in the Autonomous Region of Murcia, accounting for over 2% of the region's GDP. Too much money, too much interest to stop the frantic race to the void, to the collapse.

The farms where the tuna are concentrated have added problems. The first is that large quantities of fish (sardines, herring, anchovies) are needed to feed them. In a year, 5,300 tonnes of tuna need about 40,000 tonnes of sardine to reach 9,500 tonnes in weight. To produce a kilo of tuna, about 20–40 kg of fish are needed, depending on the species, with a survival rate of 30% in the case of very young catches or considerably greater if more mature specimens are caught. The fish that they consume (as we will see below) could be traded directly for human consumption, yet are used for tuna rearing. Much of this fish comes from assorted areas, and its bacteriological and virological control is far from exhaustive, so that diseases can be unwittingly introduced into the environment. Moreover, these huge rearing farms (usually located far from the coast where the bottom is usually between 40 and 50 m deep) create a permanent rain of organic matter, dissolved nutrients and heavy metals. On an optimized farm (many are not), only 3% of the fish supplied is 'lost'. That is to say, with the figures given above, there are some 1,500 wasted tonnes that are partly used by other scavenging organisms, but which can become a serious problem for the benthic communities that they smother (see the chapter on mariculture). The investment in catch and rearing cages has been too high to abandon it suddenly, but this is likely to happen in many places simply because of a lack of resources, due to the economic and ecological extinction of the species. This is why this animal was proposed for inclusion in no less than Annex I of CITES, a supranational body that, if confirmed (and FAO supported the data presented by the experts), would ban its commercialization across countries. But political pressures, poor formulation and the lack of external 'data' (which exist but do not fit in with CITES ANNEX I) have meant that this decision was never taken, and the species is once again at the mercy of fishing interests. 'The problem is that if there is no rigorous way to approach the issue and everyone does what they can, the credibility of the (CITES) label can be more damaging than beneficial to the sector, creating conflicts and legal nooks and crannies that are exploited to the minimum by those who want to continue with uncontrolled exploitation,' says Dr Alice Miller of the Environmental Policy Group at Wageningen University in the Netherlands.

Tuna has long been exploited in such a way that other species may also be harmed. One example is use of longlines, the miles and miles of lines on which thousands of hooks are baited to catch these and other large pelagic fish. It has been proven that, on average, the number of fish has gone from 6–12 per 100 hooks in just twenty years to only 0.5–2. Birds, turtles, sharks or dolphins are caught by both them and the endless drift nets, representing 'discards'. Much of the data we have on stocks unfortunately comes from these giant longlines, but these data greatly underestimate the catches and fish in the area, as many unbaited hooks are taken by the fish while they are in the water or just before they are taken on board, not just baited ones, yet the catch is calculated on the basis of just the number of baited hooks.

Much of the information that we have about the species comes from this type of fishing that can tell us the locations, frequency, sizes and weight of various organisms, so the reliability of the data from these and other sources of fishing is essential. Over the last few decades, these and other industrial fishing practices have reduced (according to the catch figures provided to FAO) the pre-industrial populations of the world's major predators, including tuna, to less than 10%. As mentioned earlier, fast-growing species are increasing very sharply, although their new predator (humans) may also lead them to collapse in many cases. 'The lack of predators may be affecting different ecosystems very much', said Ransom Myers of the Department of Biology at Dalhousie University in Canada. 'You have to think that about three-quarters of the fishing is close to the coast and on the continental shelf,' Myers added, 'and we have no empirical evidence of how this will affect the system if the "top predators" are removed.' We just don't know where we're going (Fig. 7.1).

This system is already showing signs of exhaustion. In the case of tuna (and many other fish), the fishing effort skyrockets. This effort is proportional to the catch made divided by the money invested (crew, diesel, insurance, etc.) and the days at sea. To think that, on average, since 1990, the consumption of diesel oil per tonne of fish caught, in general, is about 640 litres. It takes them longer every day to find a catch that is sparse, spending an enormous amount of time and fuel on a task carried out by a fleet of powerful vessels.

Fig. 7.1 Industrial capture and feeding of red tuna in the Mediterranean sea (and in other areas of the Atlantic and Pacific Oceans). Once encircled the whole school is captured; part is immediately frozen and the other is maintained and fed for future exploitation. *Credits* ADOBE STOCK

The systems are becoming more sophisticated, but the catch is decreasing and decreasing. A fishing vessel spends around 22,000 litres of diesel oil on the whole bluefin tuna catching operation (from the time it leaves until it returns to port), a fuel that is constantly increasing in price and of which, even with subsidies, a huge quantity is used just running the fleet. With an emerging energy problem, this point is becoming increasingly worrying. In Spain, for example, some 2 million litres of diesel were used in this type of fishing in 2007, and though Turkey has one of the highest legal quotas it gained an advantage by using far more fuel—an order of magnitude more: no less than 20 million litres of diesel. It is not necessary to quote statistics to understand that either the business is profitable (in part due to the illegal catch, especially in countries such as Turkey) or that the quotas are not enough to cover the millions of euros that it takes to operate the business. Meanwhile, catches of traditional species have fallen by up to 80% across the Mediterranean.

'Many species will become economically or ecologically extinct in the coming decades,' predicted Ransom Myers. 'We need to reduce catches by 40–80% now if there is to be hope.' I, too, am of the opinion that industrial fishing has changed the functioning of ecosystems in a very far-reaching way. It is not just a question of a problem for the local or global economy, nor of the supply of protein from the sea. It's something else. As I have been insisting, taking predators out of context is fundamentally changing the way that the systems work, and will certainly have an impact on the so-called biogeochemical cycles, on the carbon and nutrient balances of the planet. Tuna fish is one of those clear examples in which a key element seems about to be replaced by other organisms with a shorter life span, therefore a prompt return of their biomass to the system. This means less carbon dioxide retention, and a system less able to diversify in the short and medium term, becoming poorer and adapted to increasingly unstable situations. It will not be possible to come up with a concept of 'sustainable' fishing until the resilience of the stocks and their spawning parameters are well known, factors which, in the case of tuna, are still largely unknown. It is therefore to be hoped that someone will switch on the light bulb and begin to think about a more selective process for fishing, harming their species and others who accompany them less, in this forthcoming collapse. Almost as brutal as pelagic fishing is trawling, one of the biggest ecological aberrations our tables enjoy today. But that's a theme for the next chapter.

Tuna and Traps

The trap is the traditional way of fishing for bluefin tuna and other scombrids. Since the time of the Phoenicians, but in a much more systematic way since the time of the Roman Empire, the areas where the 'labyrinth of death' was located have not changed in many centuries. It is a cunning and at the same time primitive hunt in which deep nets are set. It takes a long time, at the end of spring, before the coveted fish swim in at high speed towards the end, a huge funnel where they are trapped. The skipper, the person in charge of the coordination of the whole operation, stays silent in the middle of the 'flake' or end of the route. He can't give the signal until most of the bewildered tuna are in the trap that they have swum into. Then he has to tell the rest of the fishermen to close the exit both at the entrance and below, and start the slaughter, in which dozens of fishermen work hard to bring the fish up to the surface of the boats with pike poles and hooks. The sea is stained red, and hundreds of tuna are caught using this ancient fishing technique. Or they were.

These ancient arts have not been able to compete with industrial methods for decades, and are therefore in danger of extinction. In spite of being controversial due to its ostentation and violence, the trap method is still a way of fishing for tuna selectively, and in which human effort and expertise, through trial and error, through patience and coordination and teamwork, manage to catch a large number of tuna without the total eradication of a whole shoal, as in current fishing methods. The collapse of tuna stocks has aggravated the reduced catches in tuna traps and also smaller traps. To give an example, in a set of traps in the Strait of Gibraltar in 2000 a thousand tuna were caught; in 2005, the same effort caught only 22 fish, and in 2006 only one entered the trap.

Historical studies of the traps have allowed us to see interesting fluctuations in the populations of this precious scombrid. For example, the Unai Ganzedo working group of the University of the Basque Country explains that between 1525 and 1756 the traps' catch fell in southern Spain. There were several specific reasons, such as outbreaks of piracy, political changes or epidemics among the population who fished this coast, but a long time series like this one allows us to discriminate between factors that affected the fishing, in this case, of Atlantic bluefin tuna. The climate seems to be behind this decline, with the fall in catches coinciding with the Little Ice Age between 1640 and 1715. During this time, rivers and lakes throughout Europe (including the Ebro and Thames) froze in winter, and northern seas such as the Baltic could be skated on near the coast. As mentioned above, tuna depend on food and temperature for optimal recruitment, and in their case this increases when the temperature is higher. Another interesting fact is the fact that in the same system of traps there was a peak of 99,000 tuna in the mid-1500s, falling to just over 6,000 in the mid-twentieth century. Once again, therefore, there has been a fall in the population to no more than 10% of what they were in the pre-industrial era of 1950. Just as at that time, the fall in catches may be partly due to the climate, yet the current drop is undoubtedly due to a dire resource management policy: we cannot blame 'climate change'.

The Forgotten Sharks

Sometimes, although unintended, human acts may have an impact on a species. In a grisly way sharks have been renowned throughout human history, yet in 1975 a film director managed to increase people's terror and aversion to these animals further. They cannot exactly be blamed for being at the top of the food chain in various oceans. Other films followed, including sequels and *Orca, the Killer Whale*, in which a male specimen turned into a cunning killing machine off the coast of Alaska. Like many others I enjoyed these films, but in large part this type of 'marketing' has made it more difficult to protect sharks, an animal that has been badly mistreated by fishing. It is deeply unknown as it is not a specific target species for fishing, thus its biology, customs, migrations, population size, and so on have not been of commercial interest. What we do know is that their numbers have declined alarmingly, especially in the last three decades. Some illustrative examples: in areas such as the north-east Atlantic (including the Caribbean), hammerhead populations have declined by 89% since 1986 (only 20 years), and white shark and tiger shark populations by 79 and 65% respectively. But for some species it has been even worse. Catches of the Mediterranean and Atlantic porbeagle (*Lamna nasus*) have declined to just 1% of what they were, from around 4,000 tonnes per year in 1930–1940 in some catch areas to only 10–30 tonnes in 2000–2007. The shortfin mako shark has the ill fortune to have flesh that is highly appreciated by the consumer—with most sharks, only the fin is used. Sharks such as blue sharks account for up to 17–19% of the fin market in Hong Kong, with more than ten million individuals fished annually (about 360,000 tonnes) using the 'finning' method, which consists simply of cutting the fin and throwing the carcass back into the sea. In a market in Hong Kong for typical local products, I saw with my own eyes sacks full of dry fins, ready to be exported to places in China or the Far East. This practice, banned in more than seventy countries, has a key advantage for the trade, among its other aberrations: no one can tell what species it was, how much it weighed, its sex or whether it was reproducing. One of the most effective ways to implement the ban is to force fishermen to bring all sharks to ports, where they can be identified, measured and weighed. But slicing the fin and discarding the rest is a more cost-effective way of doing business, as more can be stored and less time is wasted. Shark fin is worth about €60–€100 per kilo (fresh weight, then dried), in a growing market (especially in China) that demands more and more for use in traditional dishes.

Countries such as Indonesia now account for 60–80% of catches, with others such as Brazil, Venezuela and the Azores lagging far behind in catching and marketing them. But the worst part is that around half of the catches of these chondrichthyans are accidental; that is, they are part of the famous by-catch (which we will talk about more in the next chapter), thus does not even appear in the catch lists or in the cargoes that arrive on land. John Stevens of Commonwealth Scientific and Industrial Research Organisation (CSIRO) in Tasmania, Australia, estimates that in 2000 some 760 thousand tonnes of sharks were caught, but in fact more than 1.5 million tonnes were taken from the sea if including discards or by-catch in various types of fisheries where sharks are not a priority target (especially tuna). 'The number of sharks caught in longlines of kilometres in length and other large pelagic fish is terrifying,' says Stevens, 'and in the overwhelming majority of cases we don't even have data on the species or group to which they belong.' One of the biggest problems, of course, is various specialists' perceptions of the problem, which can be quite different. 'This doesn't help,' grieves David Shiffman of the Leonard and Jayne Abess Center for Ecosystem Science and Policy at the University of Miami: 'specialists are very aware of the problem, but many are reluctant to apply more restrictive measures, depending on where you're interviewing; there's certainly room for improvement and a lot of communication channels.'

There are many areas from which sharks and rays have completely disappeared, such as the Irish Sea or the Mediterranean, where it was once common to find a number of these benthic species that are now simply not there. In the last ten to fifteen years, more and more species that live at great depth have been appearing in fishmarkets, and it is assumed (because we know even less about these than surface species) that they are slower growing and to have more dispersed populations. But not all of them are doing so badly; some species, especially those with a faster breeding cycle and scavenging habits, have seen their numbers increase in some areas by up to 30% in recent decades. There are more remains and more offal to be consumed, which helps them to reproduce. It has been observed that some species of rays and benthic sharks identify trawlers with food, as do gulls, and follow them until the crew release the discards, even taking advantage of the presence of injured catch by raiding the nets at the bottom.

We return to the same line of questioning used in earlier chapters with other species. What is the importance of sharks? What role do they play at sea? Taking this one element out of the system affects the entire food chain, once again making it dominated by small, fast-living animals, mostly invertebrates. We now know that sharks are sensitive to over-exploitation, perhaps more so than many animals in the ocean. Many species are now protected in one way or another, yet if it is considered to be essential to know more organisms before exploiting them, sharks are our prime candidates because, as we mentioned before, little is known about them. Once their number diminishes to that critical to their livelihood, it will be decades or hundreds of years before they regain dominance over their former domains. There will always be sharks left but, after them having survived more than 400 million years of evolution, might we bring them to the brink of collapse?

Protecting the Open Ocean

If it is difficult to protect the coast, it is even harder to protect the open sea, especially areas where there is no national jurisdiction (beyond the famous 200 miles). The sea, as has been stressed throughout the book, is considered by many to be for 'the good of all'; that is to say, common property, belonging to no one, and experience confirms the drama. No one respects it, and everyone exploits it without measure. However, all the specialists agree that it is now urgent to create protected pelagic areas, sanctuary areas where fishing is regulated or prohibited, areas where no fish can be taken either at certain times of the year or during the entire annual cycle.

The issue of pelagic reserves, of protected areas in the open sea, is highly complex. There is no doubt that, if there is anything left to protect on this planet, it is the pelagic zones. These are what we can term the 'lost dimension' in the equation of the conservation of our oceans. No less than 99% of the volume of the biosphere is made up of this pelagic space, of which the open sea (outside the continental shelf) comprises the majority—and only 0.1% is protected (including coastal or shelf areas).

The role of this 'no man's land' is not only fishing profit. It is also fundamental as a climate regulator and an essential part of the planet's primary production. Creating a protected pelagic area forces regulation or control of the land pollution that affects it, maritime traffic (and the waste it produces) and the species that inhabit it. Unfortunately, most of the pelagic areas to be protected are in international waters (64% of the total), with highly mobile species that may be protected in one place yet over-exploited in another. It is difficult to achieve consensus when it is not understood that erecting barriers to the use of the sea is complicated, if not absurd in many cases, if the different actors (politicians, managers, technicians and scientists) cannot agree on how to legislate, regulate or monitor it. 'We need much more data on the biology and ecology of species,' writes Anthony Richardson of the Ecology Centre and Centre for Applied Environmental Decision Analysis at the University of Queensland in Australia, 'but, despite the mobility of many of the protected species, the migratory routes are gradually becoming known, as are the breeding and feeding grounds that should be protected.' Perhaps it is difficult for us to capture the concept of pelagic areas, as these should be mobile spaces, flexible to the needs and changes in the migratory habits of the species, which demands rigorous and continuous monitoring of the various zones. For example, it has been shown that closing certain well-studied areas can prevent up to 33–40% of the by-catch of sharks and other elasmobranchs in areas of the central eastern Pacific, with an impact of only 12% of the tuna catch. Other species benefit from this measure, such as turtles, birds and other fish, and is the most important thing when it comes to deciding not only what we are going to 'lose' on a given fishery, but whether the closure will lead to far greater pressure on other fishing areas.

Under no circumstances can a protected area be created for its own sake. Choosing an area is complicated, as economic, social and ecological interests must converge, and it must be effectively monitored. Using remote technology, this step could be conducted relatively easily by satellite. The problem is who sets the sanctions and which legislative body operates in the area. International waters are just that: waters for all, as is the case of Antarctica, for example. So they are either left in the hands of a competent international body, with the possibility of sanctions, or we seek (highly complicated) compromise formulas. Ideally, an international conservation law would be generated, although it might be more feasible to form multilateral pacts between the countries bordering the international waters to be protected. I do not believe that such protected areas in the open ocean are a panacea, but I am convinced that they require our attention with a certain urgency so that we do not find ourselves with nothing to fish for and nothing to protect in the near future.

8

Ploughing the Sea: The Destruction of the Marine Forest

A beautiful, untouched, primary oak woodland stretches between fields of grass near a remote village. In the trees, squirrels, birds and dormice search for food while other animals such as wild boars, deer or foxes prowl around the trunks. Not far away, in the meadows, bustards and partridges peck at the soil rich in life, closely watched by eagles. Suddenly, an iron plate falls from more than a hundred metres up in the sky, followed by another about half a kilometre or so away. The plates are connected by a steel cable that supports a huge net. The plates move, and the animals, at first bewildered and then terrified, flee as best they can from the hungry mouth of their prey. The net takes everything it finds: deer, wild boar, bustards… that's what it came for. But, incidentally, it also takes the whole woodland with it. All the oaks, bushes and other trees over a certain size are either taken into the net, uprooted or knocked down by the fury and force of the machinery that comes from the sky. When the plates and cables disappear after many kilometres of dragging, nothing is intact: the century-old oaks are damaged or have gone, and only a few animals have survived the insatiable net. The whole woodland has been destroyed just to catch a few deer, wild boar and bustards. The other animals taken in are the by-catch—the discards that will be returned, dead or dying, to the floor of the destroyed wood.

This highly dramatized and exaggerated image was given to me several years ago by my great friend Toni Polo, journalist, writer and companion on literary adventures. I can certainly reaffirm it, and I am not alone. More and more specialists are joining the ranks to denounce this absurd type of fishing. The problem is that something unthinkable on land goes completely unnoticed under the sea: we just do not see it. In 1904, an extensive scientific article by Walter Garstang disagreed with the claims of eminent colleagues

© Springer Nature Switzerland AG 2019
S. Rossi, *Oceans in Decline*,
https://doi.org/10.1007/978-3-030-02514-4_8

that there was no harm in trawling the inexhaustible depths around Britain. The author acknowledges that there have been occasional fluctuations in our observations due to the weather, the very meteorology that affects the reproduction and survival of species, but warns that at that time and in that area the decline was already evident, and that the seas were becoming impoverished by extremely aggressive fishing.

More recently, authors such as Les Watling of the Darling Marine Center at the University of Maine in the United States have compared trawling to logging. A forest being dragged away by a net is an outrage on land, but this is what happens underwater when a trawler drags the seabed's complex system: the marine forest. At the very bottom of the sea, apart from fleshy macroalgae and aquatic plants, there is a living forest of life, rich in species. Here, the dominant forms are calcareous algae, down to a hundred metres deep, needing less light, and especially all animals such as sponges, gorgonians, corals and bryozoans, all of them living, three-dimensional structures or biostructures. The forest trees that we mentioned earlier are perhaps a good analogy: large, vertical structures that accommodate plant and animal life, providing shelter and food for the large numbers of organisms around them.

In the case of coral reefs, gorgonian forests or extensive sponge fields, the biological principle is the same but it is marine animals that are dominant. They are capable of creating three-dimensional structures that allow them to provide the space, nutrients and food that their tenants (crabs, fish, molluscs, worms, etc.) will use as shelter, mating places and a food source. The more branched and diverse the structures, the more complex they are, providing more options. This description of the marine forest that I have developed throughout my works gives us a wide vision of the whole, slotting highly different systems into the same concept to help us to understand that there's more than just rocks and mud down there.

Comparing animal and plant forests will, in my opinion, help us to understand better the strategy of each component and why that tactic has been so successful in nature. While trees use their leaves to capture light and their roots to absorb nutrients and water, animals feed directly on suspended particles (hence the word 'suspensive': eats suspended food). Both create forests to cover as much space as possible, but the big difference is that plants can live on light, rainwater and nutrients dissolved in the soil alone, while animals have no real roots and seek out places that are as exposed as much as possible to currents to intercept the food they need to live. In water, these particles are abundant and available due to their molecular properties and viscosity. Depending on their physiology, in areas or the environmental conditions in which they are most comfortable, we find massive, soft, branched animals (Fig. 8.1).

Fig. 8.1 Top image: marine animal forests before and after trawling the seafloor. Once an area is destroyed, organisms may not recolonize it for centuries, drastically diminishing the biodiversity and complexity of the system, changing it from a 'forest' to mere 'grassland'. Bottom image: what would happen if the land were trawled: the imagined scenario in the African savanna. *Source* Rossi et al. (2017). Images courtesy of the Bloom Association (https://www.bloomassociation.org)

One reason why these marine forests are of key importance to understanding the whole system is the diversity that they can support. Not only does the large number of shelters and cavities increase the range of species that are accommodated, but also the exchange of matter between the creatures that make up the structure and the water around them. By eating suspended particles, corals, gorgonians, sponges or bryozoans recycle the material. Some of it will be used by an endless number of bacteria and microscopic algae that will, in turn, be food for worms, small crustaceans or snails. Thus, the range of animals housed expands as the food supply increases, as does the supply of places to live in. But there is more: because they are ramified, complex and full of obstructions, animal forests retain suspended particles for a long time, creating whirlpools, turbulence and water currents into them. This means more food for the corals themselves and also for the fish, crabs and other creatures that live with them.

It is not only the complexity of the shapes that influences the diversity of fauna and flora, but also climate stability: the fewer the seasonal 'shocks', the more life can be diversified. That's why both rainforests and coral reefs have been so successful in harbouring a wide variety of species in their interiors. Not having suffered intense glacial periods, abrupt climatic changes or mortality episodes as marked as in land's temperate or subtropical zones for a long period of time, marine species have been finding increasingly specialized and diversified ecological niches. They have had time to adapt and 'refine' their way of life to an unimaginable extent. The diversity created is undoubtedly comparable to that of the rainforest. Most surprisingly, a similar diversity can also be found in the deep sea of Antarctica or in deep, cold, coral communities, where animal forests are as lush than tropical forests, or more so, and extremely fragile. Animal (and plant) forests are fundamental to the functioning of the system, but only now, in recent decades, do we seem to be realizing this.

I keep stressing that fishing is one of the main causes of disturbance to the balance of species and ecosystem energy flows, if not the chief culprit. Of all the methods of resource extraction, trawling has definitely been the most damaging in the past century, and continues to inflict damage. Why? Trawling scrapes the bottom, clearing wide areas not only of mobile fauna (fish, cephalopods, etc.) but sessile fauna and flora (gorgonians, corals, marine plants, etc.) that provide complexity in the marine system. The impact of fishing gear on the seafloor depends on its mass, the degree of contact and the speed at which it moves. In some places, such as California, the average trawl is 1.5 times per year, and in some areas up to three times per year. Near Hong Kong there may be trawling up to three times a day, and elsewhere are areas that have seven trawls per year (there are plots where the same area can be trawled 400 times a year).

I will make a conservative calculation. If we have 12,000 active trawlers (well below the true figure) with a net mouth 25 m wide, operating at about 5 km per hour for six hours on 175 days a year, the swept area will be about 1,575 million square kilometres. This is just under three times the surface area of Spain, and 5.6% of the area of the world's continental shelves. But there are many more boats that undertake dragging in many more areas, so this estimate falls far short. Some specialists estimate that about 15 million square kilometres are trawled each year, more than half of the area of the continental shelf. As an intermediate approximation (20%), we could say that the bottom of the entire global continental shelf is being disturbed once every five years. As we have said before, there are some almost undisturbed areas and places where dragging is carried out extremely frequently. The area affected by trawling is estimated to be about 150 times that of the forests cut down on land, which is 0.1% per year. Unfortunately, the truth is that no one knows for certain how much of the planet we have been disturbing by dragging. 'While in shallow areas the impact is well studied, in deep areas of our oceans it is practically unknown', says Hilmar Hinz of the Spanish Institute of Oceanography of the Balearic Islands.

What we do know is the effect. Now we can begin to understand, we appreciate that frequent 'sweeping' prevents any recovery of marine biota. One of the most pernicious effects is the homogenization of the substrate, transforming it into a uniform mass in which opportunistic species with rapid growth and a short life cycle can increase with ease, taking advantage of the frequent disturbance of the system. In marine ecosystems, small-scale natural disturbances influence communities, generating distribution discontinuities, or patchiness. At one point, it was suggested that, for this reason, disruption by fisheries could increase diversity. But that is untrue, because it does not take into account the global distribution of species, their demography or, above all, their resilience. Resilience can be considered as the capacity for regeneration and resistance to disturbances (natural or otherwise) that a species or ecosystem suffers. For example, it takes dozens of years for red coral to reach maximum size, while the entire life cycle of a small marine worm living near the mouth of a collector can be months. There will be a huge difference between the resilience of one species and another in the ecosystem. It has been argued that overfishing has virtually eliminated new teleost predators (fish), resulting in a resurgence of the Mesozoic Era, which was dominated by echinoderms and crustaceans.

The time dimension is the interval between a disturbance and recovery. Frequent disturbances make it impossible for any long-lived organisms or ecosystems (often the ones that shape the most complex systems) to recover. In this context, we must not forget the spatial variable, the relationship between the disturbed area and the actual size of the habitat. That is to say, if the disturbance does not transform the entire habitat it can be considered recoverable, but when the entire habitat is disturbed a reversal becomes more difficult and, in some cases, practically impossible, on a human timescale. However, as many of these species are seriously affected by repeated disturbances, such as trawling, they are unable to raise their heads, and any possibility of regeneration and reintegration into the system is lost. Both the size of organisms and the size of the space that they occupy are important. 'In deep waters, where corals abound but represent a fragile environment, resilience is very small', comments Murray Roberts of the Centre for Marine Biodiversity and Biotechnology at Heriot Watt University in Edinburgh, Scotland.

Another pernicious effect of trawling is that on sediments, the substrate in which the organisms live, where the particles become compacted. The 'fresh' surface is removed, as it is stirred up and goes into suspension again, losing much of its nutritional value. In addition, near the bottom there is a so-called nepheloid layer. Particles here are not deposited but 'float' in the water, and are a food for countless organisms that form the basis of the benthic trophic chain or are directly targeted by fishing (such as many species of shrimp). Persistent removal of sediment disrupts this layer, and thus affects the availability of food for many animals, affecting global cycles of matter and energy recycling at sea. 'We're talking about a movement of about 9 kg of sediment per square kilometre along the Gulf of Maine coasts,' says Cynthia Pilskaln of the School of Marine Sciences at the University of Maine in the United States, 'but the most disturbing thing is that about 6.14 billion cubic metres of interstitial water are moving and stopping working in this area.' Interstitial water is that which is found between the grains of sediment or sand, that which is in the benthic substrate and which is a fundamental part of biogeochemical cycles; that is to say, the recycling of nutrients and carbon.

We do not know what these changes mean. 'Both in the platform and in the submarine canyons, trawl fishing is bringing about changes even in the flow of sediment and nutrients', adds Pere Puig of the Institute of Marine Sciences (ICM-CSIC) in Barcelona. We have no idea what they mean for the fauna of the area and for the balance of the ecosystem, but we know even less about the real impact of seabed transformation on a global scale: carbon sequestration and recycling, influence on bacterial communities, nutrient cycling such as silica, nitrogen or phosphorus.

I honestly believe that the problem goes back a long way—too long, according to many specialists such as Michel Kaiser of the University of Wales in Bangor, Wales, who insists that even short-term disturbance studies indicate profound changes in ecosystem functioning. Trawling has been going on for decades (in some places for more than a century), yet a rigorous test of the impact of these fisheries has only been carried out over the past thirty years. 'The most disturbed areas should be identified, the worst hit areas,' says Kaiser, 'and a rigorous map should be drawn up that includes community types, fishing benefits and the degree of disturbance by trawling.' There are scientists who go further, such as Josep-María Gili, also from the CSIC in Barcelona. He believes that what needs to be preserved are the biological corridors on the seabed: 'It is not a question of preserving just for the sake of it; we must study the areas and see the connection between the shallow coastal areas and those on the edge of the platform, whether it be the slope or the edges of the canyons.' Between these two places, on the platform itself, there are lush islands where crinoids, gorgonians, corals and sponges live and could partially feed the shallower areas. These biological corridors between the deepest and shallowest areas are exposed to a continuous hammering by trawlers, and in a way act as major roads do on land, preventing the flow of wildlife such as lobsters, fish and cephalopods. In the Mediterranean, this degradation of systems has almost wiped benthic species from the map in the past fifty years, favouring communities in which food webs are in an advanced state of degradation through the wholesale simplification of their diversity. The interconnections between the links are becoming weaker and shorter, altering the flow of matter and energy in our seas, where the most precious fish, in the upper part of the food chain, are no longer functional (because they have practically disappeared).

As fishing increases, the simplification of systems increases in a linear fashion, due to the destruction of the animal and plant forests. Fishermen have begun to perceive this but find it difficult to distinguish the causes, probably because of a lack of understanding between specialists and users of the system. Politicians are not taking a firm stand on the issue, and the explanation here is simple: any drastic change in a fisheries policy costs votes. The problem is that on a human timescale we do not know where the cap is, what is the point of no return, and the awareness of the general public is very restricted simply because we do not see this destruction as we can that of the Amazon Basin or the tropical forests of Borneo.

Deep Corals

As described, one of the least known elements harmed by trawling is the marine forest of cold or deep corals (those that extend from about 100 to 200 m deep, depending on area). At the edges of underwater canyons, at the edge of continental slopes and in scattered mountains thousands of metres down, communities of these organisms are much more abundant than we had imagined. They are special communities, built mostly on hard substrate (apart from places like Antarctica), dominated entirely by animal bio-builders such as corals, gorgonians, sponges and bryozoans. These make up an authentic reef as rich or richer than many organogenic formations close to the surface. 'Only some species of scleractinian and gorgonian corals are capable of forming habitat, in this sense there is not as much diversification as in shallower areas,' says Lea-Anne Henry of Heriot Watt University, 'but the amount of life hosted in the form of polychaetes, molluscs, crustaceans or fish is unimaginable.' These communities live at a depth of more than 3,000 m and are home to a large number of fauna in and around them, thanks to the complex morphology that we have already described for other examples.

Although some of these coral species were described by Carl Linnaeus as early as the middle of the eighteenth century, our ignorance of them is as deep as their habitat. We know that they can extend to kilometres in length and form concretions of tens of metres in height, in which those at the top of the reef are the most active. They live in low or very low temperature conditions, between 4 and 13 °C, and their growth is generally quite slow. The estimates are that some species may be centuries old, although others may grow in height almost as fast as some tropical coral species. Like any marine animal forest, they are attractors of fauna, offering food and shelter. *Lophelia pertusa*, one of the most studied corals that has adapted to these environments, may host up to 92% of the species associated with the deep reef. The 'nursery effect' (see below) is therefore highly developed in these deep communities, and we are destroying it before we even understand it. 'At great depth there are gorgonians in which sharks lay their eggs', says Alessandro Cau of the Department of Life and Environmental Science of the University of Cagliari: 'Without this living three-dimensional structure, without these gorgonians, sharks could not leave a next generation, in the form of embryos.'

One of the scientific teams to have sounded the alarm was Franziska Althaus of CSIRO Wealth from Oceans Flagship in Tasmania, Australia. In an area of no more than 20 km^2 around New Zealand there are about 800 underwater promontories of various sizes. Here, with varying degrees of intensity, trawlers have been working continuously to practise what, for the past three decades, they have no longer been able to do on the platform: profitable fishing. 'We checked the condition of the coral, gorgonian and sponge communities of these promontories and found that almost 100% of the organisms in some cases were crushed, swept or buried', says Althaus: 'After five years without any activity by trawlers, we did not detect the slightest indication of recovery in the closed areas.' Many of these places collapse economically after less than ten years of intensive fishing. After being razed, not enough biomass is produced to fish profitably and the benefits of fishing become nil, yet the fishermen do not notice that the place is not recovering. 'In the first fisheries, we know that there were 13 and 15 tons of by-catch per day worked,' continues Althaus, 'the vast majority of which was made up of corals, gorgonians, sponges…'. The marine forest had been cut down and its untouched habitat had disappeared. Coral cover, the area that the colonies together occupy, has been reduced by up to two orders of magnitude in many places such as this, causing diversity rates and biomass to fall three to fourfold for certain groups of bottom-dwelling organisms. In this area of New Zealand, the fishing boats trawl between 300 and 1,300 m deep, but there is nothing interesting beyond that.

Other areas across the globe have suffered similar or worse damage. In the Pacific alone, there are an estimated 30,000 underwater promontories, but scientists have surveyed only a few hundred and managed to collect 'statistically robust' data from only a few dozen. When robots are brought into see how the community is doing, it is often found that where there should be corals, sponges and other organisms forming dense animal forests there is nothing left, or they are badly damaged. Fishing goes much faster than science. The countdown is long overdue, and unfortunately we have no idea what will happen if these communities cease to be functional. It is symptomatic that we have gone from 10 to 40% of deep-sea fish catches since 1960: we need to extract protein from deeper areas, and much of that catch comes from areas with cold coral reefs. The ability of these areas to host wildlife is beyond question, but what about their role in biogeochemical cycles?

As mentioned, these bioconstructive species are capable of sequestering carbon in the form of carbonates and organic structures, and they are active agents of recycling matter and food that sinks from the surface. But we know nothing about how it would affect the planet to obliterate them, to take them out of the global context. We now know how much manmade carbon is locked up in the forests of the Amazon or Siberia, but what about that in the planet's animal forests? Before we have that figure, we shall have destroyed or completely transformed them.

But it is not only trawling that threatens these fragile systems. Submarine cables, deep oil extraction facilities, underwater mining and even changes in climate affect species biology, distribution and mortality in various ways. Little is known about this type of human disturbance, too, and there is a huge asymmetry in the data.

In the Mediterranean, due to the shallow depth of previously surveyed deep coral reefs, disturbances that are detected elsewhere on the planet may actually be seen. The location of these reefs was almost unknown until just ten years ago, but a group of scientists from the ICM-CSIC and the IEO of the Balearic Islands promoted a series of projects to find out where they were and their status. 'The corals of the Cap de Creus submarine canyon have seen an obvious anthropogenic impact', writes Covadonga Orejas of the Spanish Institute of Oceanography of the Balearic Islands in a recent article, 'in this case the result of deep-sea lobster and longline fishing with trammel nets.' But as the authors themselves acknowledge, despite the fact that one or two lines or nets are found abandoned every 10 m by robots or submersibles, the impact on the community is unknown. It is clear that dropping the gear will have partially broken structures, but nobody knows the regenerative power of these corals, nor if they are actually capable of growing back from scattered fragments. Some authors rule out this possibility, but there are no conclusive studies on this and other topics. We don't know what's down there, but we are destroying it. As an Italian colleague from the University of Pisa, Giovanni Santangelo, told me at a congress of deep-water corals from the Pacific, 'The less we know about a fish resource, the more we catch it.' It pains me to think that to bring fish to the markets we are destroying at great speed an ecosystem of which we know very little, without even thinking about the repercussions on the functioning of the oceans in the near future.

What is the 'Nursery Effect'?

There are places where small organisms prefer to live because there they find refuge and food, and because there is not much current. At the bottom of the sea, these places include those living three-dimensional structures that we have been talking about: algae, higher plants, coral reefs, deep coral reefs, sponge fields and polychaetes, and bristle worms. Although we have a long way to go before we fully understand, it has been possible to see that the complexity and diversity of the habitat suits smaller organisms. In the remote waters of British Columbia, it has been proven that the sponge forests, which are found at a depth of 30 m, are an ideal place for young individuals, and they are among the most biodiverse places in the area. 'Small fish need these highly complex sites to develop their early life stages', explains Charles Gibbs of the Pacific Marine Life Surveys of Canada. 'In these areas we found up to 106 species of 15 different filums per square metre, compared to another adjacent area where there are only 15 species of 5 filums because they are much poorer.' All types of small crustaceans, worms, planar and echinoderms or bivalves in their first stages of growth live here, providing food for small fish, which in turn serve as food for fish and larger cephalopods.

The case of estuaries is one of the most studied. In these places, the margin between fresh and salt water enriches the environment to the extent of creating microhabitats and a large biomass that is used by other organisms, such as flatfish. 'Macrobenthic productivity greatly favours other organisms that come to enjoy the nursery effect', explains Noémie Wouters of the Centre for Oceanography at the University of Lisbon in Portugal: 'In the estuaries of the main Portuguese rivers (as in other rivers), there are areas that are particularly rich in fauna and flora that favour the concentration of predators in search of food.' By comparing areas considered to be nurseries to those that are not, we see considerable differences: more than 4,000 individuals per square metre (about 100 g dry weight of worms, crustaceans, small fish, etc.), against just 900 individuals (less than 18 g dry weight per square metre). However, as Wouters notes, pollution and mechanical aggression can mean the loss of diversity and biomass: 'At the mouth of the Tagus, 15% of what we see in similar undisturbed areas of the Mondego River.' It is therefore essential to identify these areas and protect them.

Marine spermatophytes, such as *Posidonia oceanica*, also act as a nursery and refuge for many species. It has been observed that until not too long ago these plants occupied the coastal strip uninterruptedly, in some places, but the strong pressure exerted by humans has been fragmenting this chain until entire links sometimes disappear. The fragmentation that we are now seeing is the result of an attack in recent decades. This is confirmed by the large amount of buried rhizome (dead plant). As an example of what trawling has done to the survival of this type of plant, a typical trawl can pull off between 99,000 and 363,000 beams (the unit from which the leaves emerge) per hour. This type of fishing has been confirmed to be the cause of the disappearance of more than 80% of the plant from the Gulf of Gabes (Tunisia), 50% from the coast of Alicante and 12% from Corsica. These are confirmed data, but it is known that elsewhere it may have been much more serious. Trawling has undoubtedly been one of the main culprits, but the extraction of sand to regenerate tourist beaches and the many spills or deposits of excess sediment have also been sure causes of the fragmentation, weakening and subsequent extinction of these grasslands. Until not too long ago, trawling at depths of less than 50 m was common.

In areas where there were spermatophytes and there are no longer any, the decline in fish diversity and biomass can be as much as 75%. The recovery of the seagrass is slow, as many of the problems have not disappeared and new ones have been observed in the past two decades (mainly related to climate change). The recovery of many species of fish, molluscs and crustaceans vital to coastal fisheries will also depend on this recovery (which may take more than a century). Identifying these areas has been fundamental to creating protection zones and areas with total or temporary closures. It is more difficult to achieve the same thing in deeper areas far from shore. In very deep areas, such as cold coral reefs, the decline in diversity due to trawling is similar but less studied. It has been observed that brittle stars of the so-called ofiuroid group establish themselves on young coral when they are tiny, and begin to grow. As the coral develops, the brittle stars grow with them, and they benefit from its support, from which they extend their arms to filter the particles from the water. In these areas, an enormous amount of plankton has also been observed, as we said in the previous section, which is always welcomed by the small, medium and large predators that find abundant food there.

It is clear that every time we discover more and more how the system works, we realize that we are changing it without knowing the consequences. 'There is no doubt that we are in a field that needs a lot of exploration,' says Annie Mercier of the Ocean Scientific Centre at Memorial University in Canada, 'but every time we relate the complexity of living three-dimensional structures like gorgonians or sponges to the presence of other organisms that move like sea urchins, fish or crabs and their life cycles we find a positive relationship.' In the future, once the problem areas have been mapped in terms of diversity and the nursery effect, fishing maps will have to be adapted to take advantage of them in a way that is both profitable and respectful of the system.

By-Catch

I never fully understood, not even as a child, why so much fish was thrown back into the sea in commercial fishing. When I saw pictures of the crew selecting the catch, I wondered why some fish were picked and others not. I soon discovered that, unfortunately for some species (and sizes), only the chosen few arrive at the fish market. The rest is the so-called by-catch, organisms of various types that are discarded and that will never reach homes, restaurants, children's canteens, and so on.

The first estimate of by-catch made almost two decades ago was some 12 million tonnes of fish, crustaceans and other organisms, but this was a clear underestimate of the reality. Shortly thereafter, in the mid-1990s, the estimated figure increased to 29 million tonnes by applying slightly more precise sampling methodologies and forcing many fishing vessels to provide more realistic numbers. That is, up to a third of what is caught worldwide is declared discarded. That's a great deal. 'In some places, industrial fishing, both fish and crustacean fishing, can have between 60 and 80% by-catch,' says María Esmeralda Costa of the Centre of Marine Science Algarve in Portugal. Of this 80, 25% is usually used, depending on the area and the time of year, which causes a slight fluctuation in the amount of discards. Not everything is thrown away—some species are kept for sale—but, especially in crustacean fishing, the discards can amount to eight to ten times more, in terms of biomass, than of shrimp. In the Mediterranean, it is no different. 'At least one-third of the biomass caught is by-catch, approximately 500,000 tonnes per year between the Mare Nostrum (Mediterranean) and the Black Sea', confirms Pilar Sánchez of ICM-CSIC: 'In recent studies we were able to verify that only 115 species of the 309 caught reached the market.' In the case of fish, more than half the catch goes to the fish markets, but in the case of invertebrates other than molluscs and crustaceans, the species which are discarded represent 96%—in other words, practically all of it. Among them, of course, are those gorgonians, ascidians, polychaetes and sponges that we have been talking about. Fishing takes everything, with virtually no discrimination, despite efforts to mitigate the accumulation of discards.

One of the most serious problems is that fish, molluscs or crustaceans may be caught that are of a commercial type yet are not yet of sufficient size, so they too are part of the discard. 'Discards in shallow waters are more abundant due to the fact that more life is concentrated and there are more juveniles', explains Montse Demestre of ICM-CSIC. And most by-catch goes back to the sea dead or dying: 'Starfish suffer only a 10% mortality rate, but in small fish it can exceed 90%', says Jens Prena of the Bedford Institute of Oceanography in Canada. He and his collaborators were able to see that the most affected, from the outset, are the most fragile and long-lived organisms, as we have mentioned in previous sections.

Mortality is due to several factors. To begin with, it cannot be entirely pleasant to be compressed for perhaps hours (depending on the type of fishing, the area and depth) at the bottom of a net with other creatures, and also rocks and occasionally wheels, fridges and the like. Worse still is when the trawler (also in the case of pelagic fisheries) finds a large bank of jellyfish, all ready to sting on contact. Of course, they must pass from water to air, and how they withstand that depends on the organism and the time of year—in summer, being a fish out of water is less traumatic. Moreover, many have a problem associated with their swimming bladder. This little sac is full of gas that, upon reaching the surface from a depth of, say, 300 m, expands suddenly. Underwater, the gases are under the pressure of the water column, but the closer the fish is to the surface, the less pressure there is, and the bigger the gas bubbles. That's why many fish emerge with their bladders coming out of their mouths. If they haven't dried out on deck but managed to escape, they still have to face another challenge: being eaten. Fish, crustaceans and cephalopods caught by trawling the bottom are benthic, living close to the seafloor, so they need to swim quickly to find home again, sometimes hundreds of metres down, on a dangerous journey full of predators waiting impatiently for the remains of the catch. Among these predators are birds. Curiously enough, these can also become by-catch. Albatrosses and petrels are some of the worst affected, and both pelagic fishing vessels and trawlers present a real threat. Let's not forget that 18 of the 22 albatross species worldwide are threatened in one way or another, as these figures are important when it comes to conservation. Birds see ships as a potential source of food, so 45% of the bird by-catch is from trawling. The other 55% comes from long longlines and driftnets.

The trawling method of fishing consists of dragging a net along the bottom to capture everything, or almost everything. Anything that doesn't interest us is thrown back into the sea, dead or dying. The philosophy behind this type of fishing is fundamentally flawed.

9

The Forgotten Fish

When I was between 14 and 21 years old I used to go out with the fishermen of the village of Cadaqués, in the north of Catalonia, with trammel nets, longlines and pots. On the small beach of Llané Petit they taught me how to set the nets and longlines, and what the seasons were for each fish. Little by little, they showed me the fishing grounds that they frequented, told me where to set my nets so that I wouldn't be caught and passed on to me their decades of wisdom (passed down from father to son) to make fishing more profitable. They liked having a youngster—just me—who wanted to learn the trade, and they took me out to sea with them at six or seven in the morning. I was fascinated by this contact with the sea in the early morning or late afternoon, working out what we were going to find in our gear that day and, of course, eating the fresh catch. This last experience is unique and very few people, unfortunately, can enjoy it today.

From the age of 21 onwards I devoted myself to my studies and, little by little, stopped going out to sea with these fishermen. However, I tried to keep in touch with them, as many years ago their own children had had their interest in the art of small-scale fishing displaced by real estate companies or the tourist sector. I always have a certain nostalgia about that brief but intense period in which I learned about one of the most basic forms of contact with the sea. The work may be routine (like all jobs!), but there are few professions that bring one into deeper intimacy with nature than that of the fishermen who work the sea from small boats daily.

Although small-scale fishing is largely overlooked in practically all parts of the planet, in its various forms it extracts almost half of the fish caught in inland or coastal waters. Tens of millions of people live directly or indirectly

© Springer Nature Switzerland AG 2019
S. Rossi, *Oceans in Decline*,
https://doi.org/10.1007/978-3-030-02514-4_9

from it, and in many places it is an essential way of life, especially for small, undeveloped populations in the Indian, Pacific and Caribbean tropics. These small-scale fisheries share many aspects, such as using small vessels with one to four crew members, being usually coastal and taking a catch that is usually quite discreet. The fishing systems employed in various parts of the world are closely related to the distribution of the species of interest. In all places, both the fishing resource and its exploitation are related to the geomorphology of the environment, as well as meteorology. The fishermen who practise this type of extraction feel independent and autonomous, and more than one has started out in industrial fishing on board trawlers or anchovy fishing vessels then made the move to smaller boats. However, there is usually a strong family tradition passed down from generation to generation concerning fishing grounds, tips and the best times for fishing certain areas.

Fisherman's feelings sometimes go beyond mere profit. In certain places, they have a social function. Mecki Kronen, of the Secretariat of the Pacific Community in New Caledonia, saw this in Tonga, where fishermen are still in direct contact with their neighbours, selling their catch daily. Places like this are in transition between exchange, based on self-sufficiency, and a wider market serving other communities, such as small towns or resorts. 'Fishing in Tonga is seen as far beyond pure and simple profit,' says Kronen, 'not least because it is seen as a tool for subsistence, not something to get rich from.' For others, it is the contact with nature that gives them pleasure and independence. Luciana Queiroz, from the Institute of Environmental Science and Technology of the Autonomous University of Barcelona, saw that when the fishermen of the mangrove swamp were asked what fishing meant to them, in more than 70% of cases their spontaneous response was 'Everything'. 'Everything' is a very complex concept, because it includes feeding your family, spiritual peace in a wild or semi-wild environment, leisure with your loved ones, and so on. The freedom that small-scale fishermen or gatherers experience is tremendous.

Without wishing to idealize small-scale fishing—because, as we will see, it too has an impact when we go beyond certain limits, it is possibly an activity about which we can breathe more freely. In fact, when well managed, due to its inefficiency and low productivity of this type of extraction, the entire operation is beneficial to the ecosystem and fish stocks. An example is small-scale fishing in New Caledonia, where some 312 boats catch only 170 tonnes of fish per year, compared to 25 thousand tonnes caught by the four industrial vessels in the area. 'Fishing pressure in this area is low compared to other parts of the world,' says Nicolas Guillemot of the French National Research Institute for Sustainable Development (IRD) in New Caledonia, 'just 0.26 tonnes of catch per square kilometre per year compared to up to 40 tonnes per square kilometre per year in certain places where the exploitation is exclusively industrial.' However, long ago this ceased to be the case in practically the entire Western world, although the impact of technological progress on small-scale fishing came about much later than that of industrial fishing, even in countries such as Spain or Italy. For example, the use of Global Positioning Systems (GPS) is unknown or has been rejected in many places where fishermen prefer to be guided by means of triangulation, and until recently some continued to paddle or row, not motor, to their favourite fishing grounds. Small-scale fishing has long since entered the realm of free trade, with intermediaries and production chains in places such as the Mediterranean coast. Most of the renewable resources extracted from coastal areas by small-scale concerns are fish (between 55 and 65%, depending on the area), followed by cephalopods (octopus, squid or cuttlefish, at 20%) and bivalves (between 5 and 7%). Of course, these data are very general and depend to a large extent on the area and type of specialty of each location. Thus, in some places lobster can be highly representative (up to 10–15%), while in others there is a preponderance of shark or sea urchin.

The Balearic coasts are an example of the importance of this type of fishing (close to the 50% we spoke of) and, nevertheless, of a certain helplessness where the fleet size of this type of extraction has been halved in twenty years (there are now some 500 vessels, compared to almost a thousand before, with a 25% reduction in catch). Extractions are modest, yet select and highly appreciated, such as the 20 tonnes per year of red scorpion fish (*Scorpaena scrofa*) or the 23 tonnes of red mullet (*Mullus barbatus*), the two worth more than €600,000 per year. Fishermen are exploiting a system at the limits of its carrying capacity, partly due to themselves and partly the unstoppable tourism, which they provide with food yet interferes with their coastal activities, quite apart from the threat posed by industrial fishing, recreational fishing and the very degradation of the system. In places where tourism is an important source of income, it is becoming more and more difficult to practise the art of small-scale fishing because, where you have nets between 5 and 30 m deep, there may be divers, people in boats or swimmers to hinder you.

The synergy of the factors affecting the coast means that this type of fishing also requires strict regulation. However, studies such as that by Merino and his colleagues try to balance exploitation with the viability of fishing, calculating a price increase of 12% in small-scale fishery products if fewer fish were caught yet there was a more direct chain of sale (fewer intermediaries). The Andalusian coast has seen an upward trend in the value of certain such products, but their scarcity due to the depletion of stocks has led to a sharp drop in real profits in the sector. Thus, octopus has gone from a profit of about €2 million in 2000 to less than €200,000 in 2007, while the price has increased from €2.4 to €4.8 per kilo in these years. According to a study led by Francisco Piniella of the University of Cádiz, the price of fish has been increasing not only because of inflation, but also because of the scarcity of certain species. Restaurants and fishmongers themselves prefer choice products, and these can become increasingly difficult to source.

In fact, as in any other type of fishing, excess pressure tends to cause the marine system to collapse. One of the most frequently studied examples is that of the evolution of small-scale fisheries in Jamaica. Even before Columbus, fishing from the reefs had reduced the stocks of several species considerably, but the arrival of diseases and the massacre of indigenous people following the discovery of the islands let the system recover to the levels prior to the intense exploitation of indigenous people. 'The problem,' explains Marah Hardt of the Blue Ocean Institute of the United States, 'was the stabilization of a population of some 300,000 slaves and some 100,000 settlers on this large island.' In 1680, under English rule, it was said that 'there is no other place like Jamaica to find fish in every corner of the island'.

However, by 1969 a comment was made that, in Jamaica, 'even with technological advances such as more efficient engines and fishing gear, fishermen have to travel many miles from their home ports to get a fairly decent catch'. Archaeological remains have shown that up to twelve different families (groups of species) of fish could be found in a given place around 1400, while in 2005 there were only three types of small fish in the same place. 'The first things to disappear are always the large carnivores,' Hardt continues, 'like groupers, sharks, barracudas whose meat is tasty, consistent and concentrates a large amount of protein.' According to this scientist, the recovery of this and other systems is possible, but some species have simply disappeared and are either reintroduced or the transformed system is left to evolve on its own.

Jamaica is one of those places where the synergy of several factors has radically transformed the system, although the majority of coastal fishing is certainly small-scale in origin. More dramatic is the situation on a nearby island, Haiti. 'In Haiti there are almost no fish to catch', says John Weiner of the Foundation for the Protection of Marine Biodiversity (FoProBiM): 'People fish absolutely everything, out of an urgent need to put something in their mouths; it is an example of "emptying the ocean" dramatically, but what can they do? In my country there is 60–70% unemployment and very high poverty rates.' Haiti is an example of how the entire ecosystem can be depleted but, as Weiner adds, 'There are no hungry conservationists! You have to educate people, look for alternatives… but understand that people need to eat.'

A different matter is small-scale fishing that is so highly productive, with its unsophisticated yet effective methods, that it quickly becomes industrial. Off the Pacific coast of Alaska, capture of the giant red crab (*Paralithodes camtschaticus*) was a common between the 1930s and 1960s. The catches, always moderate, maintained the stock levels in rocky areas that were difficult for large-tonnage vessels to access. However, an increase in demand led to its extraction by similar methods yet with an increase to a hitherto unimaginable extent in terms of the number of boats and workers taking this valuable crustacean from the seabed. Between 1960 and 1970 there was a fishing boom for this species, with a peak of 42,800 tonnes around Kodiak Island alone in 1965. In 1955, some 2,000 tonnes were extracted. In 1970, only 4,900 tonnes were fished (just over a tenth of a tonne in 1965), and in 1983 there was an indefinite halt to fishing because catches were not recovering and showed no signs of recovery. 'The immense fishing pressure had "managed" to unbalance the number of males and females in the area,' explains William Bechtol of Fairbanks University in Alaska, 'with many more females than males, which led to a chronic lack of regeneration.' The final blow was struck by the transformation of the ecosystems, excessive pressure on the smaller crabs by fish in the area and an increase in temperature, preventing the recovery of the giant red crab. Here, small-scale fishing had been forced to the limit, extracting the resource from a vulnerable area in an unsustainable manner.

The problem of collapsing small-scale fisheries is particularly acute in places where there is no other means of support. In the countries of East Africa, the Caribbean or the tropical Pacific, the situation can be drastic. It has been found that 50% of Kenya's fishermen would stop fishing if catches fell to half their current levels, yet more than 40% would continue that lifestyle, as they have absolutely no other way of surviving. Extreme poverty, as we mentioned earlier, makes shifts in career orientation impossible in many places. In area such as New Zealand or the East Coast of the United States, small-scale fishermen can sell their quota, or their 'licence', to someone else or resettle and convert to another business. Subsidies have been one of the biggest problems, keeping a fictitious fishery going in places where no profit is made anymore. But these subsidies and potential conversions are difficult, if not impossible, for the small-scale fisherman in Kenya, Tanzania and the Philippines.

Small-scale 'craft' fishing now has the chance of being revived on a rational basis. The widespread collapse of large-scale fisheries has led governments of various countries to agree to tackle the problem on a global scale and to try to provide alternatives so that fish production does not fall below the more than

90 million tonnes currently being produced. This includes aquaculture, a clear option and one that is being strongly promoted (see below). However, often, another is to return to low-intensity fishing methods that respect the environment and the populations of species. In many places, small-scale fishing has been displaced by industrial fishing due to the poor profitability of the former yet, little by little, as catches have declined, this mode has become relevant again in many places, as it is more selective and provides very high-quality fish. The problem is the thin dividing line between small-scale and industrial activity, depending not only on the fishing gear but also the amount of pieces of equipment, workers and boats used for. To this must be added a change of attitude to the sea by many inhabitants of the coast, especially in advanced countries, who see tourist exploitation as more profitable than maintaining a small-scale fishing concern, and without the sacrifice of family life.

We need to understand this model of life, from the socio-economic point of view, not just the biological–ecological point of view. A future model should take into account more selective fishing and an appreciation of the product, also aspects of cultural, personal, anthropological and human diversity. Small-scale fishing has the potential to be revived, with dignity, to provide a source of sea protein in a more respectful way than the current industrial fisheries, but we must work hard to make it possible.

Destroying with Dynamite

If there is one monumentally irrational and stupid way to extract fish, it is, without a doubt, with dynamite. In general, it is carried out simply by way by looking for a shoal of fish or a suitable area with a small boat, then detonating a cartridge of explosive so that some are blown up. Only a small proportion of the casualties come to the surface and can be harvested—the rest rot on the bottom or are retained among rocks, reefs and other underwater structures. It is not only the fish that die, but invertebrates, both mobile and fixed to the seabed. The carbonate structures of the corals are reduced to rubble—In fact, a crater is formed where everything is pulverized, and a circle of dead coral is left around it. It is estimated that one of the main causes of coral reef degradation along the Indian Ocean coast is this the use of dynamite, condemned by laws and regulations that are often difficult to enforce. Actually, you don't have to go so far. In Galicia and certain areas of Greece, fishing with dynamite continues to be practised, despite close surveillance and being severely punishable by law.

The damage to the coasts of Tanzania, the Philippines and Indonesia, where this type of extraction is most prevalent, goes far beyond fishing. 'It has been proven that the damage to the reef or any other benthic system and its productivity is very long term,' says Sue Wells, 'but one of the biggest damages is to tourism: nobody wants to go on holiday in places where they can hear up to 100 explosions of dynamite a day, with the consequent degradation of the reef they have gone to visit.' Paradoxically, the main destination of this catch is the tourists themselves.

Many species are directly or indirectly affected by this way of destroying the habitat, such as populations of coelacanths on the northern coast of Tanzania or shark gathering sites in the Philippines. But fishing, which fell sharply between 1997 and 2003, returned strongly, especially in certain places because of the lucrative nature of the business: 'A stick of dynamite in Tanga, Tanzania, costs about €4 on the black market,' says Wells, 'but it can make about €270–€1,200 in profit, depending on the catch.' Therefore, despite the fact that the relationship has always been indicated between poverty and the use of this type of extraction, the reality is that a few people finance youngsters to risk their lives in this dangerous practice. 'In Tanzania alone, more than 100 people die directly from the use of dynamite in fishing every year,' Wells continues, 'a practice behind which there is widespread corruption and the complicity of many administrators and police officers.' In Indonesia, at several points the cost of dynamiting the coastline has been up to four times greater than under small-scale fishing, with up to €200,000 lost per square kilometre in fishing and tourism revenues over the past twenty years. The recovery of a reef is slow, and extremely slow after being dynamited. Some estimate it to be decades, others even suggest centuries until the environment is restored to its original state.

Naturally, if the coral disappears, then the fish, crabs, starfish, and so on, disappear as well. But all is not lost. An economic system has been found that encourages the restoration of disturbed sites following repeated dynamite explosions. It consists of putting a fixed plastic mesh in the substrate at the bottom. 'We need to find a simple and inexpensive method of reaching out to local populations that will allow new recruits to survive', says Helen Fox of the Department of Integrative Biology at the University of Berkeley. The fish and invertebrates have returned to areas that had become barren expanses of coral fragments, encouraging a hope for recovery of places devastated by the stupidity and ignorance of a few for the benefit of a few. They know the consequences full well, but are not in the least concerned about this type of problem: if one place stops producing fish, they look for another, and that's it.

In the Mediterranean, we have a clear example of something similar, that of the date mussel (*Litophaga litophaga*). This tasty bivalve inhabits hollows, cracks and holes along the Mediterranean rocky coast, usually down to 10 m deep, although it can be found below 20 m. When it is small, it settles on rock and begins to secrete a corrosive product from a specific gland that allows it to pierce the rock. Gradually, it bores its way into the hard substrate, and is most abundant in areas where the rock is calcareous and thus easier to penetrate. The only way to extract it is to break the rock, literally destroying the substrate that it lives in. This means that collectors with air cylinders on their backs need to use a hammer and chisel or even a pneumatic drill, and they destroy the entire rocky coastline from the surface down to 10 m to remove it, causing a denudation of rock of up to 10–15 cm deep. They all know that they won't have any more date mussels for decades, but they don't care much: you have to take advantage of the resource until it collapses. Anyway, 'If I don't take it, someone else will.'

But What Harm Can a Fishhook Do?

Believe it or not, recreational fishing from a boat or the shore interferes greatly with commercial fishing. The sum of all the hooks from trolling lines, fishing rods or the side of a boat can become a problem for the system in areas of high pressure from tourism. In the Balearic Islands, one of the few places on the planet where this problem has been taken seriously, just over 5% of the population (some 37,000 people) fish in this way almost routinely. The coast is an ideal place, full of coves, small beaches and places to wait in a secluded corner for fish to bite. The people of the archipelago spend an average of four hours a day in contact with the sea, taking out no less than 1,200 tonnes of fish a year in this area of Spain alone. That is a great deal, and it is estimated that 10% of the overall fishing effort in the Mediterranean is from recreational fishing. 'This fishing effort cannot be underestimated, because otherwise the data on stocks and the extraction of certain species will not be reconciled when making fishing balances,' explains Beatriz Morales-Nin of IMEDEA-CSIC. In Mallorca, recreational fishing can exceed that of small-scale concerns in some places, extracting more than half of the catch, especially where there is great tourist pressure. 'In the Balearic archipelago we went from around 5,000 fishing licences in 1998 to more than 30,000 in 2004, an increase of 600% in just six years', adds Morales-Nin, 'and the number continues to grow'—to be precise, to about 44,000 tonnes in 2014. Of course, there are many people who do not possess a licence, due to either ignorance or laziness, but there is now more and more control and the authorities want to know more about the impact of this type of extraction in various locations.

In other places where sport fishing and its impact have been studied, the situation does not seem to be so dramatic: 'In the north of Portugal, about 7 tonnes of sea bass (*Dicentrarchus labrax*) and 2 tonnes of seabream (*Diplodus* sp.) are caught with this type of fishing every year', says Mafalda Rangel from IPIMAR-CRIPNORTE in Portugal. The pressure on this part of Europe is much lower, because the sea is rougher and the weather harsher. The latest estimates at country level indicate variations from 1% of the population in Belgium participating in recreational fishing to 40% in Finland. 'Tradition and favorable conditions are important in the evolution of this type of fishing', adds Rangel. What is clear is that recreational fishing has generally been seen as a lesser evil almost everywhere, but its impact is greater than one might expect.

Recreational fishing may exert selection pressure even on a species that competes with a similar species, not the one that is targeted by the angler. This is the case with sea urchins, in certain areas. Antonio Pais of the University of Sassari of Italy notes that, in the waters of northern Sardinia, tourists who fished caught only one type of sea urchin (*Paracentrothus lividus*), leaving another alone because it is inedible (*Arbacia lixula*), which meant that the latter had a stable population while the former was in clear regression. The urchin population was declining in density and, above all, in size. This phenomenon has been observed in many parts of the world where commercial and recreational fishing intertwine to interfere with the system. The problem is that, while commercial fishing has specific regulatory channels and controls, recreational fishing is often ignored or underestimated. It seems that there's nothing that you can do, in the end. In reality, the problem is that there are a great many of us, and we are accessing more and more resources and places, therefore we have to nurture what is left so that it does not end up becoming extinct.

Hunting in the Sea: Fishing with a Speargun

I've done some speargun fishing. For a few years (in vain—I never managed to catch anything worth fishing for), I swam the waters of Cap de Creus in search of fish. Actually, what I liked was the fact that I was there, in the water, doing real marathons in coves and islets, now much more complicated with so much boat traffic. I am clear that one of the most direct ways to feel the sea is, without a doubt, spearfishing. The medium of the sea, always hostile, is at its most aggressive towards whoever challenges it by going below the surface to hunt without breathing apparatus (Fig. 9.1).

Spearfishing is also one of the most selective method of extracting fish that is known, especially by those who really understand it and are not satisfied with just any prey. However, this selectivity is not compensated for, because overfishing by this method has seriously distorted the populations of certain species, especially the longest-living and most vulnerable. Already in the 1950s, when neoprene was becoming available to intrepid individuals, the photos revealed legendary captures: dozens of groupers on the deck of pleasure boats and a wide smile of satisfaction on the faces of their human protagonists.

Not for nothing is the Mediterranean grouper (*Epinephelus marginatus*) coveted—it is an exquisite meal. Little by little, the groupers became scarcer and scarcer, taking refuge where human lungs rarely permit, beyond 40–50 m deep. Just a few decades earlier its behaviour was that it did not go so deep. The grouper is a long-lived animal and can live more than 60 years, according to studies made through its otoliths, and it may measure more than 120 cm. Today, generally, the groupers are no more than 50 cm long (at about five years old), after which their growth slows down to just over 3 cm per year. Because of their over-exploitation, groupers may experience an accelerated sex change in many localities. Male groupers are usually the largest, but before that they go through a phase in which they are females. Scientists who have studied the phenomenon have found that the renewal of

Fig. 9.1 Spearfishing may be a source of local extinction of certain species that are selected by its size. These big fishes constitute the top predators of the system. *Credits* ADOBE STOCK

the grouper population is less than expected, so spearfishing and small-scale fishing must take into account the potential for breakdown in the groups formed in over-exploited areas.

It is possible to regulate and control spearfishing but, like many other hobbies, the number of participants has skyrocketed beyond the system's carrying capacity. In Mallorca, one of the few places where this type of fishing is considered to be a real interference not only to small-scale fishing along its coast but also as a cause of distortion in the ecosystem, more than 2,000 underwater fishing licences have been issued. It is an ideal place to practise this sport, due to the clear waters and temperatures between 27 and 28 °C in the height of summer, the rocky coasts and places without much maritime traffic. Besides legal fishing, there is a high number of illegal fishermen. As we noted for recreational fishing, some are just absent-minded, but a not inconsiderable number then sell their catch to local restaurants as their 'fish of the day' for summer customers.

Seabream (*Diplodus sargus*) is the most commonly caught species (30–40% of the catch), followed by mullet (20–25%), but the star product is undoubtedly grouper. Since 1987, groupers weighing more than 4 kg have declined drastically on the coast above 40 m, partly because this is the recreational fishing area for free-divers. And competitors. Competitions have been the best source of information to confirm the decline of a large number of species, some thirty of which are a target for underwater fishing in the Balearic Islands. We can see that, since the start of official records of championships, the number of fish per competitor and fish size have fallen steadily. 'We have gone from about ten fish per championship and per fisherman to less than five fish, and the weight of all the fish has gone from 6 to 8 kg to less than 3 kg', says Antoni García-Rubies of the Centre for Advanced Studies in Blanes (CSIC). But what has really changed is the effort to obtain the same number of catch: 'much more time looking for smaller prey, more hours swimming with a considerable effort to catch fish', adds Jaume Coll of the Balearic Government's Directorate General of Fisheries and collaborator of García-Rubies. The fish caught cannot weigh less than 300 g (and moray eels, conger eels and groupers must weigh at least 2 kg), but it takes about five years for seabream to achieve this weight and it is becoming increasingly difficult to find them. Today, this type of competition, where the greatest possible number of captures is sought in a period of five to six hours, no longer makes any sense.

Perhaps it is a matter of rethinking this and other recreational practices, both sporting and competitive, to allow populations to recover from such high pressure. They need to grow as the concentration of tourism and their access to all places increases due to the facilities on land and sea. Otherwise, there will be no choice but to end up simply contemplating the sea, with no chance of any further kind of interaction.

10

The Ghosts of the Sea

We look out to sea from the shore, but can't see anyone in the water. We swim far out, alone, and there are hardly any waves. It promises to be a great day, but when we glance back to the beach we see that a flag has just been hoisted, and curse those strange organisms that seem to want to ruin our holiday. The beach guard has raised that flag to indicate jellyfish and, though we cannot see them, our instinct tells us that it would be better not to stay in the water. Perhaps we can paddle a bit to wet our feet, but no more, because we all have a cousin, friend or acquaintance who has been stung by a jellyfish, and it is annoying, painful and, in some cases, deadly.

Jellyfish are a form of gelatinous plankton and are composed almost entirely of water—more than 99%, in some cases—which gives them a very peculiar semi-transparent and ghostly appearance. They are fascinating creatures: beautiful, mysterious ghosts of the sea, some of the most simple yet misunderstood on our planet. They have spent the winter offshore, but come in with the winds that blow from sea to the shore from April. Now is the time when you notice them. Their life cycle is, in general, between one to two years, depending on the species, of which there are some four thousand, from giant size (Nomura's jellyfish) to the tiny hydromedusa, whose umbel (cap) is less than a centimetre across. Speaking of 'jellyfish' is as vague as saying 'mammals', as there are about five thousand species.

In 1990, I was fascinated by a talk given by a colleague of mine, now sadly gone, Francesc Pagés, on the life and miracles of the jellyfish. Among other things, he told us that jellyfish have been on the plant for more than 600 Ma years (about 400 Ma more than mammals), thus have managed to survive all the upheavals that the planet has delivered: there have been several

© Springer Nature Switzerland AG 2019
S. Rossi, *Oceans in Decline*,
https://doi.org/10.1007/978-3-030-02514-4_10

near extinctions that have barely changing them, lacking any brain and with an extremely basic morphology. They are primitive beings that have witnessed the evolution of other without flinching, living a simple, effortless lifestyle that is nonetheless highly effective.

Why are there so many? Has there really been an increase in these organisms? Time-series studies to follow-up jellyfish are scarce. However, in places such as Villefranche (French Mediterranean), it has been shown that from 1979 to 1989 there was indeed a progressive increase in jellyfish and that since 1990 the numbers have soared. In other places as varied as the Bering Sea, Israel, San Francisco Bay, Chesapeake Bay, la Manga on the Mar Menor and the Norwegian fjords, the trend is the same, despite the different circumstances. This represents an increase in the mass of gelatinous plankton at the expense of fish and other organisms.

Scientists and authorities have gradually come to realize that this is a global problem, not just a fluke or localized to a certain area that is isolated from the rest of the world. 'We have come a long way during the last decades in the understanding of the biology and ecology of gelatinous organisms,' says Robert Condon of the Department of Biology and Marine Biology of the University of North Carolina, 'but the road ahead of why these changes are happening will be long.'

When they find highly favourable conditions of light, nutrients, food, currents, and so on, jellyfish can 'bloom', in an accelerated proliferation through reproduction or agglomeration, forming great concentrations in a specific place in a matter of days or weeks. As practically passive bodies, the creatures are carried by sea currents towards the shore, where the fresh water from rivers acts as a barrier to their advance. When it rains heavily in spring, the rivers create a front that prevents the jellyfish reaching the shore, but when there is drought, or when the rivers have reduced flow rates due to agricultural or industrial use of their water, this front is absent and the ghosts of the sea appear on beaches en masse.

The jellyfish can adapt well to various conditions of salinity and turbidity of the water, but are mostly animals that live in the sea. It is not true that the presence of jellyfish is associated with pollution such as sewage (look at the waters of Corsica or Menorca, crystal clear yet full of jellyfish). In general, they can also survive low concentrations of oxygen, which favours their presence in where there is a high concentration of organic matter due to decomposition. However, by itself, pollution never explains the presence of jellyfish. What is certain is that some species especially enjoy waters rich in inorganic nutrients or suspended matter, and this makes them proliferate. This is the case for those jellyfish that have algae in their interior, as one source of medusa's nourishment, and are at ease in such areas. 'When the waters are polluted diversity decreases, but those species of jellyfish that resist can grow without control', affirms Mary Arai of the Pacific Biological Station in British Columbia, Canada. In the Florida Everglades, the inability of the system over much of its surface to absorb the nutrients and organic matter in suspension, coming from deep degradation, has dramatically increased a species of medusa, *Cassiopea* sp., to the extent that you can spot more than 42 individuals per cubic metre. But this is nothing compared to places like some Scandinavian fjords, where the concentration of the medusa *Aurelia aurita* reaches 300 individuals per cubic metre. The increased turbidity and organic nutrients in the Adriatic seems to be one of the key factors behind the expansion of both this and species such as *Pelagia noctiluca*. Conversely, in some areas along the Japanese coast, where fish farming imposes a heavy organic load on the environment, the closure of particular fish farms has been a trigger for *Aurelia aurita*'s disappearance,.

What is the chief culprit? One of the main triggers of the expansion of jellyfish is overfishing, without a doubt. 'In the region of Bering, where there is intense fishing, 5% of all that on the entire planet, the drastic reduction of the stocks has been accompanied by a significant increase of *Chrysaora melanaster*', says Claudia Mills of the Friday Harbor Laboratories of the University of Washington. In this and in other places, the disappearance of fish has been accompanied by an increase in jellyfish and other gelatinous plankton. The major predators of these organisms are tuna, mackerel, swordfish and sunfish, and some turtles. It has been demonstrated that mackerel are an effective predator of ephyrae, the jellyfish in their first stages of life in the open sea. As we have seen in previous chapters, these and other agents are still victims of a means of extraction that has decimated their populations and diminished their role in the pelagic and benthic systems. Already in the 1980s, academic papers warned of the extinction of various types of predators, but at that time no one linked the increase in gelatinous plankton with the impoverishment of fishing grounds. This trend has been increasing, to the extent that currently there is discussion on a ban on fishing for some species so that they regenerate. The solution, undoubtedly, is a landmark for those fish that devour jellyfish. But in the case of turtles, for example, there is little hope: excessive human pressure has reduced the extent of the shore environment needed by these reptiles, which have nowhere to go. Can you imagine turtles laying their eggs on the beaches of Benidorm or Salou?

Most jellyfish are voracious feeders and some eat fish larvae, among other things, which has led to an acceleration in the collapse of fish stocks across the planet. Moreover, their diet often includes small crustaceans and detritus, which are sources of food for other fish, so jellyfish compete for food with their potential predators and prey. On the other hand, jellyfish themselves may be eaten by fish larvae and young, or be caught by fishing vessels.

Another major problem related to the proliferation of jellyfish is the introduction of invasive alien species. There are species that have largely moved from site to site in the ballast water of merchant shipping, carrying millions of organisms (algae, crustaceans, jellyfish, etc.: see chapter on invasive species). If they are comfortable in the habitat into which they have been relocated, these organisms reproduce and come to replace native species. The number of trade routes has increased with the growth in maritime traffic.

An example that has been known for decades illustrates the successive invasions of a marine system. Over the past fifty years, eutrophication, overfishing and pollution have altered the Black Sea. At the end of the 1960s, it had intense blooms of a typical of Mediterranean jellyfish, *Rhyzostoma pulmo*, with more than two or three per cubic metre, but these stopped suddenly as the sea became a more saline environment, giving way to the medusa *Aurelia aurita*. The development of this second species is related to the lack of fresh water, since agricultural irrigation had taken water from and thus reduced the flow of the major rivers flowing into the Black Sea. But the situation became dramatic when, in the 1980s, the accidental introduction as another invasive species of gelatinous plankton interfered in the already precarious balance in the trophic chains of this sea. The alien ctenophore *Mnemiopsis leidyi* (sea walnut), which tolerates high salinity, quickly proliferated in a disproportionate way, reaching a concentration of 300–500 specimens per cubic metre. This organism was devastating, since it feeds on the larvae of anchovies, a fish that, at that time, was still a significant source of wealth in Turkish and Russian waters. The introduction of another ctenophore (*Beroe ovata*), specialized in eating other ctenophores, could regulate the concentrations of the harmful *Mnemiopsis leidyi*, but the results are unknown, to date. What is clear is that, of the 26 exploitable types of fish that once swam in the Black Sea, there are now five fewer.

Mnemiopsis leidyi settled in the Mediterranean 'officially' only recently. Scientists from ICM-CSIC in Barcelona confirmed the arrival of this invasive species on Spanish coasts in summer of 2009, where it has been found in certain areas such as Denia, Salou and Mataró, sometimes in extremely high densities. It originated on America's Atlantic coast: '*Mnemiopsis* is a very adaptable to different environmental conditions, and can withstand a temperature range of 0–30 °C and a salinity of 2–38‰', says Veronica Fuentes, from the ICM-CSIC in Barcelona. 'They often live at shallow depths,' says Dr Fuentes, 'between 2 and 30 m deep. We have detected dense banks close to the surf break of beaches, from breaking waves to beyond 200 m offshore.' The ctenophore is harmless to humans but has been shown to have a serious impact on fishing stocks, especially when in the absence of predators, such as small fish, that can feed on the first stages of life of this gelatinous component. In the Caspian, Baltic and North Seas, where it has also been detected, it also has serious implications for the food chain and the survival of stocks of pelagic fish. The aggravating circumstance in this case is that it is a non-native species thus has no natural predators.

In the Mediterranean Sea we feel increasingly that the jellyfish will begin to join us on the beach. We can consider four species as the most colourful in the Mare Nostrum, as the Roman's termed it. The first, *Pelagia noctiluca* (purple-striped jellyfish), is pink and semi-transparent, with long and tentacles that are barely visible. It is the most dangerous. Then there is *Chrysaora hysoscella* (compass jellyfish), opaque with brown stripes, which can measure more than 50 cm in diameter; it, too, has long tentacles. The other two, *Rhizostoma pulmo* and *Cotylorhyza tuberculata*, are common in less clear waters and their tentacles are short. They are less dangerous. Finally, there is a siphonophore, rare but highly dangerous: *Physalia physalis* (the Portuguese man-of-war) has long, nearly invisible tentacles and its sting is fearsome. But, as we have said before, the phenomenon of jellyfish is not restricted to the Mediterranean: they have expanded, around the globe.

The examples mentioned above are just the start of it. A changing sea is the optimal breeding ground for jellyfish, and they are taking full advantage of the reduced diversity of different aquatic systems. In many of the Scandinavian fjords, fishing now catches nothing but large jellyfish, which eat the small crustaceans and other food that had sustained fish fauna. In the San Francisco Bay, a small hydromedusa that was introduced accidentally has unbalanced the system in its favour, and its polyps and its young are ubiquitous and displacing competitors that cannot grow as fast. Other instances of expansion of these simple creatures are perhaps not so visible, by virtue of being in areas around the world that are less frequented by tourists.

It has been argued that an increase in temperature, especially in shallow water, is the trigger for their uncontrolled increase. However, by itself this does not guarantee the proliferation of any animal, nor even jellyfish. 'We need to better understand the life cycles to be able to relate cause effect', says Giacomo Milisenda of the DiSTeBA of the University of Salento in Italy: '*Pelagia noctiluca*, for example, has a cycle related not only temperature but also the availability of food.' We need to consider both warmth and food, therefore, or the cycle will accelerate in a way unlike how experts believe.

Recent studies carried out by several studies are beginning to provide the keys to questions that so far have been in the air. For example, it does seem that jellyfish now appear on our shores earlier in the year. 'Data series that we have so far cannot tell us if the increase in temperature or salinity in the Mediterranean Sea has caused an earlier appearance of jellyfish', says researcher Veronica Fuentes. Temperature and salinity, along with the availability of food, are known to influence the life cycles of these planktonic organisms, but at the moment the effects are unknown. Some groups, for instance that of Professor Jennifer Purcell of the Department of Marine Biology at the Western Washington University in the United States, have already seen that increases in temperature, nutrients and sunlight, which favour the proliferation of single-celled algae, are behind the increase in benthic substrate fixed polyps, releasing small jellyfish to the environment —'but not all jellyfish species behave in the same way,' Purcell continues, 'elsewhere, the environmental factors that affect them are different and otherwise affect their reproduction and growth'. Purcell and Stefano Piraino of the DiSTeBA of the University of Salento have also researched other species and have come to the conclusion that proliferations are on the rise, especially in cosmopolitan jellyfish that can find optimal conditions that they did not enjoy previously. Wind is another factor that interacts with the presence of jellyfish. *Pelagia noctiluca* arrives on our coasts due to currents and sea breezes, from sea to shore.

Now that our accumulated information is becoming more consistent, one of the questions asked by both visitors and permanent inhabitants of coasts is whether or not there are areas where these animals tend to become concentrated. This is because one of the main victims of the jellyfish is, without a doubt, tourism. In 2008, a report indicated that 150 million people are exposed to these ghostly bodies every year. In Chesapeake Bay, more than 500 thousand people are stung by these gelatinous beings annually. In Florida it is 200 thousand people and in Australia more than 10 thousand. Taking into account the density and population of these places, this is an extremely high number. This makes for uncomfortable swimming, sometimes even dangerous. The worst aspect is that tour operators are tending to 'map' jellyfish risk, discouraging visits to certain areas, although this is sometimes completely unfounded. Many millions are lost, because beach tourism depends on the millions who generate it—the tourists.

Jellyfish effects are also present in other sectors, chiefly perhaps fisheries, either indirectly or directly, as we have seen. On the coast of Namibia in Africa, up to 90% of the catch may constitute jellyfish. Losses in aquaculture can also be high: an eloquent example is that more than 250 thousand salmon were killed by a bloom of *Pelagia noctiluca* in Ireland in just a few days. Less well known but more disturbing is that large proliferations can clog cooling ducts. In Tokyo Bay, a species of large size (Nomura's jellyfish, up to 2 m across its umbel) blocked the port, as vessels could neither leave nor enter as their refrigeration systems were put out of order due to the immense numbers of these animals. This problem has been experienced at various plants, such as thermal desalination and nuclear pumping systems, yet only in Japan has it changed in the last few decades from being an interesting issue to a 300% increase in the cost of repairing systems. More and more jellyfish can become a serious problem and pose a huge economic cost.

What can we do? The easy (and obvious) answer is to stop fishing in the way that we do, pollute less and foster a more controlled transformation of the coastal system. However, there are people who have taken the time to think laterally: if you cannot join them, take advantage of them. For hundreds of years, jellyfish have been considered a delicacy by the Chinese and the Vietnamese. The body contains only 2–5% of protein and only 0.2% of lipids, but to transform it into food is quite simple: just dehydrate it and add flavour. According to the Food and Agriculture Organization, the jellyfish market is worth more than €120 million and is growing. This is nothing compared to fishing, but it is a market that many may explore, in time. 'The jellyfish can provide many interesting products, from an economic point of view, as pharmaceuticals, beauty products or substance stabilizers for the food industry', says Antonella Leone of the Institute of Sciences of Food Production of the Italian CNR: 'We might have to start to look to these organisms as a food source rather than a nuisance.'

Yes, while we do need to think how to reduce the pressure on the potential predators and competitors of jellyfish, the market is the market, and if we now abound in jellyfishes there will always be those who won't hesitate to take full advantage. By 2050, the human population is projected to increase by 46% from current figures, meaning ever more demand for marine fish-shaped protein—cephalopods, bivalves or crustaceans—and if we continue at this rate, maybe it will have to be… medusa.

A Small Sea of Jellyfish

Perhaps one of the places where most people have gained first-hand experience of the jelly problem is la Manga, on the Mar Menor, Murcia. This huge coastal lagoon is one of the largest in the Mediterranean, at some 135 km². No more than 6.8 m deep, it has been the perfect breeding ground for three species for the past twenty years. Swimmers have had to share the 610 million cubic metres of water with over 100 million *Cotylorhyza tuberculata* ('fried egg' jellyfish) and *Rhizostoma pulmo* (barrel jellyfish). What has happened to put these two species, introduced before the mid-1990s), and *Aurelia aurita* (the native species) so much at ease?

The circulation of water, the sediment, the concentration of nutrients and organic particles and even the fauna and flora have changed dramatically in the Mar Menor in recent decades, for many reasons. These are overfishing, the opening of the inlet from the Mediterranean (the Estacio canal), a rise in the concentration of mineral salts and urban waste. Besides contributing a much-increased level of nautical activity, Murcia is now an urban sprawl along the coast with a consequent proliferation of hard surfaces (anchorages, breakwaters, artificial reefs, etc.), and these factors together create the ideal habitat for certain species. The removal of the fish has opened the way for a jellyfish increase (as noted before), especially of species whose diet includes small crustaceans, vertebrates and invertebrate larvae. One of the three species (*Cotylorhyza tuberculata*) is greatly benefited by the huge contribution of nutrients from the crowds from the developments along the coast of the Mar Menor and from agriculture in the area. The 'fried egg' jellyfish carries inside it symbiotic algae, small plant cells that convert nutrients (nitrates, phosphates and carbon dioxide) into carbohydrates, which are, in part, transferred to the jellyfish for growth, respiration and reproduction.

But the changes that favoured the proliferation of certain species in the Mar Menor did not end here. In 1970, most of the primary production of the lagoon came from a marine plant, the seagrass *Cymodocea nodosa*. When the Estacio canal was opened, a benthic alga (*Caulerpa prolifera*) was introduced, and this began to displace *Cymodocea*. Microscopic algae such as diatoms also began to increase in concentration. Only a few patches of seagrass were left, because nutrient levels had climbed and climbed, favouring algae over seagrass. The entire system was shaken, but the final insult was still to come. One of the businesses being developed in the area a few decades ago was oyster farming. Besides the wealth of phytoplankton in that closed environment, the venture rested on the convenience of operation due to the shallow waters of the lagoon. But oysters need a hard substrate for the larvae ('seeds') to germinate and become established—as do the polyps of the three species of jellyfish, if they are to become established and grow into the future generation. The developers established the oysters and many other species by erecting artificial structures. However, oyster culture had to be discontinued due to the high levels of heavy metals that were found in the prized oyster flesh, yet it had never been agreed who would remove artificial structures.

Subsequently, the construction of piers, docks, marinas, and so on, due to the development of tourism, increased the amount of substrate available for the establishment of yet more polyps.

Both the profound transformation of the habitat and its actors favoured the proliferation of jellyfish in the Mar Menor. The hire boats that went out of San Pedro de el Pinatar to drag for fish now caught gelatinous plankton: in 2000 they caught around 2,000 tonnes of jellyfish, and in 2003 about 5,000 tonnes (96% of the 'fried egg' type). The residents of the area are divided. Almost half consider that the jellyfish damage tourism, although these species are not particularly dangerous. The majority perceived urbanization, agricultural management and overfishing as the real problem in the area, but it is symptomatic that 37% confessed to having no idea why, in a few years, there had been such a spectacular increase of jellyfish in one of the country's most important tourist destinations. The jellyfish have become the actual regulatory system, as the main 'capturer' of nutrients and particles, dead or alive. In various parts of the area, the regional government has already installed 43 km of a network to prevent the arrival of gelatinous intruders. Nevertheless, as usual, there are implications due to the other agents... and you can be sure that pieces of jelly or small jellyfish will continue to reach the Mar Menor's beaches.

Giant Jellyfish (*Nomura nomurai*)

At the end of 2009, a Japanese fishing boat sank from being overloaded by a jellyfish known as *nomura* by the locals of the area. The boat was not very large, about 10 tonnes, but the story grabbed the world's attention as it was highly alarming. Apart from this shock, the giant jellyfish *Nomura nomurai* has become a real problem on the Japanese west coast. It has been estimated that in times of population explosion or bloom there may be 500 million jellyfish in the waters lapping the coast between Korea, China and the Japanese archipelago and the area south of the coast by the Yellow River. What has happened to make the presence of this species in these waters a nightmare for fishermen since the beginning of 2000? *Nomura* is perhaps the largest jellyfish ever to exist, with an umbel that can measure more than two metres across and a body that sometimes exceeds 200 kg: a queen of a medusa.

In recent decades, two main factors may have triggered its appearance. The first is overfishing (as always), in a sea where, for example, catches have been reduced to 95% in recent years off the coast of China, despite it being one of the places on the planet where the extraction is concentrated, at no less than 11% of global fishing activity (with 9–10 million declared tonnes of extracted fish). The Sea of Japan and the surrounding areas have always been highly productive, but during recent years the level of nutrients has soared due to industrialization and intensive agriculture, as well as aquaculture-related activities, especially off the Chinese coast.

Jellyfish were the first beneficiaries of this food glut, because organic particles in suspension are an important part of their diet and there are no competitors to share it with. But there are other possible causes. According to specialists, the construction of the Three Gorges Dam on the River Yang-Tse could represent another transformation, changing the dynamics of the currents in this sea and promoting the arrival of material from the mainland to the island of Japan. 'This change in currents and an increase in the surface temperature of 1.7 °C on average since 1976–2000 are factors that help the proliferation of the medusa, along with overfishing', argues Shin-Ichi Ows from Hiroshima University, Japan. When there is an explosion in jellyfish numbers, reaching densities of approximately 2.5 individuals per 1,000 cubic meters at about 10–30 m deep, their bodies become less robust and more watery, as if they had grown rapidly. 'In only 28 days they can pass from weighing barely 1.5 g to more than 29 g', comments Ows.

But there is another factor that, it is clear, has been decisive, apart from changes in temperature, currents and available food: the increase in substrate available for nomura polyps. As in the Mar Menor and other places, the transformation of the (in this case, mostly Chinese) coast in the past two decades appears to have been a key factor behind its abundance. 'The construction of large ports, dams, artificial reefs, aquaculture farms and other artificial structures could be benefiting the formation of new young from the polyp or the "flower" which settles on the bottom and is essential for its reproductive cycle step', noted Yoon's National Fisheries Research and Development Institute of South Korea. When they reach the Japanese coast, the nomura's gonads are ripe, ready to start life over again. Polyps can survive swings in temperature between 0 °C and 27 °C, which is a very wide range, and even manage to do without food for a long time. 'It is clear, however,' adds Ming Sun of the Key Laboratory of Marine Biological Resources and Ecology of the Liaoning Ocean and Fisheries Science Research Institute in China, 'that if they have temperatures between 15 and 27 °C and much food, proliferate uncontrollably and form large blooms, that can be a huge problem year after year'.

The year 2005 was a particularly difficult year for Japanese fishermen due to an invasion of *Nomura nomurai*. There were more than 100 thousand formal complaints, losses of up to 80% of income from inshore fishing catches and considerable public alarm. There have been many selective screens devised to catch fish but not let the immense gelatinous creatures past. There are desperate measures to use nomura for food, as the Chinese did hundreds of years ago. In 2007 the Japanese government promoted a book on ways to prepare jellyfish—biscuits, salad dressing, fries—as a counter-attack to an invasion against which it felt powerless but which, together with those of other countries of the environment, it was first responsible.

Carybdea marsupialis, Mediterranean Cubomedusa

The dreaded cubomedusa is typical of tropical seas, such as around the Australian Great Barrier Reef or the Philippines. Cubozoans, or 'box jellyfish', are considered to be the cnidarian group's most dangerous, with an extremely painful sting that has changed the lives of dozens of Australians every year. It is mostly found in shallow areas near estuaries and lagoons, and the largest does not usually measure more than 20 cm by 20 cm across. Its umbel is a curious cube with tentacles of about three metres long coming out of each corner. These corners feature an optical system that is well developed for such a primitive animal, so it is able to perceive changes in both light and form, despite lacking a brain (as do all jellyfish). The venom of many cubomedusas acts immediately and may be deadly: the pain is so intense that it can cause shock, preventing the swimmer from reaching shore, and neoprene is not always sufficient protection against their sting. There are only about forty species known, and they are apparently restricted to tropical seas. Until now.

During the summer of 2008 there was a great abundance of a cubomedusa along the Spanish coast, spotted off the beaches of Denia, Alicante: *Carybdea marsupialis*, a small species. It is rare in the Mediterranean Sea, so had never been considered to be a species that would form a major proliferation, yet during the summer of 2008 the Red Cross reported a high number of stinging incidents in this area due, no doubt, to this almost imperceptible, transparent and seemingly harmless jellyfish forming dense swarms in the breakers. 'During that summer we could confirm that such assistance was due to the presence of these small jellyfish,' says Dacha Atienza of the ICM-CSIC, 'and repeatedly during the summer we made surveys and collected samples, both alive and dead.' Their sting is more serious than that of *Pelagia noctiluca*, but less so than its feared Australian relatives. It is possible that the species may reappear on another occasions, because nothing is known about its distribution or of what factors influence in its life cycle. Atienza concludes, 'Its recovery is slower than that of other common species in our waters'. Like many other species of gelatinous plankton, its life cycle has not yet been described, and much remains to be understood of its biology and ecology.

11

Precious Coral

It is just past eight o'clock in the morning in Port Lligat, Girona. Three *coraleros*, or coral fishers, are deciding where they will take their boat today, now that it is the authorized season for extracting red coral (*Corallium rubrum*) in this area of the Mediterranean. It is a difficult decision, because the best areas where they are (Natural Park of Cap de Creus) are scarce and subject to the moods of the weather: tramontana winds with stormy, unpredictable twists. They have to take the risk. It seems that the tramontana, the feared north wind that sweeps the coast and blasts this area with special intensity, will not arrive until the afternoon, so they decide to try off Farallons, an exposed headland. They have already located the rock on which this precious coral is found, at not too great a depth, between 35 and 50 m. Although some *coraleros* dive to between 80 and 160 m, here most do not usually go down more than 60 m. I can confirm that in this area there is coral from 15 m below the surface. The Costa Brava *coraleros* know that unless someone has killed the coral on this rock near Farallons in the past four years (increasingly likely, due to fierce competition between legal and illegal exploitation of the coral patches), they won't need to dive lower to extract enough to make their foray profitable.

Elsewhere, the coasts of the Alghero north of Sardinia, the view is somewhat different. Three *coraleros* are prepared to go out, but here coral extraction is not authorized at less than 80 m. This is the standard, and the law imposes far tougher rules on the Costa Brava bathymetric ranges and some other places in the Mediterranean. 'Red coral is here in abundance between 90 and 110 m of depth,' say Massimo Scarpati and Massimo Ciliberto, two bronzed, professional *coraleros*, about the coral off the region's coast. The

© Springer Nature Switzerland AG 2019
S. Rossi, *Oceans in Decline*,
https://doi.org/10.1007/978-3-030-02514-4_11

population here, near Capo Caccia and the mouths of the River Bonifacio, is quite different from those near Cap de Creus. The coral is more dispersed, yet it is of a far better size. However, in each and every place where extraction is or has been undertaken, the sentiment is the same: 'Red coral fishing turns into a kind of unabated gold fever', says a *coralero* who has searched for it the length and breadth of the Mediterranean throughout his life. It's like an itch —something that becomes a necessity, a desire. Find the largest branch, the most perfect, beautiful shape to allow jewellers to make the necklace, statue or ornament that they who work with it crave. It is a profitable worldwide business, involving several types of precious coral, worth €500 million each year (counting just the raw materials, not the finished jewellery that may contain precious metals).

However, increasingly it takes a long time to locate red coral that is both worthwhile and of a legal size—I don't need to explain. I have seen with my own eyes, and in fact it is now history. At the beginning of the 1980s there were sharp falls in the coral stocks of the Mediterranean, from about 100 tonnes extracted annually from the entire Mediterranean to only about 40 tonnes. The situation stabilized downward, with slight peaks when new 'streaks' were found in Morocco, Algeria and Turkey. Everyone recognizes that the coral that was observed in the 1970s and 1980s on undersea cliffs and in caves is no longer in existence, even at great depth: as thick or thicker than a thumb, heavily branched and ancient, like vermilion trees. Likewise, an example from Sardinia is cited by Basilio Liverino, one of the greatest connoisseurs of red coral and an expert on its history: 'At Capo Caccia, in 1956, red coral was fished to about 35–40 m deep; yet in 1958, shortly after, the *coraleros* in the same area had already to descend to about 40–45 m.' By 1964, only a few professionals, by diving to depths of more than 70 m, were able to find profitable coral there. The same story can be told about the coasts of Campania in Italy, Sicily, Cap de Creus and many other locations.

It is known that in earlier times there were huge amounts of red coral, because on the Amalfi coast, in the Italian region of Campania, 1,200 *coralines* boats, each with about four or five crew, operated between the mid-nineteenth and early twentieth centuries. Coral grew everywhere, not hiding in crevices and caves, as now, out of reach of the rudimentary gear that has ruined the supply in the Mediterranean. And when it had been exploited in this way, we turned to scuba equipment to violate its last few places of refuge, getting in everywhere, into all the cracks, over the cave roofs and along the tunnels. Such was the mania that even children of eight or nine years were supplied with air cylinders so that they get into places where adults couldn't, to remove the long-awaited reward, the red gold. Some children were killed by their parents' greed, and for decades several countries, including Spain, prohibited children from diving using air cylinders (Fig. 11.1).

Today, in all places, the red coral that is found has gone from a forest of trees to a meadow of grass: 'the underwater landscape has really changed in the past two decades, and not only because of the coral... an impoverishment has been detected that certainly it is not attributable only to the *escafandristas* (scuba divers)', says red coral historian Arnald Plujà, who knows very well the area in the North of Spain where coral was extracted for many decades. In deep areas, the decline in the populations may have left room for other, faster-growing organisms: 'From 100 m down, we find *Lophelia pertusa* and other corals known as "cold", which grow faster than the red coral', says Covadonga Orejas of the Institute of Spanish Oceanography in Santander. She and other specialists suggest that in certain areas the red coral may have been superseded by species that have monopolized the space.

Fig. 11.1 Red coral, probably centuries old, is now actually extinct across the entire Mediterranean at depths of less than 60 m. Bottom right: for comparison, coral that may be found at depth in many locations. The amount of carbon that is stored in the first example is several magnitudes greater than in the latter, severely affecting the role of the species. *Source* Georgios Tsounis

The lack of scientific rigour in the coral extraction operation (despite extensive publication by several specialists) and its high market value make it a very appealing product, not only in the legal but on the organized black market, and the management of fisheries and the environment includes a focus on poaching. Victoria Riera, the Director of the Cap de Creus Natural Park (source of more than 90% of Catalonian coral) is explicit: 'A number of areas have been created in which extraction is prohibited permanently within the park; it makes no sense to create reservations if we cannot regulate effectively extraction or fishing for the organisms which inhabit it.' Unfortunately, there is a lack of legal control to regulate illegal or abusive extraction: 'It is not for want of vigilance that coral is still being extracted everywhere', says a warden. A key issue in the penal code is the state of the coral: 'If it is removed illegally at an immature size, this is regarded as administrative misconduct, not a criminal act.' Therefore, while the extraction of the species is regulated, when it comes to awarding a penalty only a fine is imposed.

The situation is critical in many ways. A study funded by the Generalitat de Catalunya, carried out by myself and other scientists from the CSIC between 2001 and 2006, came to the conclusion that coral, at 20–60 m deep, was present along almost all of the Costa Brava, but of an average height of about 3 cm and a stem width of about 4–5 mm. What else measures 3 cm? A £2 coin measures 2.8 cm across, and a €2 coin 2.5 cm. Compared with the data from twenty years ago, the difference is significant, because at that same depth in that area the coral was about 12 cm high with a diameter of nearly 9 mm. Deeper, where the *coraleros* often already fail, red coral is recovering slowly in Cap de Creus. Below 60 m we find branches of over 10 cm in height and 1.5 cm across. There is still a long way to go before we see a recovery of the populations. 'Effective surveillance supported by legislation that actually imposes penalties for poaching and a reduction in the number of licences will considerably relieve the situation,' said Georgios Tsounis, of the University of California, Northridge, 'but above all coupled with biological data on the management of coral.' It is has been proved that a well-branched sprig of coral 3 cm high is capable of releasing 90 larvae each year. Remember that the larvae are coral 'babies', or sexual products, after the egg is fertilized. By contrast, a well-branched coral of 12 cm in height produces some 3,000 larvae each year. This is a big, big difference. It tells us that coral's ability to sustain itself, to keep regenerating young forms, is dwindling because large branches are becoming scarcer.

To make matters worse, years ago another problem emerged to check the recovery of the various populations of red coral around the Mediterranean.

Poachers collect coral when it is small, so the coral cannot regenerate. One of the properties of this always fascinating animal is that if you break the 'tree' off at the trunk, half the time it will grow back from the base. But the problem is that coral that was before only considered marginally suitable for the market now has buyers (up to €300 per kg) who make it into either small coral objects or a paste for preparing aphrodisiacs and homeopathic remedies that you can purchase online. And there is no longer any capacity for its special kind of regrowth, as those who undertake this type of indiscriminate extraction take not only the coral but the rock and substrate—along with all the other long-lived organisms. One of the professional *coraleros* of Cap de Creus complains about poachers: 'They are a blight. There is no control and there is no real punishment. Poachers roam freely without anyone doing anything.' Therefore, coral is not extracted in a selective manner.

There is a second problem, one that is more dramatic and is under little or no control. For ten years, thermal anomalies have been detected in the waters of the Western Mediterranean. Temperatures in the top 30 m increase dramatically in August, keeping up intense waves of heat for weeks. The water is made stagnant by the intolerable temperatures, thus currents are reduced, killing many organisms (not only corals) that rely on the movement of water to feed on suspended particles.

The recent massive deaths of red coral (and other organisms) due to strong thermal anomalies in shallow waters are becoming more frequent and have had a serious impact on to surface populations, which may become locally extinct. If we combine overfishing and the mining of immature coral with this thermal phenomenon, we can see that the resource may soon be economically unviable. A meeting of experts in September 2009 to discuss both overfishing and mass mortality strongly recommended the cessation of coral extraction from less than 80 m deep, so 'that coral would give more benefits to the world of tourism and the business of the *escafandristas* (scuba divers), who delight in seeing the coral in all its glory live on the cliffs', said Georgios Tsounis. However, Tsounis sees another danger looming ever clearer: the use of robots for coral fishing. 'It's logical to think that being a trade of risk, especially at depth, a robot would be give a good location of the resource and an effective support tool if there are problems', Tsounis said. 'The problem is that they might be used without control, and there is little real control anywhere; they could spend hours taking coral but, very possibly, we would not know.'

We have spoken extensively of red coral, because this is the main species within the group of so-called precious corals. But there are several other species that are being exploited, in some cases more aggressively. Pink coral, gold coral and similar terms are applied to a group of species generally living at the bottom of the Pacific and Indian Oceans. These corals, living deeper, are extracted by trawling. Quite simply, it is an ecological disaster to be totally indiscriminate and destroy all the organisms that thrive at these depths (between 300 and 2,000 m). It is true that attempts had been made to extract coral more accurately using the robots that we spoke of, but the venture was shortlived, as the cost of extraction rose too high at such depths: the robot was too expensive (buy it, keep it, move it with the boat) for the benefits obtained. For that reason, it was back to using bars and deep-fishing trawlers, and razing still deeper communities with yet slower rates of growth to extract Mediterranean red coral. These corals, in some cases, could have been centuries old, according to experts such as Brendan Roark of the University of California, who has studied coral growth rings by a sophisticated technique of stable isotopes.

That is why the idea was revived to put all species of the genus *Corallium* and *Paracorallium* on the list drawn up by the Convention on International Trade in Endangered Species of Wild Fauna and Flora (CITES). In 2009 the Americans again wanted to include them on the list of protected species, after not achieving this in 2007. Two meetings were convened in Hong Kong and Naples in 2009 among biologists, coral industry professionals, managers of resources and CITES specialists to see if it was possible (and appropriate) to include them in Annex II, which regulates trade between countries. The United States is one of the leading importers of red coral red (more than half of world production), both of Pacific origin (70% of the total of precious coral) and the Mediterranean (30% of the total). The Italian coral trade (the most important in the world) has a turnover of about $230 million a year, although this figure could be much higher since much of the coral is 'diluted' in other jewellery, perhaps forming part of a necklace, a pair of earrings or a bracelet made of precious metal, in which case the coral is not taken into account when valuing the piece. 'We are not ready to follow this irrational plundering of the resource', said the promoter of the meetings, Andy Bruckner, the American NOAA.

CITES or not CITES… in the end, the coral species did not make it onto the list in 2010, and it was left until 2016 for consideration. The reality is that coral is one of our most overlooked renewable resources, without a management plan anywhere on the globe and with little knowledge of stocks anywhere, especially the Pacific. 'In Taiwan, only 2% that is removed is alive, the rest is either dead or in very poor condition', says Dr Chin-Shin Chen, of Taiwan's Department of Fishing. Fishing in the Pacific is inefficient, as trawling destroys the majority of coral and leaves it on the seabed unharvested. Only a small proportion reaches the boat because, among other things, in some cases it has to go up to the surface from more than a thousand metres deep, so it mostly falls off uncollected. Dead coral reaches the boat, because its population has been trawled over and over again, leaving many colonies destroyed. Not too long ago, this was the situation that we had in the Mediterranean using St Andrew's cross gear and the Italian bar.

It is difficult to estimate how much red coral there is, because there have been few serious studies. Areas of the Pacific, on which there is practically no information, are currently the worst in this respect: 'We know nothing of what there is, especially at depth', says Giovanni Santangelo, lecturer at the University of Pisa: 'It is one of the most neglected resources of the fishing plan in the world.' How can specialists call for the imposition of a precautionary principle to prevent further exploitation when they do not know for certain how much coral there is? Moreover, not everyone agrees that going on the CITES list is the solution: 'with the CITES the management problem cannot be solved', complained Professor Richard Grigg of the University of Hawaii: 'We have experience of *Anthipates* (black coral) being included in Annex II since the mid-1980s. It has only served to increase the bureaucracy, but what is needed is a plan tailored to each region's stock management.'

These plans fail, and the first to worry has been regional and national government, both in the Mediterranean and in the Pacific countries that exploit this precious resource. The sector employs more than 5,000 people in some 270 factories in Torre del Greco, Naples, alone. 'The inclusion in CITES of precious corals would be a catastrophe for the industry', laments Ciro Condito, President of ASSOCORAL, the *coraleros'* union in Naples: 'People would not buy coral on account of the stigma of being on the CITES list, as they will believe that it is a species in danger of extinction, and the business will suffer much.' The president also complains of the bureaucracy that it would generate, and problems in identifying the coral genera in customs. Others support its inclusion on this list for the above-mentioned principle of caution: 'If we don't know how much there is, we must regulate its trade and do everything we can to manage the resource', says Andy Bruckner. In the end, whatever happens, 'probably very few countries will decide to make comprehensive maps of the stocks of one of our most valuable marine resources, from a monetary and cultural point of view', concludes Professor Santangelo: 'I have been hearing this same old story for more than thirty years.'

But a door opens to hope. A European collaborative project between the CSIC and the University of Pisa is installing marble slabs to transplant adult colonies direct to where they have become extinct. 'This is an ambitious project but one that has already been proven to work', says Lorenzo Bramanti, a researcher at the Italian University now living in the French CNRS (see in detail in chapter on restoration of benthic ecosystems). But perhaps the most interesting venture is to cultivate red coral. Although slow, its growth can be promoted by several methods that are now well understood to raise colonies for produce coral for powders and paste. This involves removing young forms from small colonies (highly risky) and placing them on a coral 'pasture' under controlled conditions. There is always the danger that these natural young forms would not in fact become extinct, otherwise, but might grow into huge branches of 'luxury' coral. What is clear is that the coral above 70 m must be managed, for both tourists and for anyone else, such as divers, to appreciate a well-preserved landscape and an endemic species of unquestionable beauty that, when allowed to grow at its own pace, creates shapes and colours that are second best to none, including the tropical coral reefs. A change of mentality that would be of great benefit is the growing shift to ecotourism and the rational exploitation of our shores in a less aggressive manner. What is clear is that coral such as was seen by the Greeks and the Romans more than two thousand years ago, beautiful and omnipresent, will not now seen for generations.

A Bit of History

As discussed in the first chapter of this book, red coral has been used by humans since prehistory. The artefacts found in Chamblades, Switzerland, date back to Palaeolithic times, about 25 thousand years ago, and were supposedly amulets, and coral is also found in the necropolis of Vinica, Slovenia. The spread of articles from that time made of this precious substance indicates a burgeoning trade the length and breadth of the Mediterranean. However, it was in the Neolithic when coral started to be used as currency and be exploited in a more systematic way in certain areas. Already by the time of Hellenic hegemony in the Eastern Mediterranean, Theophrastus (370–288 BC) asked about the nature of this 'red stone', so different under and out of the water, 'similar to a red root that grows in the water'. We now know that coral at the time formed dense forests, some probably just a few metres deep. Red coral branches were authentic trees, about 20–30 cm in height on average, with some reaching a metre.

The coral business had a major boost from the tenth century onwards, and the Catalan coast (Begur, to be precise) came to be an important source from then up to the thirteenth to sixteenth centuries. Trade was extended by the Silk Road, which also became the coral route. In a sense, the business remained closely tied to Mediterranean waters but in changing localities, as the centres of trade may have depended on the greater or lesser influence of various Mediterranean cities.

Catalonia was one of the places where the coral business flourished between the fourteenth and fifteenth centuries, coinciding with the region's mastery of the Mediterranean. From the sixteenth century there began to be more difficulties with piracy of the precious loot, coveted for both its value as a raw material and as currency. By the end of the seventeenth century the coral on the coast of Begur and further to the north seemed to have been exhausted, and the *coraleros* moved with their gear to places such as Sardinia and Corsica to continue exploitation. At the end of the eighteenth century the centre of trade became Torre del Greco, near Naples, where the business became firmly established in the nineteenth century.

At all times, red coral has been considered to be a precious commodity, as both a product to trade and a jewel of great value, but with different meanings. In India, coral is associated with the deity of Mangala and with Mars, while further to the north, in the villages of the Himalayas, it is a symbol of protection against the vastness of the landscape, the weather and the harsh climate. Buddhists considered coral and turquoise to be symbols of the perfect harmony between the forces of nature. A sign of fortune and protection from the tenth century BC in Yemen, coral also indicated status and prestige, as in West Africa. The spread of coral did not stop at Europe, Asia and America, but crossed the Atlantic and was incorporated into the culture of native Americans, especially in the north, where curiously it was taken as a symbol similar to that of the people of the Himalayas: protection against inclement weather, and the landscape of planet Earth being in communion with nature.

What is really strange is that this precious organism was regarded as a plant but also recognized as an animal, something that many people, even today, cannot believe. The confusion is reasonable, because an animal 'kills, becomes red and hard to get out of the water' according to Pliny the Elder, so coral does not exactly possess the typical profile and, between the seventeenth and eighteenth centuries, the leading sages of the time had not decided. There was a quarrel between those who considered it an animal, a plant or even a mineral, until one of the greatest naturalists of the mid-1700s, Peyssonnel, catalogued it as definitely as animal. After an arduous exchange of letters, he persuaded the leading members, and Carl Linnaeus, the greatest classifier of all time, catalogued red coral as *Madrepora rubra*. There were continued disputes among scholars. Today, people still believe that it is really a plant: a beautiful vermilion tree, the result of the decapitation of Medusa.

Emperor Sea Mountains

One of the most spectacular underwater mountain ranges on the planet lies in the middle of the Pacific, north of the Hawaiian Islands: the Emperor Sea Mountains, stretching more than 5,800 km from the Aleutian arc to Kure atoll. The peculiarity is that this vast region is mostly submerged, the skirts of the mountains extending several thousand metres under the sea. There are gorges which climb from 1,000 to 400 m, and their unique position allows them to host almost unknown wildlife. In 1965, the Japanese fleet that specialized in the extraction of precious corals discovered vast colonies of these and other cnidarians of great value, and began systematic plunder. For this purpose, in areas such as the banks of Milaaukee on the seamounts of Koko, tens of boats were equipped with gear consisting of nets hung with heavy weights to drag in everything before them. The Russians and Taiwanese, taking advantage of the deep banks area in the midst of international waters, followed the example of the Japanese. In 1966 the catch was 375 tonnes of precious corals, especially *Corallium* spp. In just three years (1965–1967), more than a thousand tonnes of very long-lived coral were taken, and some parts might have been centuries old. The catch has fallen sharply, not only because the coral was taken from many places but because the price dropped sharply due to the volume of the product on the market. During peak production more than a hundred large-tonnage boats were sweeping banks that had remained undisturbed for thousands of years. In 2006, a Japanese expedition made several transects in the area using a special robot able to descend to more than 400 m. The report mentions sparse fauna, just a gorgonian or coral branch here and there. We will never know about the forest of coral in that area prior to the looting, because the systematic removal of all sessile organisms has done so much damage that their distribution and numbers are unclear. Some places must have been beautiful, filled with all kinds of organisms taking advantage of the currents and the shelter of this marine forest in areas where no one would have imagined so much exuberance. It will be hundreds, if not thousands, of years in the future before the area recovers from this devastation for the sake of items of jewellery.

An Example to Follow

It has not all been bad management, in the case of precious corals, as there is an exception. In the Hawaiian Islands, for nearly four decades people have commissioned a young researcher to undertake an evaluation of the status of black coral to (mainly *Anthipates dichotoma*) and draw up a model for fishing. Richard Grigg was especially interested in work, because as well as being a marine biologist in Hawaii he was a collector of black coral in that part of the world. The first thing he did was to assess the minimum capture size; that is, as mentioned in the text, the minimum heights and diameters that could be extracted to allow the entire coral population to survive. The study came to the conclusion that coral should not be less than 1.20 m high and about 2–5 cm in diameter. The fishermen of the area adhered strictly to these guidelines, in part because they identified with Dr Grigg. With other people, Grigg promoted the creation of exclusion zones where coral (and other organisms) could grow and reproduce without human interference.

In 1998, the demand for black coral in the market rose by 50%, so the minimum sizes were relaxed, and colonies of less than a metre in height and of a diameter markedly less than the threshold size could be gathered. Technological advances allowed the populations to be pinpointed by GPS, leaving nothing to chance. In just a few years the biomass of the population fell by 25%. This sudden decline was due to an invasive species, *Carijoa riisei*, which began to interfere with the biological cycles of the precious coral. This species tends to grow on top of other species of arboreal corals, interfering with their food capture. Fortunately, Grigg's work was reviewed to devise a way to remove it, following approaches based on the biology of the species and the conditions in the ecosystem. This is an exemplary model that takes into account the evolution of the system, the advances in technology that can damage the resource and the biological basis of the exploited species. Richard Grigg, now deceased, managed to combine his knowledge as a marine biologist with his expertise in the extraction of a renewable resource to maintain something precious that would otherwise have already gone.

Part III

Effects of Pollution

12

The Silent Threat

When, at the beginning of December 2008, a maritime accident off the coast of South Korea caused a spill from the tanker, the *Hebei Spirit*, I little imagined that just two weeks later I would be travelling there to try to assist in the emergency. With no experience of maritime oil tanker accidents, with two other colleagues who had extensive knowledge of contamination by petroleum products I was invited by the regional government of the affected area in South Korea to shed some light on the spill's impact on the bottom of the sea and the potential for the recovery of its communities,. At the beginning, a little dazed and trying in the shortest time to absorb the most that I could of the protocols, methodologies and past experience in this field, I understood how my colleagues had felt at the *Prestige* oil spill, an emergency that had also demanded rapid response and immediate actions.

When we arrived on the affected coast, there were hundreds of people in small, well-organized and disciplined gangs, cleaning off the oil that had arrived, stone by stone, with unshakeable determination. We saw them working away relentlessly on the beaches, cliffs and bays. In a short time, the greater part of the spill had been collected up, so much so that it was hard for us to find visible signs of the catastrophe at various places that we visited. It was then that I understood something that threw me: only 4% of the oil that enters to the marine environment is from accidents such as the *Hebei Spirit*.

It is estimated that more than half of the petroleum oil that goes into our seas each year is natural, from springs located in various sources the length and breadth of the planet (especially in the Gulf of Mexico). The remaining 46% is the result of refuelling, bilge cleaning and other routine operations. However, a report promoted by several associations, through the Oil Spill

© Springer Nature Switzerland AG 2019
S. Rossi, *Oceans in Decline*,
https://doi.org/10.1007/978-3-030-02514-4_12

Conference, notes other, lesser-known potential sources: at the bottom of the sea are 8,569 wrecks of ships that are considered to be of a medium or large tonnage, of which 1,583 are oil tankers and other 6,986 transport materials and people, or are warships. More than 75% of these vessels sank due to clashes or attacks in the Second World War, thus are what the experts consider 'mature', so that in many cases their structures have corroded and have begun to leak the oil that was their cargo and in their fuel tanks, derived from crude oil. After more than sixty years of waiting, some areas where subsidence has been especially intense may be affected by this deterioration, especially in shallow, closed seas such as the Baltic and around the Philippines.

The most affected areas of the planet are Southeast Asia, the Persian Gulf, the east coast of the United States, the North Sea and Baltic Europe. 'More alarming estimates tell us about 70–80% of ships with their tank, or tanks, filled', reports Jacqueline Michel of the Research Planning Incorporated of the United States. It is not easy to tell exactly what quantity of fuel had been carried by either tankers or freighters, and there are vessels for which there is no reliable information. In any case, even the more optimistic estimates speak of 2.5 million tonnes of submerged crude oil, while the most pessimistic put it at over 20 million. It is believed that transport tankers may each contain between 30,000 and 40,000 tonnes of crude oil, and other cargo or military ships have between 700 and 3,000 tonnes of fuel.

The issue is to find a solution to the silent threat. 'Recovering crude oil or refined fuel from ships is very expensive', says Dagmar Schmidt of the Environmental Research Consulting in New York: 'The majority cannot be made without underwater robots (ROVs) and few companies have the appropriate technology.' The *Prestige* case demonstrated that, with the appropriate technology, it is feasible to recover fuel from a ship that is thousands of metres down: after the accident, and more than 14,000 tonnes were pumped up by one company, REPSOL. However, fuel recovery from more than 250 m deep is considered to be a highly complicated and expensive operation, 'and the worst thing is who is willing to pay for it. Many sunken ships are due to acts of war, thus there are legally prescribed responsibilities', concludes Schmidt. In the study there is nothing about other contaminants, such as chemical weapons, hazardous or toxic cargoes or radioactive materials. Logistical and legal gibberish that does not facilitate the planning of an adequate clean up acts to the detriment of a system that hosts such a large number of potential threats to the balance of the submarine world.

The problem, once again, is that we can see the untimely discharge if it involves something striking, such as an oil rig accident. For instance, the inhabitants around the Gulf of Mexico witnessed the *Deepwater Horizon* explosion and subsequent inshore spill of crude oil (see below). The black tide had an immediate, scandalous effect: fish and birds die, and beaches and rocky areas are stained with crude oil, which quickly spreads over large tracts, prompting many people to clean the area to restore the site, trying to return it to its 'pristine' condition.

The effects of such a discharge are many and varied, and may be both immediate and postponed for a long time. One of the first things to go is the light, in some instances decreasing photosynthesis in plants (microscopic, multicellular algae and marine). In open areas, with strong currents and swells, this effect is minimal, but in more enclosed areas where the water circulation is less it can be dramatic. Just before the end of the Gulf War, more than 700 km of coastline were affected by the discharge of millions of tonnes of crude oil, forming in some places a continuous and impenetrable black or dark brown film. One community in the column of water, phyto-plankton, was unaffected by this lack of temporal light, due to its regenerative power and great resistance. However, for the shallow communities of sper-matophytes and areas of coral reefs (which have symbiotic algae inside), the immediate and prolonged lack of light caused by the fuel slick in certain areas resulted in mechanical asphyxia.

In 2001, ten years after the oil discharges in the war, only about 20% of the marshes affected by the sabotage by Saddam Hussein's forces—between 90 and 95% of them—had recovered from the asphyxia caused by the oily coating on water, sediment and plants. In areas like these marshes, where the flow is weak and there is a low water renewal rate, the effects can last for centuries, according to some experts. There are still some areas that are not fully recovered. But in more exposed areas, beaten by strong waves or wit-nessing daily a strong tidal motion, the effect is much less prolonged. After the crash of the *Exxon Valdez* off the coast of Alaska, the more exposed areas fared better than the backwaters or secluded coves, with macroalgae, where the water barely circulates.

The methods for cleaning up spills, as we have learned much from recent disasters, are vigorous: in the case of the *Exxon Valdez* in 1989, involving more than 40,000 tonnes of crude oil (260,000 barrels), the cleansing was so conscientious that it sterilized an important part of the rocky areas, making it take far longer to recover. Several methods were used: hot and cold pressure, injections of sand, water steam, pulverized dry ice… Paradoxically, the more intensively washed areas recovered slower those areas than had been treated less vigorously, the worst being where pressurized hot water hoses had been used. This had literally sterilized the substrate, killing any organisms. That's why, in the case of the *Hebei Spirit*, the conscientious volunteers were cleaning the area stone by stone, with cloths and biodegradable products. Those areas least accessible by land or by sea were left unattended, yet the sea, typically more aggressive in such places, achieved this patient washing without the intervention of man. Waste fuel, coating rocks, sand, gravel and everything in its path in an oily mass, quite apart from the obvious ecological damage due to suffocation and mechanical traction involves visual and landscape damage that affects us in a special way. As we have said before, wild or rural landscapes affected by the oil slicks incite the population to remove as soon as possible the stain that both affects us and makes us feel so guilty.

But there is another component, much less visible and therefore overlooked, which is many times more damaging to the ecosystem if there is a spill, either by accident or by a continuous toxic drip. Dumping fuel in any form, or other contaminants such as radioactive or toxic chemical residues, causes changes in marine ecosystems. We've all heard how an area is closed to fishing following a disaster. Along the Galician coast and part of the French coast after the *Prestige* incident, it was months before fishermen could go out again, though the closure was, in some cases, only a precaution. After the first few months, much of the crude oil enters a latent phase, as while the oil has disappeared from our field of vision it has not gone from the affected area. This can continue for a short or longer time and be either intense or less so, depending on the type of discharge, yet at this stage the oil tends to take its toll, especially on certain links in the food chain. The most vulnerable are those organisms in which toxins accumulate, as they lack the opportunistic nature of other species and their way of reproducing. In one way or another this decreases, through morphological change, such as genetic mutation, or reduced immunity, their progeny and their ability to survive.

After the disaster of 2002, on the Galician beaches, for example, between the high and low tide marks there was a major reduction in the numbers of creatures fixed to the rock. In 1995 there were more than 2,500 in an optimal state of health, but this had been cut to less than 800 by 2005, and they were in a precarious state of health. The recovery in this area had been more rapid than in many others, depending on the substrate, the type of crude oil and the hydrological and geographical characteristics of the area. It also depended on the affected community. Plankton did not appear to have been much affected by pollution from the *Prestige*, due largely to their short life cycle and its wide fluctuations from year to year. The intertidal zone was affected in a quite different way, yet it is also one of the communities that is recovering, thanks to the continuous purifying action of the sea, concentrated due to the area's high-energy tides, currents and waves.

In terms of deferred effects, the outlook is less well known for those areas of platform that lie tens or hundreds of metres deep and attract trawler fishing: 'We know that the type of fuel that the *Prestige*, due to its low solubility, did not affect the respiratory system of the fish,' explains Francisco Sánchez of Institute Spanish Oceanography of Santander, 'but much fuel has been brought into the system in the form of marine snow (aggregates of organic matter that float in the water), and enter the food chain mostly through its predatory organisms.' Much of the crude oil, when fragmented, can enter the mud and sand on which many commercial species feed.

In the clean up after the *Prestige* incident, the largest crude oil slicks were converted into smaller droplets, then tiny particles. Once ingested by an organism, the particles' effects depend on the type of hydrocarbon. For example, it is known that after the *Exxon Valdez* disaster the eggs and larvae of fish and invertebrates such as mussels and sea urchins were malformed, causing higher mortality rates and a much lower than expected subsequent reproductive rates. Most of the studies after the spill focused on the early stages of life of the species that were of ecological or economic interest, being more fragile, thus potentially more affected by the accumulation of contaminants. There were also studies that focused on adult humans, for example research into cancer, infections, reduced immunity and other effects that can last for generations in places where a pollutant is present for a long time. The recent BP disaster will have long-term implications, and today we are seeing only the tip of the iceberg. Probably in ten years we may be able to make an objective assessment of the real impact of a release crude oil almost twenty times greater than that of the *Exon Valdez* spill.

Most hydrocarbons are degraded by certain bacteria, fungi, protozoa and phytoplankton. But not all. Some—the most recalcitrant—can go years without being affected in the least. The worst is when they come to be part of the sediment, where the biodegradation processes slow down tremendously. Depending on their concentrations and the reactivity of the metabolic processes in the organisms' cells, the damage may be major or minor. The worst thing is that effects spread gradually from one organism to another causing diseases, vulnerability to parasites or death.

The quantities of oil spilled during the last forty years have been declining dramatically (apart from catastrophes such as BP's *Deepwater Horizon*). If we collect the total discharges in this period, 56% were in the 1970s, and in the first decade of 2000 we have seen less than 3%. There has been progress both in the ecological sense and in the irretrievable and useless waste of a valuable product. What remains clear after gaining much experience is that we should conduct long-term monitoring. Only in this way may we assert with confidence trends in any ecosystems that are exposed to pollution from the discharge of oil or other contaminants in future, compared to their initial condition, and quantified in the form required. Logically, we ideally need some notion of how things were before an accident, yet this has not always been possible and we must employ a certain degree of realism regarding the affected areas. A series of indices of the state of health of ecosystems (integrating data chemical, physical, biological and ecological data, and also those from the engineering, management and the economy of the area) will allow us to deal with the silent threat, that which is hidden behind the phenomena of contamination by organic and inorganic agents, in a consistent and well-structured way. Unfortunately, as we have said before, the problem goes beyond oil being spilled by accident, due to both widespread practices in both maritime traffic and the disposal of hulls of tankers that have transported crude oil (which should be all mandatory, by now). From leakage along the way, continuous discharges of all kinds and this huge number of wrecks waiting in latency will make things even worse if nothing is done. Let us remember once again that our oceans are the synergistic sum of effects; we are adding to the equation more variables that are detrimental to their proper operation. The future impact of radioactive drums being opened up by the erosive action of the sea, shipments of bombs or chemicals from previous wars, now about to 'hatch', sunken ships, or those that have lost part of their payload and that have ended up at the bottom of the sea may be felt by the living populations of the seas in just one locality or spread around the globe.

BP Disaster

On 20 April 2010, eleven people lost their lives in an explosion on an oil rig that extracted oil from a depth of more than 1,500 m. A series of mechanical and human failures led to what is considered to be the greatest ever oil catastrophe involving oil extraction. The *Deepwater Horizon*, in something less than 90 days, poured out some 4.9 million barrels (780,000 cubic metres). To understand this magnitude, just remember that the *Exon Valdez*, in one of the largest oil spills in history, discharged 'only' 260,000 barrels.

As we will see in a later chapter, underwater mining and the extraction of oil from the deeps is a reality. In August 2010, two other platforms were about to become operational at more than 4,000 m down. And so the practice continues, in the Arctic, deep and remote zones, prospecting in unexpected places… We need oil, and we need it now. So, although experts continue to say that what happened was 'virtually impossible', experience shows us again and again that impossibilities are precisely what we must take into account, as humans are capable of creating the most advanced technology and at the same time making the biggest mistakes. 'Despite being a very rare accident, we cannot fail to pass over it', says Paul Bonner, Professor of Petroleum Engineering at the University of Texas.

It's always a bad time for something like this, but it is agreed that many economic, social, physical and ecological variables made things even worse: it was the time of migration for birds, the hurricane season, the fishing season and beginning of the second main wave of tourism in the area. The oil slick expanded from 1,500 to 10,000 km^2 in just five days (from 25 to 30 April), penetrating the marshes, beaches and delta of Florida, Alabama and Louisiana. But, as insists Prosanta Chakrabarty, a Louisiana University ichthyologist, it was 'not only pelicans' that were affected by a disaster of this magnitude. Unlike other accidents this occurred at depth, and the dynamics of the spill is very different. 'The stain will introduce oil and gas at medium and great depths', says Samantha Joye of the University of Georgia. 'Part of the oil evaporates, but another will enter into the pelagic and benthic system with yet unknown consequences.'

In fact, a state of anoxia that is both harmful and little understood has been added to the consequences that this area suffers, year after year, thanks to the 'dead zones' (see next chapter), water with little or no oxygen generation due to algal production at the surface and bacterial decomposition. We now know that this has affected many communities. 'One of the worst affected areas on the border between the land and sea are the marshes', says Brian Silliman of the Department of Biology of the University of Florida. After two years these wetlands were still coated in crude oil, and the system still does not function well. 'Even though time has passed, the erosive effects remain incalculably devastating, with loss of biodiversity and biomass', insists Silliman. The fact is that part of what often wears down those communities we do not see, as it takes place at depth. 'Deep corals, of slow growth, have been very affected', says Nancy Prouty of the US Geological Survey. Flocculent material covers these organisms so they cannot eat, breathe or perform in these conditions. 'These organisms can grow in diameter about 15–30 µ, very little,' adds Prouty 'implying that a disturbance like this can leave out a lot of colonies.' In adjacent areas, more than 45% of corals were damaged. 'More than 90% of colonies were affected, with what that implies for its recovery', witnessed Helen White, a chemist at Haverford College.

After the first onslaught of the *Deepwater Horizon* accident, some of the solutions were worthy of *Marvel* comic, such as proposing closing every channel with stones to prevent the entry of oil, or to send a fast and running barrier of sand down the coast, involving some 68 million cubic metres of sand, at an estimated a cost of $350 million. 'It was simply awesome', said Joseph Kelle, a geophysicist from the University of Maine: 'Those who proposed this have simply not have thought about the consequences.' Bacteria have begun to break down the oil, but the spill will affect the system for decades.

What is clear is that we are a long way off mastering this type of exploitation. Once it has happened, a catastrophe at these depths is exceedingly complicated to combat efficiently. But it is not enough to look at what is happening today with the *Prestige*. A study carried out between March and October 2006 found that the *Prestige* was still pouring out fuel. Even after the initial spill and recovery by REPSOL the ship had more than 50,000 tonnes of fuel onboard, and this has continued to escape from its tanks despite efforts to seal the leaks. The study, carried out by Saioa Elordui-Zapatarietxe of the Institute of Science and Environmental Technology of the Autonomous University of Barcelona, focused on the dispersion of the fuel in the various layers of water. 'In this area we find a very complex structure of oceanic masses, with surface waters, Mediterranean and deeps', the scientist says. 'Fuel discharges were especially intense in October 2006, four years after its collapse.'

These discharges could be observed from the surface, but their concentration was greatest in the layers closest to the wreck. Bearing in mind that in this and other areas more fish from very deep zones are being extracted all the time (an increase from 5% in 1970 to 35% in 2007), the release of fuel that had been circulating in just the remotest layers of the ocean may end up interfering with fishing.

The Tragedy of Plastics

We have gone from generating about 30 million tonnes of plastic per year in the 1970s to more than 200 million at the start of the new century. Bottles, containers, dividers, packaging balls to protect products, bags and a long et cetera form a long production chain in which the various forms, compositions and textures of plastic have become part of the technological and economic success of our society. Plastic is cheap and useful, and a kilo of packaging costs no more than €1.3 and occupies a considerable volume. Far from being optimal for recycling, a part of these products (around 10% of plastic produced, depending on the area of the world) goes into the sea via sewers, rivers or floods. In fact, around 80% of floating debris comes from the mainland. Surface currents and tides carry it and tend to make it accumulate in the centre of the bounding oceanic currents. In the middle of these systems, like in the Sargasso Sea in the North Atlantic, and in the North Pacific, the winds are slack and currents weak. 'Plastics accumulate in areas with little traffic by navigation, away from the coast', says Rei Yamashita of the Japanese University of Mie Kuroshio: 'In one of the most important in the world, the current moves large quantities of floating objects from the coasts of China, Korea and Japan, strongly industrialized countries, to the rotation of the central Pacific.' In this area, plastics may form in excess of a million elements per square kilometre. And waste, in large quantities, also arrives here from the coast of California and Oregon. In this and other areas, as in the North Sea, the accumulation is a very worrying fact: 'Only in the restricted zone between Great Britain, the Netherlands, Germany and Norway has it been quantified, by pelagic fishing, underwater robots and visual transects, that there is up to 600,000 tonnes of floating and submerged waste', says Gerard Cadée of the Netherlands Institute of Marine Research: 'More than 80% of the floating debris of medium and large size bears marks of having been bitten by sea-birds, which indicates that they take it to be part of their diet.'

The problem is much more than aesthetic. Plastics (the main culprits) of various shapes, textures, buoyancies and sizes are interfering in the components of marine ecosystems. David Shaw reports that 'The intake by marine creatures such as whales, birds, turtles, fish and squid bodies is very high.' Shaw works in the Institute of Marine Science at the University of Alaska, and has seen that, especially as chicks, species such as albatross and petrels are the hardest hit: 'Adults see the small pieces of plastic, including plugs or balls of Polyexpan (gel-forming reactant, used in waterproofing), as potential prey, ingest them, head to the nest and regurgitate.' The intake is selective, since textures, shapes and certain colours confuse the birds, which identify the plastic with potential prey. If they have not already died, the birds regurgitate these objects, affecting the lives of their progeny, their future generation. Unfortunately, it has recently been discovered that many of these small- and medium-sized plastics, moreover, are capable of accumulating toxins such as DDT and PCBs: 'Toxic build-up in the floating plastic has been proven,' says Dr Lorena Rios of the University of the Pacific, 'and the worst thing is that 44% of the species of marine birds consume them routinely.' The problem would be merely an interesting fact if the amounts of plastic that floats were insignificant, but in certain areas of accumulation, as in the Sargasso Sea and the North Pacific twist, there is a huge, highly variable mass several times more extensive than the Iberian peninsula. 'We were able to see that in these areas the relationship between the weight of the plankton and the weight of the plastic and other floating objects was one to six: that is, per kilogram of plankton, there are six of waste', reports Charles Moore, along with retired Professor Curtis Ellesmeyer of the Algalita Marine Research Foundation, both researchers who have been involved for more than a decade on this problem. 'We do not know the extent to which so much material has affected marine ecosystems, pelagic or benthic, but it is clear that they are interfering in the food chain and the proper functioning of the system.' In some areas, it has been calculated that the 'cloud' of objects can reach between 20 and 30 m deep. Recent research has estimated between 5 and 13 million tonnes of plastic entered the marine system in 2010. 'The worst thing is that the prospect for 2025; if we continue at this rate,' says Jenna Jambeck of the University of Georgia in the United States, 'it may be up to an order of magnitude greater.'

The problem is exacerbated when we consider that the plastics become fragmented. Visible plastic does not interfere so much as microscopic bits. 'We don't know very well the parameters such as the speed of collapse or the effect on real organisms', says Andres Cortazar of the Faculty of Sciences of the Sea at the University of Cádiz in Spain. 'Although the concentration may not be as high as is thought in many places, it is essential to understand the impact on food chains, especially that of small pieces'. In just one cubic metre there may be 100,000 tiny bits of plastic. 'These sometimes-microscopic pieces of plastic cause abrasion, ulcers, blocking of the digestive system by toxins, metabolic disorders, malformations... shall I continue?', says Stephanie Wright, in an extensive review of the subject.

In addition, we have the problem of degradation. Items break down slower at sea than on land, partly because the temperature is constant and low and partly because at sea there is more protection from the action of the sun's ultraviolet rays. While studies vary in terms of their conclusions, estimates of degradation of most solid waste of artificial origin (bottles, bags, etc.) indicate that its disappearance exceeds the scale of a human lifetime. While a bus ticket takes only two to six weeks to vanish, a plastic bag can spend fifty years in circulation, and a PVC bottle more than five hundred.

The Mediterranean, due to the great urban, agricultural and industrial concentrations on its shores, is one of the places most affected by this phenomenon. In the Catalan coast there is between 1,000 and 1,200 cubic metres of floating waste every year just in the area near the coast (in the summer of course, due to the crush of tourism). 'The other problem is that these objects have demonstrated that they can be bearers of invasive species, such as the algae that form red tides,' says Mercedes Masó of the Institute of Marine Science-CSIC, 'polyps that release young jellyfish or small thalli of macroalgae, for example; this partly explains the rapid expansion of some algal blooms.' In Catalonia, a flotilla of dozens of ultralight boats records pollution spots and collects floating solids when they pose a problem for tourism. But tourism itself produces much of the tide of waste, especially in certain areas. 'In Mallorca there can be up to 36 objects per metre along on the beaches, more in the summer, up to twice that of the winter; and there are no major rivers to blame', notes Martínez-Ribes Lorraine of the Mediterranean Institute of Advanced Studies, CSIC. The amount of waste always increases in summer, when antisocial people decide to use both beach and sea for their trash. Overall, what can we do for the sea? The issue of plastics is one that is only now beginning to be studied rigorously. Until recently, only a handful of specialists have been trying to draw attention to the possible effects of the huge discharge of solid waste that our seas have to endure. It seems that we are now taking it a little more seriously… but maybe the effects have already been so damaging that we cannot reverse them in the short term.

Mercury in the Arctic

As oil derivatives, many other compounds enter the marine system and can harm both organisms and mankind. Their tendency to accumulate as they remain in the food chain has been known about and tested for several decades, when studies on DDT and mercury made it clear that higher organisms (eagles, dolphins, seals, bears and humans) were the ones to suffer the aftermath through foetal malformation, disease and metabolic deficiency. For this reason, the following example is of mercury and the latest findings in the Arctic, a system particularly vulnerable to the accumulation of toxic substances.

While extracting the last ice cores to check the accumulation of pollutants, Dr Feiye Wang's team, from the University of Manitoba (Canada), confirmed that mercury levels seemed to have stabilized in the environment yet not in the animals of the Arctic. Emissions of mercury during the last decade have decreased: the regulation of waste and alternative treatments in various industrial processes seem to have had an effect. However, during the same period of time, in animals like beluga whales and polar bears the levels of so-called organic mercury, or methyl mercury, have increased by an order of magnitude.

This apparent contradiction may be due to one of the side effects of climate change: 'everything that is happening in the Arctic is very complex; people see above all changes in the level of the sea or open navigation steps, but biologists see the outstanding impact on flora and fauna.' Methyl mercury is a highly volatile substance, and it travels long distances by air from temperate places, where there is more industrial activity, to the poles, due to air currents. This liquid metal is absorbed by contact with seawater and may form molecules of methyl mercury, easily assimilated by the tissues of creatures forming the various links of the food chain. It seems that times of thaw, when the crystal structure of water changes, are the most delicate, since this element is then absorbed more easily.

Unfortunately, methyl mercury has a long history of negative effects on wild flora and fauna, and humans. Its accumulation caused scores of deaths and hundreds of affected people in the Minamata Bay in Japan, but its victims are spread all over the planet, from the Amazon to the Inuit of Canada and people of Greenland. The places where the population is most vulnerable are those where a large amount of fish is consumed, and people tend to have levels of mercury in the blood far above the average. 'When the water is liquid, there is no greater absorption of this element', says Wang, 'but due to climate change water at the North Pole stays frozen for less time, so the amount of mercury absorbed by the system tends to increase.' In fact, the mercury that is trapped in the ice is retained, and does not circulate through the marine environment until the ice thaws. 'We have stopped issuing mercury in an uncontrolled manner,' says Dr Gary Stern, Department of Fisheries and Oceans, again at the University of Manitoba, 'but agencies such as belugas continue its tendency to accumulate in the body.' Dr Stern has made a scrupulous follow-up of these cetaceans since 1981, and says that we 'are aware that there is a tendency to build up from the most basic parts of the food chain (microscopic algae) to the highest (beluga whales and polar bears).' This is normal, because as we rise in this chain toxic molecules accumulate in our tissues, and also have an effect on the various types of prey taken by a single individual. 'The build-up that we detected is insufficient to explain the gap between what is emitted and what accumulates. There has to be something else.' According to this expert, it could be centuries before there is control of this metal entering the environment. This is one of those little-known effects of climate change, called 'wicked' by some specialists, because they are not expected and not easily visible. Some scientists predicted that areas of the Arctic would be free from ice in 2015. Today, in 2018, there is a quasi-permanent passage in which the ice is weak or non-existent. It is the freezing and the thawing of polar water that seem to be favouring the entry of mercury into the marine system, so that, although the emission source has decreased, much of the mercury currently circulating in our atmosphere could, through the poles, end up in our oceans.

13

The Swamped Shoreline

There are so many of us! It is a well-known topic, much repeated, and is something that seems almost banal, yet it is an inescapable reality. We consume a great deal of energy and products, we need resources of all kinds and generate mountains of waste. One of the biggest problems is that more than half of the world's population lives less than 200 km from the coast, and all this activity, all that movement, has a direct impact on the sea. Coastal areas are, undoubtedly, one of the more pressured and threatened ecosystems on the planet, as a group of experts warned in 1990 in Kobe. One of the direct effects on our agricultural, industrial and urban areas is the production of nutrients (especially nitrogen and phosphorus), artificial organic contaminants (mainly derived from oil) and heavy metals. When there are nutrients in excess, some species are capable of assimilating them rapidly, multiplying tremendously, such as in single-celled algae's rapid development. These algae prevent the growth of organisms whose life cycle is slower, and swamp anything that cannot compete in the race to reproduce. In the end, a huge quantity of organic matter is created that then dies and is taken over by bacteria, resulting in hypoxia or anoxia (see below). The cost of reducing nutrients is very high, especially nitrogen and phosphorus. In the United States alone, wastewater treatments across the country in the last thirty years have cost some €360,000 million, a considerable amount.

As we shall see in the following examples, this effort is working, but is it enough? Unfortunately, there are few ventures that can claim with confidence that the conditions are gradually returning to 'normal': we would need a long time series study to tell us if the conditions were equivalent to a return to the pre-industrial state experienced in a particular town in the mid- to

© Springer Nature Switzerland AG 2019
S. Rossi, *Oceans in Decline*,
https://doi.org/10.1007/978-3-030-02514-4_13

late-nineteenth century. It is not that the specialists (including myself) want to return to those times. This is impossible, for obvious reasons of demography, expenditure of energy and materials processing, but that model uses a baseline towards which must aim. In the great city of Oslo, sediments have been studied using a few specific organisms as indicators of changes over time: 'Sediments are an excellent time integrator if you know how to read and take into account variables such as rates of sedimentation, temperature, currents, etc.', said Elizabeth Albe of the University of Oslo. The organisms used are the foraminifera, protists that live both in the water column and on the seafloor. Looking at various places in Oslo Fjord where sediment had been deposited each year, Albe and her colleagues were able to see that, over a century, there had been profound changes in the waters of this area of Scandinavia. 'The anthropogenic pressure had reduced the number of species of foraminifera as much as 73%', adds Albe. Eutrophication and heavy metals had been the main cause of the impoverishment of the fjord, also the reduction in the diversity of the system both at the bottom of the sea and in the column, due to intensive fishing. Despite thorough controls that now remain in this area, the current populations of these organisms are far below those found in 1900, the starting point of this study. Predictably, in following the time series through the sediment cores that were studied they saw that the most drastic change was in the 1960s, when industrial and agricultural expansion influenced decisively this and other areas of the world.

Many degrees of longitude and latitude distant, across the planet, there is another example of a short time series study, also one with major implications for the trends in many developed countries. In California, a rigorous study of over thirty years long has demonstrated unequivocally an improvement in many aspects of the state's marine ecosystems. The volume of solids in suspension has improved drastically in some cases, leading to the recovery of biodiversity in coastal areas and on the continental shelf. 'At the beginning of the 1970s, 75% of flatfish on the Central California coast were affected by a malformation of their fins, today that percentage to past to be less than 5%', explains Eric Stein of the Southern California Coastal Research Project of the United States. A reduction of 80% of waste derived from mineral and vegetable oils, or the virtual disappearance of PCBs and DDT, has to have had a concrete and palpable effect. It is not the total quantity of heavy metals and PCBs in the environment so much as its availability for pelagic and benthic organisms. However, a good environmental policy can bring about recovery even in badly punished environments. We must not forget that the sea has a capacity for recovery greater than that of terrestrial ecosystems, in many respects. Although the recovery of the fundamental pillars of predators, such as large ecosystems or coral forests, is much slower, the system is capable of responding promptly to certain improvements introduced by humans if we follow guidelines, apply the supporting science and treat the chance of a recovery seriously (Fig. 13.1).

But human stupidity is great, and our memory of history and ability to learn almost non-existent. Let's explore a part of the planet in full development: the sea of the Arabian Gulf. Charles Sheppard, of the Department of Biological Sciences of the University of Warwick in the United Kingdom, and his colleagues conducted an in-depth analysis of the latest outrage to human progress: the runaway expansion along of the coasts of this region. Due to uncontrolled growth, every shore in Qatar, United Arab Emirates and Kuwait is undergoing a rapid degradation of its ecosystems. We must bear in mind that the waters of this sea, mostly off the Arabian peninsula, are very shallow and some areas have been flooded for only three or four thousand years, due to the ingress of the sea caused by the melting of glaciers and polar ice caps after the last ice age about fifteen thousand years ago. We are therefore facing coral reefs, mangroves, seagrass beds and quite recent macroalgae. Coastal ecosystems are areas where the sheets of shallow waters are very large. Since the beginning of the 1990s, when pressure began to be exerted for industrialization, the creation of ports, airports, canals or desalination plants, as well as residential complexes or luxury business centres, there is no more than

Fig. 13.1 Example of urban overpressure on the Mediterranean coast: Algarrobico Hotel in Andalusia, Spain. *Source* GreenPeace

40% of the coastal area left. In some cases, there has been such drastic alteration that it has led to changes in currents and areas of increased evaporation of water in an area of the world where the sea is already more saline due its hydrography.

In certain areas, such as the famous artificial islands off Dubai, the coastline has been increased by 11% (more than 90 km^2), gaining ground on the sea and degrading the surrounding areas by adding sand. More than 150 km^2 have been lost, where the turbidity, the addition of nutrients and changes in currents already reflect the drastic changes in the dynamics of systems. Such is the lack of foresight by the engineers and biologists of the region that there are already recurrent eutrophic algae blooms and an absence of hydrodynamics. In the houses on the artificial islands (priced on average between €2 and €3 million), the bathrooms are impracticable due to the concentrated dirt and pestilence. Environmental impact reports are opaque or confidential, giving the main actors (government or real estate and large infrastructure businesses) recourse to commercial or strategic protection. But the reality is that the changes in this place have been extremely fast. The area has gone from having 60% of the seafloor covered by live coral to less than 1%, and gravel and dead reef areas have gone from 10% to as much as 60% along the most developed western coasts. There, the second population of dugongs (*Dugong dugon*) on the planet and one of the most vulnerable populations of green turtles (*Chelonia mydas*) no longer find healthy meadows of flowering plants, as these have been replaced by fleshy algae's rapid growth, in the most fortunate cases. Systems of high complexity and difficult recovery, such as mangrove swamps, have gone into reversal at an alarming rate.

We must not forget oil exploitation, as the area's ports offshore installations (more than 800 throughout the coastal area) handle 60% of the world's oil, and with more than 25,000 tankers full of crude oil and derivatives. The area acts synergistically, adding fuel to the fire. We cannot ignore that it is one of the places in the world where desalination plants are most concentrated, extracting some 11 million cubic metres a day to meet the need for water in one of the driest areas of the planet.

Two mega projects may yet be completed that will affect this area irreversibly (always in our time scale, of course). The first is the construction of a mega dam of the Aswan type, in the south of Turkey, which would retain the waters of the Tigris and Euphrates. The retention of the waters and their sediments would decrease the water balance in a sea of very high evaporation (as we have said before, one of the most saline of the world). But that would not be the most devastating effect. Due to the conversion of the area to other sources of income for when the oil business ceases, there is a serious proposal to build a new type of dam: to close the Strait of Hormuz. The purpose would be to create a hydroelectric plant with a waterfall from the Indian Ocean to the sea of the Gulf, of between twenty and forty metres high. Inside this dam, the sea water would evaporate. This project, if it takes place, would take an indicative budget of €44 thousand million and 30% of the annual production of cement and concrete on the planet. And it would close a sea. The traffic of oil and other goods is already achieved in other ways, but what is certain is that no one has stopped to think about what would happen to the coastal and oceanic ecosystems in the area, nor the serious consequences of a kind of gigantic Aral Sea, all to create energy. And no one has thought of people, of course—of the millions of inhabitants of the coasts here. The same engineering approach has been on the table for the Red Sea and the Mediterranean, although, in these two cases, I doubt that it will ever be progressed.

There are two further examples of nefarious coastal management: Cancun and the Riviera Maya, Mexico. Tourist expansion here has been the fastest and least planned ever known. Cancun alone has about four million tourists every year, spending more than $4 thousand million dollars. It is a very desirable cake and, for the sake of tourism, the regional and state government removes any real obstacle to unbridled expansion, even changing the opening hours to accommodate tourists from the east coast of the United States and Canada. Beaches have been transformed and mangrove swamps destroyed, and there has been strong eutrophication and erosion of the reef, which has gone from 40% live coral cover in the 1880s to less than 5% in 2010. There is no control, and the resident population, without proper services, has grown from about 60,000 inhabitants almost 800,000 in just two decades. As an example, the last stand of mangroves under the influence of the city of Cancun, the Tajamar, disappeared under bulldozers and trucks overnight (literally) to offer more expansion for the construction of hotels and apartments. 'Cancun lagoon has gone from being a precious emerald blue to a disturbing brown', explains Ernesto Arias of the Laboratory of the Ecology of Ecosystems of Marine Reefs, CINVESTAV, in Mérida, Mexico. 'Everything has changed, the chemical flows, availability of drinking water, biodiversity, the complexity of ecosystems', added Arias: 'The problem is that the waters smell ever worse, the beaches are no longer transparent and idyllic, the reef is poor and devoid of the life that there used to be, and our tourists feel more and more disenchanted.' In 2012, visiting its most iconic beaches, I found myself at six in the morning, with the resorts' employees, digging holes to dump the algae that grew out of control and that wreck the landscape. These algae are now rotting under the sand and leaving a worrying stench.

But the worst thing is that they are also out on the reef, which is lethally poisoned by pollution from uncontrolled nutrients and stifled by the lack of fish that had controlled both them and the feast of phosphates, nitrites and nitrates released by faulty hotel control system. But, no problem! This sector is already eagerly looking north to the Yucatan Peninsula, in the area of Holbox, planning the next expansion phase. Bread for today (for a few), hungry tomorrow…

There is no need to go so far afield to witness chronic coastal deterioration. The Mediterranean is the sea that has most suffered most from human action by all those who has swum here since time immemorial. But in recent decades it has experienced a series of profound transformations, especially in European countries such as France, Spain, Italy and Greece. On top of over 130 million coastal inhabitants of the Mediterranean, double that number of visitors migrates to its shores in summer in search of fun in the sun. By 2020 it will be a global movement of more than 345 million tourists, concentrated in the developed countries.

The Spanish case is especially striking, due to its poor coastal management. Spain is one of the main tourist destinations on the planet, and the sector earns about 11% of its gross domestic product, supports more than 2.7 million jobs, permanent and temporary, year-round and seasonal, and is worth approximately €40 billion every year, thanks to the nearly 60 million tourists who visit us—or come quasi-permanently, in the case of the pensioners from northern countries—each year. But that has a cost. Between 1987 and 2000 alone, there was an increase in artificial surfaces (marinas, docks, etc.) of more than 30%, which resulted in a total change for more than a third of the coastal strip from what it was at the beginning of the last century. No less than 34% of the first kilometre from the beach is now completely transformed, and our beaches have become, mostly, a constant press of people. Only 31% of the coast possesses what, according to European directives, can be considered a 'healthy' beach in terms of the concentration of people per square metre (one person every 6 square metres). In any case, this is not a superfluous measure. People value an absence of overcrowding, and appreciate the quality of the landscape and the identity of places. Has the transformation managed to deform both, making places unrecognizable? The problem is similar across Western Europe, even in areas where the tourism is not yet not as bad as in the Strait of la Mancha. The French and English coasts are also busy, with excessive building, problems of coastal transformation, pollution and the depletion of natural resources that that entails. However, we should recognize that exploitation has been uneven, even within developed countries, depending on regional policy.

From experience, I can say that the coasts of Lazio or Liguria in Italy, Tuscany or Sardinia have not experienced the same uncontrolled expansion as on the French Côte D'Azur, and the as yet undisturbed Corsica is the same. Driving from the marina at Bosa to the town of Alghero in Sardinia, I had to stop in disbelief at what I saw: not a single building for 41 km of coastline. Nothing. In Spain, the disparity occurs primarily between the areas on the Valencian, Murcia or Catalan coasts, and the coasts of Menorca, Ibiza and parts of Andalusia. However, the pressure is also immense in these places, and it continues to be subject to economic projections based on the building of second homes. While the impact of cement is year-round, the tourist is seasonal. A population like Roses can muster about 20 thousand inhabitants in winter yet rise to more than 200 thousand in summer, which involves a huge expenditure of energy and water and the production of waste, among other things. Just think of the gradient of disturbance at the bottom of the sea, an unmistakable imprint of our direct presence in the area. 'If we look at the seafloor of the Natural Park of Cap de Creus,' says Rafael Sarda of the Center of Advanced Studies of Blanes (CSIC), 'we see that 23% of the rubbish (mostly string, tubes, cables, bottles, etc.) is near the port of Roses, while on the same cape, in the most remote area of all the urban centres, there is just 7%'. Fishing products tend to be found more frequently near the shore and ports. In places like the seafloor or the platform of California, in every two hundred metres there are between one and twenty-four objects. This is not surprising, as it is estimated that up to 630,000 tonnes of rubbish are dropped from boats. Considering that only 20% of the solid waste is from boats and that the other 80% comes direct from the land, the amounts are vast. And everything remains there, in the background.

Everything including a shopping trolley, refrigerators and a bath, which I have seen with my own eyes at the bottom of the sea. And a bulldozer. It's huge. What on earth happened? Was it abandoned after its day's work to collect on the insurance? Many of these objects provide a new substrate for the establishment and shelter of certain organisms (part of the solid substrate), yet the majority may be toxic, can shatter ecosystems and be harmful for or affect fish, mammals and aquatic reptiles.

So much construction, so much use of the place as both an anchorage for pleasure boats and fishing, and so much waste have tangible consequences for fragile communities like that of the spermatophytes. As we have already seen, trawling has been one of the main culprits behind the marine system's regression in various places, but not far behind are the proliferation of ports, dredging of sand and waste, and the contamination of water by large urban conglomerates or by intensive fish farming. 'Loss of flowering plants is due mostly to coastal development, in all its senses', says Charles Boudouresque, of the Marseilles Centre of Oceanology, France. The issue is that grasslands, which as we have seen are essential for a useful (for us) operating system, are declining widely. 'Many meadows of *Posidonia oceanica* have disappeared, but they have been replaced by others of *Cymodocea nodosa*', explains Boudouresque. What happens is that these others grow faster and are better adapted to unstable environments, but do not accumulate the complexity and diversity of *Posidonia*.

Between the 1980s and 1990s, over 2,900 km^2 of spermatophytes were lost, according to documented records, but it is estimated that this figure could be more than 12,000 km^2. Other estimates speak of up to 33,000 km^2 up to the beginning of 2000, implying that, if we calculated a surface of 177,000 km^2 of these plants around the world, we are talking about losses of between 7 and 19%. On the French coasts of the Mediterranean it is considered that there is no less than 19% of water below 10 m deep with a degree of anthropization. This depth is vital. All beds have a very narrow bathymetric range. They are plants; they cannot live below certain depths because they need light, finding a home at perhaps between 3 and 5 m and perhaps between 15 and 20 m, depending on the clarity of the water. Of course, in certain places they are found deeper, but the water clarity must then be greater, for it is the depth to which light reaches that determines whether the plant can create enough oxygen to compensate for its respiration (plants produce oxygen through photosynthesis). This problem, from the point of compensation, could also aggravated in a short time by the potential rise in the sea level, as the Intergovernmental Panel on Climate Change (IPCC) models predict. Plants have coped with continuous rises in sea level, but if a rise is too fast the conquest of new territory is difficult for a slow-growing organism. So, we must bear in mind that, on a human time-scale, the regression of the seagrass *Posidonia oceanica* and other spermato-phytes is irreversible. I always insist on the human timescale because, in the very long run, communities can always return to the previous balance or find new paths of complexity.

We need to integrate concepts to achieve viable management of our coasts. It seems that we have started to understand this point and, in many respects, the management of our waters, our sea bottoms and our coast have started to improve. The case for a much more rational management involvement remains urgent. Tourist exploitation, for example, has to strike a balance between the benefits that bring visitors with their needs and the preservation of landscapes, materials and resources for local people. It has to be considered holistically, and not only as the water needs of a golf course and the estates surrounding pools in areas that have a chronic lack of potable water. And is not a matter of building advanced desalination plants or increasing the energy available for air-conditioning, but achieving a good integration of human and natural factors to arrive at the best solution.

We also have very robust systems and long time series health status indicators. There are specific indicators to help us to understand this human impact. 'We cannot assign coastal management randomly or to the wrong hands', says Adriana Brizon, of Portugal LABOMAR in Brazil's Federal University of Ceará: 'We have to conduct rigorous monitoring of coastal communities and bring out a number of problems so that future generations do not have a system that is marginalized, deprived of the diversity that it had a few decades ago.'

Part of my work is a search for this type of indicator in coastal systems and platform. All elements must be considered: plankton, benthos, seaweed, grass, fish… We always have to refer to the needs of the area, its socio-economy and the vital structure that it will support. 'If it is proven that coastal ecosystems, both on land and at sea, help us to survive, why treat them so badly?' asks Mark Spalding of the Global Marine Team at the University of Cambridge in England. 'It is clear that maintaining these ecosystems in good health will help us to have a dignified life. We must not waste this last chance of doing things properly.' It is a vision of the future, in the long run. You can, as we have suggested before, see that the problem is no longer developing little by little in the advanced countries. It is now in the emerging countries, which see in tourism, in large cities and giant infrastructures a great business opportunity, and whose respect for the environment and vision can be so short-sighted, just has ours been for many decades.

Dead Zones

The surprise of divers must have been immense when great shoals of fish from deeper areas invaded the shallows (about 20–25 m) off the Californian coast in the summer of 2002. For them, it must have been an impressive and pleasing spectacle, but the fish were fleeing. Crabs (*Cancer magister*) that had already been caught in fishermen's pots were not so lucky, and when they were taken out it was found that more than 75% were dead due to lack of air.

The reason for the flight and lack of air was the ascent of a mass of oxygen-poor water from the depths—a 'dead zone'. Dead zones are fast changing from something rather interesting to a serious problem for the health of ecosystems and fisheries. They are bodies of water in which, due to an excess of decomposing organic matter, the oxygen level becomes very low. This is termed hypoxia, and it can progress to anoxia (a total absence of oxygen). It has become more common since the 1960s, with double the number and extent every decade since then. In 2007, there were more than 400 dead zones across in the world, covering just under a total of 250,000 km^2 (half of the surface of Spain). Some, like that of the Gulf of Mexico, fluctuate in size from year to year and place to place, growing off the coast of Florida from 5,000 to more than 15,000 km^2 in just a few months. In reality, this does not seem a huge figure, but it conveys to us the importance of what has been lost in terms of fishing and the functioning of the ecosystems. In Chesapeake Bay, which began to register acute hypoxia in the 1930s, there may have been a loss of around 5% of secondary production, which would mean less food for fish and an estimated loss at catch of 10,000 tonnes, especially in deep areas.

There are more dramatic examples. 'In the Baltic Sea, in a hypoxic deep sea in certain areas, no less than 264,000 tonnes of carbon are being lost in secondary production (zooplankton); that is, 30% of the potential production of fish in the area', says Rutger Rosenberg of the Department of Marine Ecology of the University of Goteborg in Sweden: 'This would mean more than 100,000 tonnes per year of fish lost to fishing,' The worst thing is that dead zones continue advancing, reaching increasingly shallower areas, which are traditionally more productive. Altogether, experts in the field estimate the losses of secondary production as between 350,000 and 730,000 tonnes in the whole of dead zone. 'If we extrapolate the consumption of 60% of that secondary production (copepods, cladocerans, etc.) to fish, we lose between 210,000 and 438,000 tonnes of fish a year on those approximately 250,000 km^2, Rosenberg concludes.

Dead zones are due in most cases to an increase in the amount of nutrients in the sea. In fact, the main areas where they are concentrated (the Baltic Sea, the Kattegat between Sweden and Denmark, the Gulf of Mexico, the Black Sea and the China Sea) have intensive agricultural, industrial and urban development. Nutrients (especially nitrogen and phosphorus in various molecular versions) reach the sea mainly from rivers and human activity. 'The increase in the quality of life and the cost of energy and materials are an important part of the equation', explains Robert Diaz of the Virginia Institute of Marine Science of United States.

Hypoxia is generated by bacterial respiration. This process is due to the consumption of organic matter from algae that has not been assimilated by other agencies. An excess will stay at the bottom, where it is consumed by aerobic bacteria. These bacteria need to consume oxygen to perform their metabolic processes, and they take up oxygen from the water that surrounds them. Many species of bacteria are capable of withstanding very low concentrations of oxygen, but not animals, as a rule. In the Bay of New York, already in 1976 there was a mass mortality of organisms such as fish, crustaceans and other invertebrates in an area of over a thousand square kilometres, while other fish managed to avoid the undesirable area, migrating to deep waters.

This adds fuel to the fire. Scientists such as Brian Grantham of the Washington State Department of Ecology's United States study other types of hypoxia—seasonal, physical upwellings. These are distributed around the globe and are in the most productive areas of the planet. The winds that run parallel to the coasts of the Atacama Desert, Namibia and areas like Oregon and California move the surface waters, causing upwellings of deep, nutrient-laden waters. These nutrients naturally fertilize the surface waters where phytoplankton grow in abundance throughout the year, but especially in spring and summer. Despite supporting great secondary production (as we have seen in previous chapters), not everything is consumed and part goes to the bottom, where our bacteria are waiting. There is therefore a body of water in these areas that is hypoxic, usually at between 50 and 600 m deep.

Climate change could be forcing a different situation. On the shores near these upwellings, the difference in temperature between the landmass and the water may be increasing the intensity and frequency of these winds, causing upwellings that are more intense. In this case, the hypoxic area will extend upwards to shallower depths and be more persistent: 'While temporary wind records are poor, everything indicates that off the coast of Namibia there has been an increase in the intensity of the winds during the last decades', says Grantham. Indeed, the disruption off California and Oregon in 2002 coincided with a greater intensity and persistence of these winds. And, in Namibia, lobsters are found increasingly in waters that are shallower, fleeing from the dead zone.

But there's more. It is estimated that if 94% of these dead zones are in the coastal area, there will be an increase in the air temperature of more than 2 °C. Water will be heated proportionally (never to that extent, of course), and that will reduce the already poor availability of oxygen. 'Many factors are acting synergistically in these places,' says Andrew Altieri of the Smithsonian Tropical Research Institute in Panama, 'and we need to understand in what way they will impact the poorest and the services that they provide to us. We have gone from a few dead zones—10 to 15—in 1910 to over 400 in 2007; that we would have to reconsider.' Altieri estimates that up to 90% of giant clams disappeared from those coral reefs affected by this phenomenon between 2004 and 2012. 'More than 3% of the coral reefs are affected by this problem, and the thing is increasing', concludes Altieri. Less certain is the impact that it will have the acidification of the oceans (see later sections on these dead zones). 'Due to the relationship between the amount of carbon dioxide increases and hypoxia, we know that the chemical balances might be altered in a very profound way in the near future in dead areas', says Frank Melzer of the Helmholtz Centre for Ocean, Jiel Research in Germany: 'The magnitude of these changes in coastal areas could be much greater than we thought.'

Paleoecology tells us that hypoxia is not a recurring phenomenon in many areas: in areas such as Kattegat and the Baltic Sea, it has no precedent and is human-induced. The majority of our seas' hypoxic areas are recent creations, but above all they are intense and recent. By making our ecosystems unstructured, allowing the more accelerated life agencies to dominate and turning the sea into a kingdom of microbes, it is possible that we are accelerating and perpetuating a human timescale process. Everything moves in a synergistic manner to a same destination of a 'marginalized' ecosystem.

The Destruction of Coasts

I've worked for a long time in the area between Estartit and La Escala, in Montgrí on the north-eastern Spanish coast (the Costa Brava). If you disembark at the first port, you see a fairly broad unspoilt space, preserved in part as military territory in the past and in part by the expertise of the various actors in the city, non-government organizations, scientists and the Departament de Medi Ambient de la Generalitat, who have managed to avoid the construction of houses, breakwaters, and so on, in that area. But it becomes puzzling when you arrive at Cala Montgó. On the right, seen from the sea, the Mediterranean forest still reaches the sea. On the left, the municipality has sold everything and you couldn't fit a single extra house in. There are no more places on which to build, and an invisible but tangible line of division between municipalities is evidence of this outrage, as a paradigm of human greed and lack of vision for the future. The same could said of coastal urbanization policy on ports and the maintenance of public spaces such as beaches, breakwaters and containment dykes.

The number of people who have embarked (never better said) on nautical leisure activities has grown almost exponentially in the past three decades, which has involved meeting the need for basic infrastructure to allow mobility along the coast for boats of limited range. In Spain, we have 325 leisure facilities designed to accommodate boats, with more than 107,000 moorings (of which two-thirds are on the Mediterranean). Worldwide, these facilities began in estuaries, fjords, inlets and industrial ports, then between 1960 and 1970 there was a boom, especially in the Mediterranean, the north of France, Great Britain and United States. In Spain, the number of licences for sailing vessels rose from some 30,000 in 1990 to more than 43,000 in 2006, and after the economic crisis it decreased to some 37,000 in 2015. Sport fishing licences rose from 40,000 to more than 71,000 in the same period, decreasing to 57,000 in 2013. Despite the fact that the crisis slowed it down a little, gradually the place is being monopolized by a proliferation of these structures and causing alarm about the excessive numbers of users in these sectors. The latest crisis (people don't see what is happening at all) is once again talk of expanding ports and jetties, without learning from the past.

What risks are involved in the construction of a marina? Is it always necessary to build one? Many coastal towns have wanted to have one to promote the area and to ensure more permanent tourism throughout the year, as tourists with a boat and a second home do not stay only a month for their summer vacation. But have marinas managed to attract sufficient vessels to avoid going into deficit?

The first effect of a marina is an increase of the number of boats in the coves and beaches in the vicinity. This increases the impact of tourists on the ecosystem in a more or less controlled manner, depending on what measures have been adopted by the relevant administration in terms of the flow of people in these areas: more boats, more impact. While the number of marinas needed is questionable in terms of real need, problems resulting from their construction are generally fairly concrete (the environmental management usually is). Forming barriers and jetties along a section of a bay or beach requires a thorough study of the ocean currents, the grains of sediment (size, shape and composition) and the communities that are established in the area to avoid causing an accumulation of sand at the mouth of the port, the displacement of sediments and formation of sludge, or the destruction of communities of ecological interest if they become trapped by the changed current dynamics.

A port also produces an increase of waste of all kinds, from organic to hydrocarbons and heavy metals from the bilges, varnish, use of patent preparations made from copper or lead as antifouling, transfers of fuel, and so on. All this must be disposed of and will impact the areas adjacent to the port. Both sport and non-sport fishing increases, within the limits of the port and also in the most remote areas, due to the increase in population, occurring mostly in summer in that part of the coast.

The second part is the coast policy designed to maintain beach tourism. The beach is one of the main attractions on the Mediterranean, especially along the coast of the Iberian Peninsula. From late spring to the beginning of autumn, hundreds of thousands of people seek entertainment on the sand, which seems ever more difficult to retain. What is happening to the beaches on our coast? The proliferation of ports and jetties to a large extent has caused a double effect: an uneven accumulation of sand on the beach (more on one side than the other) and a failure to recover after major storms. These storms may be shortlived, maybe now a little more frequent and more virulent, heading inland from the sea and carrying off the finer sands, leaving the shore naked or with only a coarse gravel beach. The main currents, which are responsible for slowly building up the sand again after strong gusts have dispersed it, have been often displaced by the action of man: now, to keep the tourism, you have to produce a sandy beach artificially. Another constant is that the most abundant fluvial contributions decrease once there is a marina: industrial, agricultural and urban consumption reduce the flow in the rivers, and this is another fundamental blow to the accumulation of sand and mud. Indeed, the regeneration of beaches is a measure of erosion, but its effectiveness will depend much on the problems that arise.

Conducting a good study of its impact is essential before proceeding with beach regeneration, because it is certain that we will affect both the site of sand extraction (especially if it is taken from the bottom of the sea) and the place in which we are going to pour it (feedback is especially important). There will be an impoverishment of organisms where the sand is sucked from, and it is important if the place is surrounded by prairies of spermatophytes or rocky substrate with animals and plants. We have to remember that a significant proportion of what we extract is too fine to make it into the boat, and this falls down to the bottom to create a layer of a very soft mud of up to 30 cm thick on the seafloor. This slurry of inorganic particles can lead to serious problems for the organisms that remain at the bottom or near the site of extraction, among other things reducing the flow of oxygen in the water column. Another problem is that there have been instances when toxic substances that were trapped in the sediments or sand have been put into circulation, thus harming the environment. The effects, although local, can affect many fishery resources in the area, such as clams and other molluscs, which can disappear not to make another appearance for years. If the affected community is a meadow of flowering plants fed by currents along the ocean floor, the recovery can take more than a century.

In fact, my question is whether beaches should be regenerated. Is it worth make a big effort every year to lose it all in a few months? Some propose 'hard' or more drastic measures, proposing the construction of submarine concrete breakwaters (which would help mobile substrate retention) or the destruction of a large number of poorly erected breakwaters that hinder natural currents. We have both colonized and urbanized our surroundings; we are now looking for artificial corrective measures to amend our mistakes. Of course, I am the first to acknowledge that we need to enjoy our beaches and have marinas, but we should demand more rigour and at the same time more imagination from our authorities in managing our coast. We should not be afraid to propose solutions that are in keeping with the viability of the environment, not just our comfort.

Too Many Visitors

When a natural area becomes protected and is seen as being of ecological interest, it triggers a justifiable desire in people to visit the place and redis-cover a part of nature that we lose if we live in the city. However, due to the fragility of these places, an excessive influx of individuals may damage them. We can see the influence of human beings on the flora in a nature reserve such as the Medes Islands, in Spain.

In place like the Medes Islands, what harm can come from people who love the sea and intend simply to enjoy the beauty of the seabed for a while by scuba diving? The bottom of the sea can be rich in species, shapes and colours. The most beautiful places are largely composed of slow-growing structures made by often fragile organisms (corals, gorgonians, bryozoans, sponges, etc.). These structures, as well as the large number of fish, obviously attract divers. To study their impact, a team of scientists studied a species of fragile bryozoan (*Pentapora fascialis*) at various points on the coast of Estartit and the Medes Islands. They basically measured its population density, diameter and height and noted its location in space (exposed, in crevices, on gorgonians). They considered the less frequented and the more frequented types of area (about ten divers in the first area and ten thousand in the second, each year). 'We looked for areas of similar composition, depth and similar biological community to find how to attribute the results to these parame-ters.' 'Over the years, in the most frequented places the density was lower, the height and diameter were lower (at the break) and the position of the colonies was more hidden away (cracks, holes or slopes) than in less frequented areas', says Enric Sala, marine ecologist for the National Geographic. In the little-frequented spots the bryozoan colonies were in the most exposed positions, and of larger diameters and heights. It showed the negative effect of subaqua diving; and there is no reason to assume that the rest of the community is not equally affected.

To study the impact of dives, a team of scientists analysed the case of the purple gorgonian, *Paramuricea clavata*, a species that it has been considered necessary to protect due to its restricted distribution, slow growth and reluctance to recover. It has been established that the presence of humans on outcrops where this gorgonian lives causes significant damage to the population as they are inadvertently torn or broken by the force of flippers and aqualungs or by leaning on these fragile structures. The bryozoans would be depleted even by moderate frequentation by subaqua divers with some experience, due to these effects.

It is clear that forming a protected zone is always something positive, giving nature some respite from pressure imposed by humans, and providing what have been proved to be beneficial effects on the development of many species (see the chapter on this subject). However, there are always disputes—it never rains to everyone's taste! Conscientious follow-up of the various aspects of the biological, ecological and social (in this case marine) systems of prevention, tailored to each situation, is absolutely essential in carrying out a serious management project. We must not forget that a protected area is to be enjoyed, as well as preserved.

14

The Stained Sea

Here comes spring, and with it heat and light. It is the time for algae, utilizing the abundant nutrients that are dissolved in the seawater to proliferate, forming large populations that will become the basis of the food chain in every ocean. But not all proliferations are good for or edible by the organisms that live in the sea. Sometimes the algae proliferate abnormally, forming dense populations, from a few hundred cells per litre to tens of millions in the same volume. The worst is when, in addition, these algae have associated natural toxins and become what the specialists call harmful algal blooms, after exponential growth. They are the famous red tides, also known as discolourations (red, white, green), absolutely nothing to do with physical tides involving the rise or fall of the sea level. In some places, what was once a sporadic natural phenomenon has become a recurring nightmare for coastal residents, the creatures that inhabit the waters as well as the humans who live near the shore. Some species of algae produce toxins that enter the food chain in these organisms, killing or causing serious damage to those that consume them.

On the Florida peninsula red tides are already common, but the worst thing is their tremendous expansion. '2005 had one of the worst red tides there have been ever', says Bob Weisberg of the University of South Florida. Fish, manatees, turtles and dolphins died, painting a bleak prospect, and clouds of airborne algae formed by the sea foam carried the algal toxins inshore and affected the villagers, causing allergies and irritations. This was nothing new, actually, because in this same area the effects of a very intense red tide in 1996 killed up to 10% of the population of manatees—150. However, the frequency and virulence of this phenomenon are, in many

© Springer Nature Switzerland AG 2019
S. Rossi, *Oceans in Decline*,
https://doi.org/10.1007/978-3-030-02514-4_14

places, becoming greater. There are many more cases of collapse of entire areas due to the proliferation of these microscopic organisms. In 1991, hundreds of birds died in Monterey Bay in California from eating anchovies with extremely high levels of domoic acid, and by the mid-1990s more than a dozen humpback whales had died off Cape Cod, Massachusetts, from eating poisoned juvenile mackerel. Algae produce a toxin that accumulates at different trophic levels: the higher level in the food chain, the greater accumulation of toxin, as we have seen in the case of mercury in another chapter, and therefore it is more likely that the biomolecule will be lethal to an organism. Barbara Kirkpatrick of the Mote Marine Laboratory in Florida says, 'After more than eleven years of recurring red tides, people still know almost nothing about this phenomenon; only those who suffer from asthma are reported properly and the only people paying attention to whether you can or not eat seafood.'

We don't yet have to hand information that tells us what causes a bloom or what sustains it, and what is the mechanism that makes one disappear. We have several approaches, however, always different because each alga requires different conditions for sustained growth into a toxic bloom. 'Phytoplankton show different strategies to minimize losses among their own population and maximize the use of resources such as light or nutrients', indicates Willem Stolte of the Ecology Marine Group at the University Linnaeus in Sweden. These are the conditions of temperature, salinity, irradiance and concentration of nutrients that lead to one or another alga becoming utterly dominant in the column of water. But one of the most important features of these algae is their ability to form what are called cysts of resistance (fundamental to many organizations facing adverse conditions) in the background. They may be around for months, years or decades without being activated until there is an explicit signal to stimulate them. This signal is likely to be a particular combination of the water's physical and chemical conditions that we mentioned. 'It is possible that the cysts have an internal clock to make them hatch,' writes Anderson of the Woods Hole Oceanographic Institute, 'but they must have some flexibility to "catch the moment" that is most suitable for its population explosion.' The cysts are 'sleepy' and remain passive when the conditions are adverse, waiting for the appropriate conditions for the optimal growth of the species. Explains Anderson, 'When they are transported offshore, the cysts may respond to the physical changes in the water as a trigger to form in the coastal zone', thus avoiding any dilution of the population in the area further offshore.

But why are there more red tides now than before? Why are they happening more and more frequently? There have always been red tides. These discolourations would not be as interesting if, as we have said before, they were not so extensive and their proliferation was not so persistent. It is clear that one of the factors is that they are now detected more, because many more countries than before are undertaking specific sampling of the water column: there are more observers, more monitoring, more management of coasts. But that is not the most important explanation. 'We have been able to detect an unstoppable advance of some species that have become cosmopolitan', says Ester Garcés, a researcher at the Institute of Sciences of the Sea: 'many of these species are not our own, but for instance from the Mediterranean, yet they have adapted and now they are spreading rapidly.' There is no doubt that there are two main factors that help these algae to conquer new places. On the one hand, vessels unceasingly carrying goods the length and breadth of the planet have led to species movement, thanks to the so-called liquid ballast (as in the following chapter). This enables large ships to balance their load, among other things carrying algae and other organisms that stay in a lethargic state until they find the optimal growth conditions, perhaps on the other side of the world. However, many species need fairly calm water in order to activate the process of growth. Then the second factor comes into play: port development is taking place in many countries (see below), especially in the first world. The extension of shorelines' physical profile (currents, coastal dynamics, retention of springs and containment dams) and chemistry (higher concentrations of dissolved nutrients) are key factors in the increase in ever more intense algae blooms. In certain areas in the sea the concentrations of dissolved phosphorus, a nutrient essential to the growth of algae, has increased up to three times, and the nitrogen up to four times since the pre-industrial era. In some places such as the Baltic or the southern North Sea, such increases have reached ten times the pre-industrial concentrations. Some specialists have directed their investigations at the increase in concentrations of carbon dioxide and temperature in certain areas and for certain species, but perhaps one of the most neglected factors, yet which some claim to be fundamental, is the disappearance of our seas' filter feeders (oysters, mussels, and so on). As we've noted repeatedly, the elimination of creatures that are able to 'clean' seawater by filtering it to feed is a factor to consider when it comes to algae, as filter feeders control their level near the coast, along with that of zooplankton.

Many animals die due from the toxins inside algae, yet others are able to survive and to proliferate without any problem. As we have said before, the toxins of some species can be deadly, even to human beings. So the appearance of red tides is much more than an interesting story; it can leave fishermen and the seafood trade without a product for entire weeks. When I was in Chipana, in the North of Chile, an intense red tide of a virulent species

(*Protoceratium reticulatum*) was well established along hundreds of kilometres of the coastline. For weeks its proliferation continued uncontrolled, yet in some areas the shellfish gatherers and inhabitants indicated that it was 'better not to talk about red tides' aloud, otherwise any type of extraction would automatically be closed down. They commented that the algal blooms were becoming more frequent and that they had not known them before to be so vigorous and persistent.

Gradually, we can begin to understand that the destruction of habitats, even on the land, may be a factor that stimulates the growth of these algae. 'We're checking that the disappearance of the Everglades marsh in the Florida area is directly related to the persistence of these algal blooms', says Bob Weisberg. The complex system of canals and the vegetation that lives there are natural sewage treatment plants, removing a large quantity of nutrients so that the nutrients do not enter the sea. Without this natural barrier, the nutrients arrive in the sea in large quantities and the algae find optimal conditions. In this place, the concentration of nutrients has multiplied by 15 since the 1950s. Nature's kidney is no longer working, here, and excess human, agricultural and cattle wastes reach the sea as they are no longer being retained, filtered or reused. Algal proliferation has important economic consequences. South Korea, for example, has lost more than $120 million in the last three decades due to this type of phenomenon: 'We need protocols to respond to emergencies, particularly in aquaculture zones where there are bivalves', says Tae Gyu Park of the Southeast Sea Fisheries Research Institute of South Korea: 'Water oxygenation and nutrient retention have so many times been ignored, and they both require more and more economic effort.'

What is not clear is why toxins form inside this type of single-celled organism. One of the reasons may be related to the fact that having this so-called secondary metabolism could partly prevent predation, for example by zooplankton such as copepods. But these views about negative relationships between some species do not tend to be universal. Another possible explanation is that excess growth due to high concentrations of nutrients and optimal conditions of light, together with the right water turbulence, is conducive to the formation of a series of complex molecules. There is a high energy cost if these cells' photosynthetic machinery is stopped. For all plants, photosynthesis is a continuous process, and if matter production cannot be stopped due to the bonanza, why not create molecules with a high energetic value to divert part of the material produced by this continuous process of photosynthesis? Another explanation is the creation, by phytoplankton, of a kind of 'pheromone' whose purpose is unclear. In any case, the number of algae that produce it are few in comparison to the rest. Of more than four thousand species of algae in the waters of our planet, only 200 are toxic and barely 80 produce diarrhoeal, amnesic or paralyzing toxins. And of course, not all red tides or discolourations kill or are harmful. Many are unpleasant to the eye or form a marine mucus-like slime, but they are not necessarily a cause for concern. What is clear is that those which can give rise to long-term changes in the composition of ecosystems; toxicity can force some organisms to adapt and others disappear or move in search of places free of these pernicious 'tides'.

Several projects have been designed to investigate the still not fully understood mechanisms of red tides' operation. What most experts agree on is that the profound changes suffered by both land-based and marine systems are taking their toll, and proliferations of algae will continue their expansion and interference in ocean ecosystems. 'Knowing when and how there is going to be one of these uncontrolled proliferations, which may last weeks, will benefit us. In 1500 there were already marine discolourations, as described by Spanish sailors, but now they are continual, and their frequency and extension have run riot,' says Cyndi Heil of the Florida Fish and Wildlife Research Institute, 'so we did a comprehensive monitoring programme: temperature, salinity, nutrients, concentration of cells in the water… information that every time helps us to better interpret the phenomenon of the sudden increase of these toxic algae.'

The data in the models are applied to know where the red tide is going to go, not only physical data like currents, winds and saline fronts (the difference between salt and fresh water at the surface), but the biology of the organisms that form the discolourations (growth rate, affected by the turbulence and the balance of the preferred nutrients). But Florida is not the only place where these blooms are taken seriously, keeping track of this phenomenon. Spain, Italy, New Zealand, the United Kingdom… a long list of countries has understood that if you want to preserve the public's health greater efforts have become necessary, not only to control but to follow up blooms in real time. In one of these centres of continuous monitoring (the Institute of Sciences of the Sea), Esther Garcés says: 'Algal blooms are becoming a real problem for the tourism sector in areas like the Mediterranean: it is the toxic algae themselves, which simply form a bulky mass of plankton, and also because they affect not only the seafood but the quality of the beaches.' The water becomes unpleasant for those who want to have their two weeks swimming on the coast, as is looks dirty and is unappealing. Efforts to understand algal blooms are growing, as it is now understood that they are a phenomenon that will continue to harm the world's health and economy. Their increase is a fact in all the oceans, and involves an impoverishment of the ecosystem and a new balance of species. Our response will trigger a feedback mechanism: this will make them occur more, and thus there will be greater economic cost; thus it will be trickier to reduce them, and every time the blooms occur they make the quality of life more precarious in many parts of the world.

Red Tide and Invasions Through Ports

As we saw in the previous chapter, for years major changes have taken place in both the density of the summer population and the infrastructure—the buildings, urban developments, marinas, reconstruction and regeneration of beaches, and so on—for those who come to enjoy the Spanish coast (mainly the Mediterranean).

Besides the problems already discussed, one of the most serious has been left for this section: the dispersal of red tides. Near the port of Barcelona in 1998, each litre of seawater had over sixty million cells of a toxic algae, *Alexandrium catenella*. This dinoflagellate had occasionally been seen in various parts of the Spanish coast, but only as something mildly interesting to report. Considering the usual moderate to low concentrations of this and other organisms, fluctuating between 300 and 10,000 cells per litre, the explosion in numbers of this little creature alerted scientists for some time that the evolution of the species continues. 'The first proliferations appeared in 1996 in the port of Barcelona,' reports Magda Vila of the Institute of Sciences of the sea of Barcelona (CSIC), 'but the most interesting thing is that they had spread northwards and southwards from Calella to Coma-Ruga (more than 100 km) in just three years.' Why had it spread so fast? The answer to this and the spread of other proliferations is the number of ports— in some areas of the Costa Brava, for example, there is one every 7 km.

In the port of Barcelona, every month tens of thousands of cubic metres of water are discharged within a radius of twelve miles of the port, and only 60% of this water is of Mediterranean origin. In 85–90% of cases, the water has been less than thirty days at sea. In this way, surviving in the darkness of the bilge, a species could arrive en masse in the ballast water of any freighter. Other likely mechanisms are wind dispersion, as for plastics (as mentioned) above, and vessels loading and discharging surface water. In the port there are nutrients in abundance and mainly calm water. 'The port conditions are favorable for many species of phytoplankton, but those that require a certain degree of calm waters and lots of nutrients can get ahead of others requiring moving water', says Magda Vila.

In the sediment in ports the cysts of resistance mentioned earlier build up, where a cell in a 'sleeper' state is quiescent until it finds conditions conducive to growth. A port, once 'infected', has the cells to recommence a proliferation again and again, favouring a spread along an extremely developed coast. This is a serious management problem, because such species, as we have seen, can be highly detrimental to health, both directly or indirectly.

We make matters even worse by allowing uncontrolled urbanization that is exempt from planning regulations. On the shores of the Arabian Sea, the explosive development has been accompanied by a sudden proliferation in algae (toxic and otherwise). 'Marine systems have undergone major changes recently', says Maryam al Shehhi of the Center Institute for Water and Environment of Abu Dhabi in the United Arab Emirates: 'There is greater salinity, less fresh water and more nutrients, and the construction of levees and ports that retain water means that there is much less circulation.' These algal blooms may reduce the productivity of desalination plants by up to 40%, 'and can pose a risk to people's health from toxins that seep into water for human consumption', adds al Shehhi: 'Proper planning may be impacted by the incidence of these blooms, and many economic sectors are affected; its impact, however, is difficult to quantify.'

Dark Slime

In the spring of 2009 along the coasts of Montgrí (Costa Brava), you could see a thick layer of algae, especially the species *Dyctiota dichotoma*, so dark that it was difficult to see the other creatures. Years before (1998) the area had witnessed another large algal explosion at Cap de Creus. I discussed it with my friend Manel González, who put me in contact with the authorities to comment on these occasional anomalies in which a large amount of algae proliferates in a disproportionate manner for various reasons. Algal explosions are recurrent in spring in the Mediterranean. Conditions of luminosity, temperature and availability of essential nutrients for their growth make this season the most conducive to their proliferation. However, more and more frequently we see a massive presence of slimes and species of algae mass (not unicellular), macrophytes sometimes of 5–20 m and longer, for months. In the case of the Costa Brava it is the first 15 m down that are most affected, creating a typically green blanket that smothers all the other organisms.

Back in 2009, the rains that had continued since the previous autumn may have put into circulation a large amount of nutrients from rivers and floods. However, proliferations of this type of algae are poorly studied. 'We have found a relationship between the proliferation of mucilage algae with the high surface temperatures detected in 2003', says Stefano Schiapparelli of the University of Genoa of an earlier event. His team of scientists evaluated the covering and its impact on the community of the alga *Actinospora crinita*, observing it off the Ligurian coast in the south of Italy, at 'the point of the boot' (Calabria). There are not many species that form these slimes, rich in polysaccharides, forming a true marine mucus. They include diatoms and dinoflagellates, but also other species. The algae began the proliferation at between 5 and 15 m deep, reaching a maximum expansion in July 2003, coinciding with a heatwave that apparently favoured its growth. That year, in surface waters, the temperature reached 28 °C in July, when the average is usually around 25 °C. A strong and sudden storm took all the slime to a greater depth of 20–40 m, especially in sandy areas. The storm could be what saved the creatures that had been coated in thick carpets of algae, which are very dense and cause the filter structures of animals such as gorgonians, sponges or bryozoans to collapse. In addition, the storms may have been responsible for between 40 and 50% of the algae in the first few metres being torn off by the force of currents.

Slimes seem to have been becoming more frequent since the 1980s, and some research has already quantified not only the covering but the mortality of organisms such as gorgonians, so essential to the operation of the coast. 'Up to 80% of gorgonians can be covered by the mucilage in different proportions, and in case of lasting months may kill them partially or totally', says Simone Giuliani of the ICRAM in Rome. Gorgonians and other benthic suspensives (the aforementioned eco-engineers) favour the complexity, diversity and increase in biomass of other species, so if they are severely affected the environment in which they live may be transformed and its diversity simplified. Another research group has recently shown that these mucilages (some of whom may last more than a year in coastal waters) have a high bacterial and viral diversity: 'Up to 90% of bacteria and viruses from the mucilage are not present in the surrounding water', explains Antonio Pusceddu of the Polytechnic University of Marche, in Italy. Many of these bacteria and viruses are infectious for various marine species, and the spread through the mucilage and its extensions can be fast; it's like a kind of 'itinerant flu' about which we know very little. In more than sixty locations studied, these mucilages have increased in frequency and extent over the past two decades. Mortalities in benthic organisms of the Mediterranean and other issues (such as heatwaves, changes in food, excessive pressure from tourism and divers, etc.) add to this singular phenomenon whose frequency appears to be increasing in many locations.

15

The True Alien Nightmare

The look on the face of the first Cuban or Florida peninsula fisherman to find a lion fish (*Pterois volitans*) in his nets must have been worth seeing, back in the mid-1990s. This fish had never been spotted here before, so it was a focus of attention, especially for subaqua tourists in the Western Pacific. While it is a beautiful fish its venom can be deadly, yet it is even more harmful to the ecosystem, to other fish and crustaceans, due to its voracity. Its introduction seems to come directly from marine aquarology, possibly inadvertently, and is damaging. Rapid migration has already established this invader in the tropical waters of Venezuela and further south, 25 years after the first official record in 1992. I saw it myself in the waters of the Riviera Maya in 2013–2015. The flesh could be bought at fish stalls—and its consumption was encouraged. It is one of the many cases of an alien species that has managed to be introduced in a very particular type of pollution, that of biological invasion or contamination.

Here we need to introduce a term that is widely used in ecology as is highly relevant to the topic that we are now going to deal with: an allochthonous species. This refers to a species that, by various means (for example, the disappearance of a mountain range, the joining of a strait such as that of Gibraltar or human trafficking from one place to another between distant points) penetrates an ecosystem that is not its own, in which it is 'foreign'. The natives will be those who have long since adapted to the environmental conditions of the area. Species that are introduced into a new ecosystem may react in different ways, either simply disappearing because they cannot adapt to the environment, becoming aggressive because they do not have predators

© Springer Nature Switzerland AG 2019
S. Rossi, *Oceans in Decline*,
https://doi.org/10.1007/978-3-030-02514-4_15

or competitors in the new space, or displacing other species that have similar requirements in terms of space, food, and so on,

A harmless species at its source may become invasive elsewhere. If it tolerates the environment, has the ability to multiply faster, can compete for food efficiently and has no predators, competitors, parasites or diseases to limit it, it can displace the native species. Of course, not all species that are introduced are successful; quite the contrary in the case of algae. Most remain as rarities, relics confined to small sites that are sometimes difficult to locate by a non-expert eye. But in some instances this has not been the case, and certain species have become real pests. Sessile organisms, such as algae, ascidians or corals, usually start at a few highly specific and scattered points; nearby, we will definitely find their place of entry, such as a port, a drain or an aquaculture facility, that had helped the species to introduce itself. Then, if successful, it reproduces quickly and its density increases. From then on, if conditions favour, it can begin to spread to other places, where it may or may not have a stable development.

When the presence of an alien organism becomes an invasion, it's no joke. In many places, it has been found that 80–90% of extinctions of local fauna and flora are due to the introduction and subsequent success of non-native species (especially on land, where the topic has been advanced by the scientific groups that study it).

It may be the case that the native species may encounter a tenant who takes up space and food or nutrient resources. In this case, the aggression is not straightforward, but having someone more efficient than you at eating or colonizing the rock or sand that you live in may mean that you have to give in more and more to your competitors until you are reduced to nothing. For example, Marc Rius of the Department of Animal Biology at the University of Barcelona found that invasive marine invertebrates, a species of ascidia, were capable of displacing their indigenous competitors from the earliest stage of their lives, as larvae. 'In the early stages of life, we found that the alien ascidian was able to accumulate space faster and inhibit the settlement of local species', writes Rius. But if these effects are serious or very serious, the potential change to the habitat is even more so. Invasive species can sometimes alter the habitat, the place where other species live, impoverishing it or making it sterile for the life cycles of plants and animals that had depended on its structures, holes, recesses, shelters or exchanges of nutrients, which change due to the arrival of the new tenant. This is especially true for algae species and other sessile bioengineering organisms; that is, those capable of creating living three-dimensional structures on which other species base their existence. Therefore, invasive species can and do have effects on the food chain and ecosystem structure by attacking the complex web of ecosystems at several points at once.

Other effects, perhaps less understood in the marine sphere, are those concerning the introduction of disease (such as smallpox and other epidemics among the natives of Mesoamerica in the sixteenth and seventeenth centuries), or the hybridization of species (the arrival of species that cross with native ones, diluting the so-called genetic acerbic). All these and other effects have made biological invasions one of the greatest problems for the economy, health and biodiversity of marine (and terrestrial) ecosystems.

Just consider something shocking: while we have been reducing pollution by oil, mercury, artificial substances or organic waste, at the same time we have been increasing pollution through invasive species of all kinds. An example of this is the ctenophore that was discussed in previous chapters and that is beginning to form dense banks on our Mediterranean coasts (*Mnemiopsis leidyi*). This gelatinous animal represents an estimated loss of more than €150 million per year, mainly due to its harmful effects on fishing. Unfortunately, the effects of marine invasions (fishing, health, tourism, etc.) are far less quantified than those on land, due to our eternal ignorance of the marine environment. And all taxa can become invasive species, going from seemingly harmless to a serious social problem. On South Atlantic island coasts, goats have come to incorporate in their basic diet algae that are exposed to the tides...

Although we know that the movements of species from one side of the planet to the other are totally inadvisable, some have been premeditated, voluntary and 'studied'. In the overwhelming majority of cases, they have ended up as an ecosystem fiasco. For example, researchers from Thiagarajar College in south-east India conducted a study showing how harmful an invasive alga can be to coral reefs in one of the biosphere reserves in this country. At the end of the last century, people looked into the possibility of importing for industrial treatment this seaweed (*Kappaphycus alvarezii*) from its place of origin, the Philippine archipelago. In 2001, following studies by the Marine Chemical Research Institute of the Gujarat region, Pepsi-Cola cultivated it for use in a food stabilizing product. For some time, it had been thought that this alga was 'ecosystem friendly', meaning not only was it non-aggressive to other species but promoted biodiversity.

However, an accidental introduction of *Kappaphycus alvarezii* into the Hawaiian archipelago in 1976 had already had negative consequences. The alga proved to be harmful to the development and survival of the corals in the area: by growing on top of the calcareous structures, it prevents the arrival of light and food to polyps, choking the animals that form the reef. For the time being, its spread in Indian waters is asexual, but if it starts to produce spores it could spread to other reefs and invade the surrounding ecosystems in this part of the country. Pepsi-Cola has disengaged, saying that the origin of the invasion is probably the Marine Chemical Research Institute that brought it in, in the late 1990s. The Chemical Institute has replied that it stopped experimenting in 2003 and that the propagation sites are in the plantation area of the food company. The reality is that the cultivation of an alien species had been promoted in full knowledge that it could be harmful to a native ecosystem (Fig. 15.1).

However, not all invasions can be considered harmful to humans or the ecosystem. In South Africa, for example, ten non-native species have been officially detected, yet one of them has had an apparent benefit for the country's ecosystem and economy: the European mussel. Since its introduction in the mid-1970s, *Mytilus galloprovincialis* has formed dense, fast-growing mats, displacing the native mussel. It is capable of forming several layers where the native one creates only one, and a stock of some 35,400 tonnes has been counted along more than two thousand kilometres of coastline, especially in the western part of the country, where production is

INVASIVE ALIEN SPECIES

Jellyfish, bacteria, crabs, starfish, etc.

Fig. 15.1 Ships' ballast water is the main means of entry for invasive alien species into our oceans

concentrated. Faster to grow, better able to withstand desiccation and more capable of efficiently filling space, native mussels in South Africa have been left standing, in many areas. But the most interesting thing is that the fauna and flora associated with the new tenant are at ease in their new home, being more abundant, with more diversity per square metre and finding more food to eat. Due to the large pulses of larval production (up to 20,000 per square metre), there have been no predators to control it, neither sea snails such as *Nucella* species, nor birds such as the oystercatcher, which however are seen in greater numbers thanks in part to the abundance of food offered by the new inhabitant of the hard substrate of these coasts. The cultivation of this bivalve, as is to be expected, has increased and the European mussel has become a popular meal from a high-quality catch.

Interestingly enough, while in this part of the world the European mussel triumphs, thousands of kilometres to the north, in the Wadden Sea, it languishes. There, an Asian variety of oyster (*Crassostrea gigas*) has invaded the area, thanks to intensive aquaculture. In this case, the apparent effects on local biota have been as good or better than those in South Africa. Algae, bryozoa, hydrozoa, ascidians and barnacles have increased in number, biomass and diversity, and also moving organisms, which enjoy the greater particle retention of sediments by these rougher and better-oriented organisms than by their twin-shelled mussel relatives. In this case, there is something that makes oysters better than mussels: they clean the water of plankton in suspension. The more than two thousand mouths per square metre are powerful vacuum cleaners that suck better, not only because they are more numerous than their mussel colleagues (whose density is about 1,300 individuals per square metre) but because they suck faster. Therefore, in the case of the coasts of the Netherlands, Germany and Denmark, as well as those of South Africa, invasion has not only adapted but has apparently meant no loss to the community, marine or human.

How do invasions enter the area? The main vector of invasions is undoubtedly the ballast water of ships. Since the end of the nineteenth century, all ships have had to carry a certain amount of water either in their interior or outside to compensate for the load that they carry and thus sail safely. This water may be taken in at the port of Hong Kong but, when the consignments of teddy bears, cars or soybeans arrive at the port of Marseilles, for instance, it needs to be evacuated, and there may be as much as 120,000 cubic metres (and more) of foreign salt water. This is a very large volume in which all kinds of organisms enter and leave: microscopic algae, jellyfish, crustacean zooplankton, larvae of various species of invertebrates… the problem is not so much being taken on with the liquid ballast as surviving the weeks or months of passage from one port to another across the planet. But many organisms can cope with this, as they enter a quiescent state, waiting for better conditions, as we spoke about in the chapter on red tides. It may also be that organisms are attached to the outside of the hulls of vessels, as fouling or even, as we saw two chapters ago, in plastic bags that can drift around from anywhere on the globe, making long 'migrations'. Aquatics and mariculture are also common means of entry into ecosystems.

Algae are some of the most passively proliferating groups on the bottom of the planet, coming from ships, aquariums or marine cultures. Of all the algae, one of the most successful has been *Sargassum muticum*, an ugly seaweed from Japan where, in its native habitat, it grows to a moderate size (about 1.5 m long) and has a limited niche. However, it has been exported to areas of the East Atlantic, probably through both oysters and maritime traffic, and is now highly developed in some coastal areas (growing up to 10 m long in some parts of the French coast). It can develop a monopoly on space, excluding native species, and can become dangerous when it is entangled in ships' propellers. There are many more examples (some are discussed in the following sections) in the various seas of the planet.

The next question is how to stop the invader? How to end the invasion? There is no magic formula. And, at sea, the issue is more complex than on land because of our greater ignorance of how the system works and, in many cases, because the species can move more quickly, given the natural and artificial connectivity of the oceans. The first step is clear: avoid the main entrance, which is intense maritime navigation. The ships' liquid ballast has to be treated, and all kinds of techniques have been thought out (and applied) in this respect. Mechanical separation, by filtering, and ultraviolet treatment are used to kill the organisms. Heat is applied, ultrasound and biocides are used, the water is deoxygenated... yet all these treatments have problems. For example, ultraviolet radiation is incapable of killing all the organisms, cavitation (ultrasound) is very expensive, deoxygenation is complicated, heat is not efficient for all species and is cumbersome (and expensive in terms of the energy required) and mechanical separation does not retain everything—on the contrary, a large number of organisms are left. Combining several techniques is one solution, but some of the most effective treatments cost at least €280.000 (treating about 3,500 cubic metres per hour) for normal ships with about 100,000 cubic metres of ballast water. That's a lot of money. The worst thing is that there has to be consensus, otherwise biological pollution can be controlled in one port yet go unnoticed in another, because there are no mechanisms to regulate it. In July 2009, only eighteen countries in the world had signed an agreement on the treatment of ballast water, which represents only 15% of merchant countries.

Once the invasion has become a reality, we move on to the second phase: killing the alien. On land, this is still difficult, despite decades of experience (in some cases centuries), yet at sea it is far harder. Everyone agrees that early eradication is the most effective method: find it, exterminate it. But this requires specialists in the field, operational public and private bodies and money, sometimes a great deal of money. Moreover, there are various proposals for the eradication of pests at sea. Examples? For algal blooms, one is the introduction of viruses or pathogenic bacteria that cause cell lysis and population decline, but the apparent lack of specificity has made this measure dangerous for other, non-hazardous, algae. In the case of the *Carcinus maenas* crab that has invaded the western Atlantic coasts from the east of the Atlantic, as well as the seas of Asia, Africa and Australia, one remedy would be a small, single-celled organism or a multi-celled parasite; in both cases there would be a 'castration' of the voracious crabs that eat everything on the bottom. But the possibility of mutation of the parasites is an eternal problem. Will they change and become less specific? Will they attack native species? The remedy may be as bad as the complaint.

The same possible disadvantage of biological control applies to algae and their specific herbivores. *Caulerpa taxifolia* (see below) is an alga that herbivores shy away from in their 'artificial' environment; the sea urchins, fish and sea slugs in the places that they invade will have little to do with it. When the alga spreads, there is a possibility of introducing a sea slug (nudibranch) that has been found to be highly effective as a biological controller in its place of origin, denuding it. The problem is that no one knows if it would continue to be specific to that alga or, worse, how it would affect the fragmentation of the plant, which had been shown to be a good method of dispersing the invader. Biological controls are too new at sea, far too new. There is little or no knowledge of the consequences, and on land they have often proved ineffective. There are several proposals, but actual implementation is almost nil.

In the end, it all comes down to one thing: we know too little, we have too little information and we lack the perspective to assess its consequences. Meanwhile, invasions continue to transform ecosystems, altering the cycles of matter and energy, depriving native fauna of space and resources. There have always been biological 'invasions'; there have always been alterations in systems due to alien species. What is not clear is whether at any other point in our natural history there have been so many at the same time and in so many places on the planet in just a few decades.

The Mediterranean: A Sea of Invasions

If the Mediterranean is the sea with the most history on the planet, it is also the one with most invasions of alien species. Already in the sixteenth century the arrival of the first allochthonous organisms began to be travel with the Spanish galleons full of cocoa, gold and silver. Sessile organisms such as algae, ascidians and bivalves were undoubtedly the first tenants of these immense, rigid, floating substrates that arrived mainly from the Americas. But we would be mistaken if we believed that everything was unintentional, since the so-called Portuguese oyster (*Crassostrea angulata*) was introduced knowingly at the end of the same century, especially to southern Spain and, of course, Portugal. But without a doubt, the main impetus did not come until the end of the nineteenth century when the Suez Canal was opened and the so-called lessepsian invasions began. From 1869 onwards, something occurred that rarely happens in the history of the planet: bringing together two of the most distant bodies of water in the world, those of the Red Sea and the Mediterranean. The first place officially to detect an invasion of its waters was the city of Jaffa on the Israeli coast in 1891, and in 1925 it was considered appropriate to make a biological laboratory dedicated to monitoring the species from the Red Sea. From the 1950s onwards there were around thirty or forty new invasive species per decade, until today there are more than 300. Of these, as many as 96% are from the area adjacent to the Suez Canal, the so-called erythreian species also found on the Mediterranean coast.

Some of these species are so abundant that already in the early 1900s there were fish and molluscs in the markets of Alexandria and in the Lebanon. In the 1940s, a fish in the same group as the red mullet (*Upeneus moluccensis*) contributed no less than 15% to the catch in Israel, rising to 83% in 1954–1955, a particularly warm winter. With oscillations, catches stabilized at 30% in the 1960s, confirming the displacement of the native species (red mullet, *Mullus surmuletus*). Some of these lessepsian invaders are now seen as a pest, such as the swimming crab, *Charybdis longicollis*, which can represent up to 70% of the benthic biomass on sandy seabeds along the Mediterranean Levant coast.

The increased frequency of maritime traffic from both the Suez Canal and the Strait of Gibraltar has besieged the Mediterranean, especially in recent decades. But aquaculture has also contributed to the problem; from species professed to have been introduced for exploitation, to organisms that have accompanied others, there has been a flow between quite separate areas that has caused more than one problem.

Oyster farms quickly became big business on the French Atlantic coast, especially from the nineteenth century onwards. The non-native species, those from the Pacific, seemed to be more profitable than the native species, so they were imported and conditions favoured their growth. The businesses soon saw such benefits that they decided to introduce them to the French Mediterranean coast. With the oysters came two problems. The first was the epibionts; that is, the animals and plants that live on the rough outer surface of the oyster. One example is that of the Japanese oyster *Crassostrea gigas*, which carried algae and animals on its surface that were then introduced into the system. Ten of the fifteen species of macrophyte algae carried by this bivalve have been identified as originating in the Japanese archipelago. The second problem with these oysters is that their success replaces the native ones (*Ostrea edulis*). As both filter feeders eat the same and settle on the same hard substrates, the native oysters in the area have found themselves cornered and in some places disappeared, as *Crassostrea gigas* is quicker to grow. The final blow was delivered by a plague of a parasite that affected *Ostrea edulis* more than its competitor, possibly because the origin of the parasite was from distant seas and the latter was more accustomed to it.

Problems increase with the Mediterranean invaders, as algae that have long covered the rocky substrate grow each spring. Of the sixty invasive algae, eight can be considered as a serious problem for Mediterranean benthic ecosystems. Some examples are that in April and May, *Asparagopsis armata,* an alga possibly of Australian origin, grows into huge, continuous carpets. It was first detected in Banyuls sur Mer, France, in 1922, and is an organism that has several advantages, apart from the obvious acclimatization, over other fleshy algae: (1) it is avoided by herbivores (such as sea urchins); (2) it can reproduce vegetatively (a fragment can grow into a new plant); and (3) its hooks allow it to cling to new colonizable substrates when displaced by the currents. These advantages make it more competitive, thus more suitable for expansion.

Something similar happens with another species of alga, *Caulerpa racemosa,* which officially appeared at the beginning of the 1990s to colonize the seabed at a depth of more than 70 m. It is another example of recent expansion, and it is found in places as disparate as the Island of Elba and the Mar Menor. In Tuscany alone it has grown from 3,000 square metres to more than 300 ha in two years. Once they have found the optimum conditions and are able to withstand the first seasonal onslaught, the alga spread by covering both soft and hard bottoms, with varying success, forming a complex web in which the native fauna and flora are altered in abundance and diversity. It can cover more than 60% of the available space, reducing the presence of other native algae, especially incrustations. Interestingly, it seems that it has taken up the space previously occupied by native algae, but it is tolerated by another invader that prefers deeper environments, *Womerseleyella setacea*. It is a complex system of competition.

Many other species have invaded the Mediterranean, causing real problems. But it is a sea accustomed to invasions and, although they pose a problem for its functioning and impoverish its ecosystems, many of the species that have entered will end up forming a synergistic part of its sea bottom and water. Some experts have suggested that the establishment of new tropical species may be partially favoured by global warming. In the section on climate change, we will see how some species have been conquered, in stages, in less and less time, especially mobile species such as fish and cephalopods, both from east to west and from south to north.

'Killer' Algae

Caulerpa taxifolia has become a paradigmatic alga in terms of invasive marine plant species. The first time I heard of it was in 1991 from Professor Charles Boudouresque in Marseilles and, like the data showed, we all felt that it was a growing threat along the north-west Mediterranean coast. Boudouresque told us that the most likely source of introduction was tropical water aquariums. Uncontrolled washing of an aquarium for non-native species can pose a real problem for the ecosystem. In 1984, in front of the Oceanographic Institute of Monaco, a metre-square mat of this alga was detected for the first time, and, since then, it has been observed in various places. In 2000, an estimated 6,000 ha were affected in the Mediterranean, and only two years later the estimates put its coverage at more than 30,000 ha.

This alga can withstand emersion, low temperatures and even desalination. A simple fragment of a few millimetres in size can start a small population that spreads at the expense of other native algae. Such fragmentation can be natural or manmade. Dispersion can occur by fragmentation or by reproductive elements carried on currents, animals (sea urchins, crustaceans, etc.) and by human means (anchors, fishing gear, handling). The progression has been rapid, and it has been documented at up to 99 m deep.

The main problem is that the algae does not have predators, as they all seem to shy away from it, and when they grow on top of other vegetation (in areas between one and 20 square metres) they create a network that tends to accumulate sediment and become denser, so they end up shading out and suffocating everything that is below them and needs light. The most serious case is *Posidonia oceanica*, which becomes coated and may not function properly because the light does not penetrate the covering, and the amount of nutrients available for its growth is reduced. *Caulerpa taxifolia* has it all: a wide ecological spectrum of settlement possibilities (even on soft bottoms), a well-off population persistence, high competitiveness, resistance to herbivores and strong reproduction. During the last two decades since the confirmation of its expansion, losses have been quantified in various areas both directly (damage to the ecosystem) and indirectly (damage to fishing due to the disappearance of recruitment and feeding sites, and damage to tourism due to the degradation of the coast, etc.).

After the initial alarm, however, several things gradually became apparent. The first is that the maps of algal extent and cover could be a little exaggerated: 'The algae cover is an order of magnitude smaller than was described', says Jean Jaubert of the European Oceanological Observatory of Monaco: 'With more extensive methodologies, we have been able to see that it is not as widespread as we thought.' Another of the reports that has come to light is that the algae seem to prefer disturbed, slightly eutrophic places where *Posidonia oceanica* was already in regression, dying or even dead, as Jaubert indicates. This would indicate that it is sometimes occupying ecological niches not exploited by spermatophytes. In addition, the effect on *Posidonia* does not appear to have been as severe in healthy grasslands. On the other hand, as of 2009 the algae seem to be regressing both in the place of origin (the Côte d'Azur) and in other places such as the Tuscany coast. Indeed, Monica Montefalcone of the University of Genova explains that, 'after a very sharp expansion between 1984 and 2001, *Caulerpa taxifolia* began to disappear in many places or to drastically reduce its distribution'. The system itself seems to have been adapting itself to the invading 'impetus' of the misnamed 'killer alga' that has now begun to be accepted as food by herbivores who, until recently, had not wanted anything to do with it.

This adaptation is not the case for another species, which has expanded in a more, shall we say, 'silent' way: *Caulerpa cylindracea*. This other invasive species in the Mediterranean has an abundance of one and a half times greater than *Caulerpa taxifolia*, has adapted better and spreads through the seabed, changing the diversity, complexity and sedimentation rates. 'This species also has its biological and physical controls,' writes Dr Piazzi of the University of Sassari in Sardinia, 'but each species has its dynamics, its way of proceeding.' *Caulerpa cylindracea* is actively ingested by fish and other organisms. The problem is that some of its molecules may be having repercussions on its predators and those who, in turn, ingest those predators. Recent studies by the Department of Environmental Science and Technology at the University of Salento in Italy warn of this possibility. '*Caulerpine* accumulates in the tissues of the bream,' explains Serena Fellini of this department, 'which is not surprising because we can find fragments of the alga in up to 86% of the fish's stomachs; the problem is that this accumulation causes problems in the hepatopancreas and in the species' ability to reproduce.' And it can also accumulate in the organisms that consume these fish, although the impacts are not yet proven. Ernesto Mollo of the Institute for Environmental Protection and Research of the Italian CNR speaks clearly of a new problem, that of 'invasive metabolites' or 'alien metabolites': 'It is a field that requires our attention, because we do not know how these types of molecules are influencing the food chains, the behaviour of species, their reproduction,' Mollo insists.

We have to understand what we're dealing with. In the case of *Caulerpa taxifolia*, in my opinion the problem has been magnified without being fully understood, to the extent that in some cases an ignorance of the issue on the part of environmental managers may have been a benefit. Far more agile tools must be provided to eradicate an outbreak of invasion immediately, but in this case, because of all that the 'killer alga' issue involves, I think it is another lesson to be learned by authorities and scientists. However, on the other hand, we cannot ignore that there are species that really are changing the ecosystems and the way that the organisms that inhabit them interact and survive, as in the case of *Caulerpa cylindracea*. There is a long way to go, and we cannot afford to ignore any detail.

A Problematic Crab

Species introductions are often not only deliberate but are intended for the social and economic benefit of a region. In general, these introductions are always harmful. We are going to illustrate this by a story that reflects this kind of blunder by people who obviously had little or no understanding of the functioning of ecosystems and their equilibrium. Between 1961 and 1969, Russian scientists and fishing technicians decided to introduce a valuable species of crab from the waters of the Aleutian Island arc (between Canada and the Kamchatka peninsula) south of the Barents Sea (northern Norway and Russia in Scandinavia). In that period, more than 1.5 million larvae, 10,000 juveniles and 2,600 adults were released into the fjords of the northernmost, Russian part of Scandinavia, in the hope that Russia, too, could establish and profit from the imposing king crab (*Paralithodes camtschaticus*), one of the largest arthropods on Earth.

The king crab went from about 15,000 adults in the late 1960s to more than 12 million in 2002 (between 1.2 and 2.8 tonnes per square kilometre). With a shell of more than 22 cm wide and weighing more than 10 kg, this underwater beast can be worth around €40 a kilo, and is a highly profitable catch in the Pacific area, where tonnes were extracted (until its virtual collapse in some areas, as we have already seen). Such a plan should have had great results: the fishermen in that area were to enjoy a new source of wealth, given that many fisheries were in decline due to a lack of fish. But it turns out that this little friend is a voracious predator, an insatiable consumer of everything, dead or alive, at the bottom of the sea.

Juveniles feed in shallow areas between 50 and 60 m deep, on the lamellar algae on rocky or stony bottoms near the coast. Adults migrate to greater depths to feed, to between 300 and 400 m, looking for all kinds of food. Among their preferred prey are bivalves, starfish and low-mobility benthic fish, but they make ascents to take carrion or sessile invertebrates such as alcyonarians or ascidians. In short, they wipe out everything. And they are considered, within the system, to be one of the highest links in the food chain, precisely because they are insatiable carnivores. The bivalve-rich bottoms (especially scallop-like shells) are like carpets of food, where the crabs crush the outer structures of their prey without great difficulty. After a period of growth to maturity, the crabs moult and reproduce again in the algae forests, where they prey on everything before them. In just two years, the prey can drop by up to half its biomass in the affected area (up to 450 broken shells per square metre in just 48 hours), forcing other predators to move or disappear. A single female crab can carry more than half a million eggs, which is understood to be why the expansion was so rapid. In fact, it is one of the most studied cases of marine biological invasion, because it is known exactly when it started and where it is right now: in just forty years (1961–2001) it managed to reach the west coast of Norway going around the northernmost fjords. Since 2002, it has been regularly on the market in Norway.

Although it is still not entirely clear how it will affect the functioning of the ecosystem, it is known that the fishermen are desperate because their bottom-set nets and longlines are found with less than 60% of the fish intact, the rest having been eaten by the crabs while on the bottom. The fishermen have to set their fishing gear far from the coast, fleeing from the predatory plague, but even so the crabs are invading every habitat, every corner, every fjord. The profit that the crabs make as a piece is not negligible (in 2005 they earned €12 million in Norway alone), but the price has been devalued outrageously and traditional fishing has experienced considerable regression yet has not been compensated for at all. 'The problem with the capture and impoverishment of the Atlantic is overfishing,' says Walter Courtenay of Florida Atlantic University in the United States, 'and what we need to do is to manage resources well, not to introduce new species whose consequences for the balance of the ecosystem are unpredictable and generally very negative.' He and other specialists are alarmed that a politician or manager has addressed the issue of recovering fisheries by introducing a new species. In a recent model, a 'simple' conclusion is reached: the crab must be completely exterminated if the ecosystem is to return to normal and recover. The long-term productivity of the system depends on the complexity of the food web. In other words, if such a destructive species spreads further, only short-lived organisms adapted to the crab catch rate will survive.

Meanwhile, the king crab continues its expansion, and in 2008 an Italian group found a specimen in the depths of the Western Mediterranean…

Part IV

Impact of Climate Change

16

Climate Detectives

I remember well the first time that I heard about climate change. When I lived in Montreal between 1987 and 1988, a couple of elderly men regularly visited the restaurant where I worked as a waiter. One gentleman (whose name I unfortunately cannot remember) gave me a complete dossier laying the groundwork for reducing CFCs (chlorofluorocarbons), due to the significant depletion of the ozone layer detected in polar latitudes (especially in Antarctica). In the afternoon I read it without understanding much of it, partly because I had not yet started my university studies that would give me the tools to understand the chemical reactions involved in this phenomenon, and partly because my level of English (why deny it?) was regrettable. But what I did realize was that human beings could create a problem in Russia or the United States that could be reflected in the southernmost part of the Southern Hemisphere. At that time, I remember that there was also talk of acid rain and Scandinavian forests being affected by the clouds from Great Britain. There were even futuristic films (very bad, by the way), in which there were areas where the trees were frazzled by the effect of the acid and the protagonists drove along in cars slashed by a rain halfway between radioactive and acidic (to supply extra morbid detail).

There is much talk today about climate change, and most of humanity has realized something: we are vulnerable to the climate. I believe that for the first time we are aware that the climate may no longer be what it is today, and that this transformation may affect us much more forcefully than we had ever thought possible. In fact, although some are still determined to ignore it, climate change is already changing our lives, and will be the greatest challenge facing humanity in the coming years (not decades or centuries—years). There

© Springer Nature Switzerland AG 2019
S. Rossi, *Oceans in Decline*,
https://doi.org/10.1007/978-3-030-02514-4_16

have been intense rates of exports and imports of resources, the mobilization of energy based on fossil fuels and an unstoppable advance of consumerism based on an affluent society that, deep down, does not even want to know about changes that could really affect us, both in our quality of life and in our daily lives. These have caused a unique situation that, until the last century ago, had been granted only to God: to influence the very climate dynamics of the planet.

As we explained at the beginning of this book, the Earth has been changing over hundreds of millions of years of planetary evolution. It is not new that it can change the climate. That variable, and the adaptation that goes with it, has always been there. The problem lies in the speed with which this transformation has taken place. I would like to say by this that it should make us reflect a little on the fact that in less than fifteen thousand years we have gone from 200 ppm to more than 400 ppm of carbon dioxide, with the most drastic increase in the last 150 years, closely connected with the industrial explosion and the transformation of the economy and society.

'Climate detectives', the people who work to understand past climates in order to gain a clearer picture of future climates, have recorded interglacial periods of about 280 ppm carbon dioxide throughout the planet's history, except at particular times when abrupt changes had resulted in a major shift in the Earth's climate dynamics. But even in the most exaggerated cases, this shift has never been so seemingly precipitous as it is today. For more than four billion years the Earth has been in a defined temperature range, allowing life to exist on it. Glaciations, as we understand them today, are a relatively recent phenomenon that has been affecting the planet for the last 15 million years, with special intensity and frequency in the last two-and-a-half million years. There is no doubt that the whole planet is used to these oscillations, but what about us? How will we cope with such a sudden change if one comes about?

Nature around us has seen many upheavals before. The one that we are experiencing seems to us to be particularly acute, because a series of synergistic effects are joining forces to transform our entire environment at an accelerated pace. In addition, we are the first organism to see the planet as a whole and to begin to understand the impact of, for example, a significant decline in biodiversity. No doubt the direct and indirect activities of human beings are affecting (and will affect) the number of species that inhabit the planet. But climate change will perhaps be the blow that precipitates this development already evident in our daily lives. How can climate change in the oceans affect the diversity of the planet, for example? Changes on land (desertification, land use, mobilization of pollutants, etc.) affect ocean systems, especially coastal systems where most of the production is concentrated, as we have seen before.

For example, it is estimated that about 20% of coastal wetlands may be lost by 2080, representing areas of extraordinary richness and productivity since they act as highly complex borders (much more so than beaches or rocky coasts) between continents and oceans. It will be the loss, modification or fragmentation of habitat that will induce major changes in certain areas, as some organisms will adapt without problem to the new conditions while others, restricted by their physiological or metabolic needs, will not be able to do so and will disappear. If a species has reduced mobility, it will be put under pressure by a changing habitat that will soon cause its displacement or disappearance.

The areas most affected are the poles. They always have been, because at this latitude the temperature changes are more abrupt. Many species will find a perfect breeding ground to flourish in slightly warmer waters, while others, adapted to the polar rigours, will find their existence compromised. Those that are already vulnerable today to a number of direct or indirect anthropogenic factors will bear the brunt, as they have less adaptive room for manoeuvre.

Ecosystems dominated by long-lived species can take a long time to respond to such rapid changes and their destructuring can lead to a negative feedback; that is, a precipitation of degradation by being put out of operation as a carbon sink (i.e. as an element capable of sequestering carbon and not releasing it back into the atmosphere). Throughout the following chapters we will go into more detail about the various problems in the oceans associated with climate change, but here I would like to focus a little on how these conclusions have been reached and on the basis of which we set up future projections.

From the perspective of most scientists, the complex phenomenon of climate change (added to the synergistic effect of local, regional or continental changes due to other causes of direct management of the planet) has fallen prey to what everyone saw as inevitable: politicization. Questioning science is inherent in the profession, of that there is no doubt, but creating pressure lobbies at one extreme or the other is dangerous and does not help at all. Most scientists who study the causes and effects of climate change are attempting to make progress by trying not to see the noise that they are making around them. But it is a difficult task when it affects so many people, when it compromises life as we understand it today. That's why almost everyone that I know is prudent but firm, some working step by step, even with blinkers or ear protectors, so as not to look or hear either side or let themselves be influenced.

All the people whom I have spoken to or am working with in this scientific research are prudent and have expressed to me their conviction about the change, but in no way take a catastrophic view of it. That would be counterproductive. We must accumulate truthful, rigorous and testable information so that we do not jump to erroneous or alarmist conclusions.

However, there is the evidence. We scientists who are on the frontlines of 'combat' are depressed, disoriented and less and less motivated. We're not being listened to. We see that we have a Problem, with a capital P, and we begin to resign ourselves. At congresses, you sometimes find that there are people who have thrown in the towel, others who are hopeful and younger ones who are pushing for solutions. It is not a simple issue, because up there, in the real command centres, there is considerable resistance. Now, when this book comes out, the most influential and powerful politician on the planet, the Caesar of modern times, is a climate change denier, a clear promoter of the fossil fuel-based industry and even a creationist who does not believe that evolution is proven. Donald Trump is the champion of these deniers, and at the moment I don't see a society that is seriously willing to rebel. In an open network called the Coral List (conveying opinion and information about coral reefs around the world), warning messages have begun to be posted because we know that the little that has been achieved will be eliminated, questioned or delayed. There was discussion, and some said that chat was not the place to exchange political opinions. But we are desperate, and we see that the system will soon be transformed and there will be nothing, I repeat, nothing, that can change issues such as desertification, the increase in air and water temperatures, the increase in violent phenomena and the rise in sea level.

But let us again try to understand what is happening and why. The first approach is to study the past. Over the past few decades, a large group of researchers from various nationalities have been studying past climate to better understand what the future climate will be. As we mentioned earlier, these are the 'climate detectives', people who work on the parameters that describe in the short and long term what our atmosphere was like, allowing us to know how much the temperature in different regions will increase or decrease, how this change in temperature will affect the rains or aridification of these places and whether the change in sea level or its acidification will occur in a gentle or abrupt way.

In general, we can speak of two speeds of scientist: those who focus on what has happened in recent centuries and those who look far beyond, towards horizons so far away that they stretch back millions of years. From the first group of scientists, those who study climate change in the short term, we have very precise approximations—so much so that a small mistake can ruin a well-established predictive model. This is the case in one study, where it was found that poor data recording was changing a series of predictive models of water temperature. 'We were going crazy,' says Susan Salomon of the US National Oceanic and Atmospheric Administration, 'we couldn't explain what caused an anomaly in the ocean temperature series detected in 1945.' In that year, the temperature at sea dropped by 0.3 °C, while the rest of the series of temperatures recorded up to that time were always rising. The use of two atomic bombs in Hiroshima and Nagasaki? 'In the final assessment of the IPCC, which came out in 2007, this figure dropped to a low point. Nothing explained the anomaly, and predictive models were affected,' says David Thomson of the University of Colorado. But the answer was quite simple: an anomaly in the collection of reliable data for that year. During the Second World War there were not many centres in the business of collecting water temperature data. The only data that came out and are considered reliable for climate models were those collected by the United States Navy. But these measurements of the temperature of the ocean water were recorded from devices located at the intake of the cooling nozzles of warships. 'We took the data from the British Navy and we could see that there was no anomaly in them: the temperature was taken directly from the sea overboard', explains Thomson. The data collected by the Americans gave slightly lower figure than the English. A banal methodological error had put a large community of scientists in check. 'He wasn't the only one. We know for example that the temperatures of the drifting buoys are also slightly different from those collected until 1970', says Richard Reynolds, also of NOAA. 'The bad news is that the models we are making now, more accurate, indicate that the warming is faster than we thought.' The tuning and harmonization of results make the climatologist's violin work better and better. The detective work of removing the data until we found the error had worked.

In a sense, it is harder for 'climate detectives' who focus on records from hundreds of thousands or millions of years ago: palaeoclimatologists. But their findings are also fundamental to understanding the changes that are taking place now. They have been able to demonstrate that the leap between the glacial and interglacial periods can take place in just decades. The tools to reach these conclusions are varied: the concentrations of carbon dioxide, methane, oxygen isotopes or organic products of plant origin that persist in sediments or ice for millions of years. It is the specialty of the scientist Alfredo Martínez-García, from the Geological Institute of Switzerland. 'We've been able to see amazing things thanks to our studies and other parallels made in the sediment or in the ice', says Martínez-García. 'Some organic molecules like n-alkanes are typical of higher plants and are carried by the wind in interglacial times. Wind has always been one of the important factors in the governance of climate on our planet. What we have seen is that the wind, which carries dust to the sea, fertilized it with iron', continues the researcher. 'In glacial times, aridification was intense, and the transport of elements such as iron to the sea in the dust was very important; it is inorganic elements such as these that cause algae to proliferate that capture carbon dioxide from the atmosphere.' Many factors are involved in the oscillation between glacial and interglacial epochs, such as the orbit of our planet around the sun or its inclination, factors that lengthen or shorten the periods between glaciations. Thanks to these advances in palaeoclimate, we understand how the current climate works and where future changes are taking us. 'It is important to contrast data from different parts of the world in order to assemble the complex puzzle of the past climate', says Frédéric Parrenin of the French CNRS Glaciology Laboratory. His data also speak, like those of the researcher at the Geological Institute in Zurich, of something disturbing: climate change is very abrupt. 'The problem is to realize that in the paleo-climatic registers the transitions are very abrupt', insists this French professor. This implies a short capacity to react. Abrupt climate change could prevent immediate adaptation, and unfortunately the models that we have are partial and imperfect. Actually, we've already embarked on transition…

The second working group focuses on future prospects. Predictive models are not really predictive models. I'll explain myself. What scientists do is to use not a crystal ball to predict the future but a simulation, in which they enter a series of data to first understand how much climate can change and, secondly, how it could affect species. These increasingly complex and accurate models present a picture based on probabilities that takes a great deal of data to work. 'The data and simulations needed to estimate the extent and nature of future changes in ecosystems and the geographical distribution of species are incomplete', explain these scientists through the 2007 IPCC reports, 'meaning that the effects of climate change can now be partially quantified.' However, at the IPCC in 2015 and 2018 the voices were much stronger: 'Almost in all probability…'. Things are already changing, but scientists need to be cautious about impacts and models.

The biggest challenge that we face in this regard is to make simulations that include ecological, biological and socio-economic data. Let's take an example. As we have already mentioned, red coral is a much-exploited species in the Mediterranean, and in certain depths, especially in the first tens of metres, it is at the limit of exploitation. In models used by ecological and mathematical experts from the University of Pisa and Trieste, using demographic data such as fertility, recruitment, growth and size structure, the potential effects of both fishing and climate change are studied. What has been proven is that, in shallow populations, a fishing event does not lead to local extinction of the species, nor to total or partial mortality due to water warming. However, 'the combination of too hot water every three or four summers and intensive coral fishing, even of legal sizes, can make it disappear locally', explains Lorenzo Bramanti of the French CNRS, a researcher involved in these simulations. These types of models, which combine data on the species with possible climatic fluctuations and direct human interference (fishing), are very scarce, but they are the ones required to understand where we are going and how we should react in order not to lose the race for the conservation of our natural heritage.

Changes in the climate transition phase in primary productivity levels (staple food) or increases/decreases in predators are still unknowns that need to be addressed urgently. It is clear that the species that make up a community will move differently in the future to other places, some of them staying behind and others managing to flee if conditions are not favourable for them, so that the variables of the future model become more and more complicated. What we do need to be clear about is that, despite not having the whole deck in our hands, the poker game has already started, and we can't go blank or try to cheat the dealer. It is a race of knowledge that affects us all and in which short- and long-term climate studies combine to understand, against the clock, where we are going.

Changes in Carbon Dioxide Balances over the Last Millennium

Changes in carbon dioxide capture and emission may well have taken place before the so-called industrial era (1850 to present day). A group of German scientists from the Max Planck Institute and the University of Hamburg have compared the carbon dioxide and isotopic data collected by ice cores and corals over the last 1,100 years (since 850 AD) with the profound changes in land use (agriculture and forestry). 'The influence on biogeochemical cycles and especially on carbon dioxide itself began much earlier than we know to be industrial,' says Julia Pongratz of the Max Planck Institute, 'but these changes did not have much effect on temperature as they did not reach critical levels like the present ones.' Almost half of the mainland has now been substantially modified in one way or another, but agricultural use and forestry have been the major processing agents. 'The carbon released in the pre-industrial era could have reached 320 gigatonnes between 850 and 1850,' according to Pongratz, 'and at that time, land uses began to change, becoming more extensive and at the same time intensive, with fast-growing plants unable to retain carbon dioxide in the form of cellulose and other molecules like trees for tens, hundreds or even thousands of years.'

Another source of imbalance came with the extent of the cities and their need for charcoal, but the final insult was paid, in many cases, by the construction of immense fleets that required top-quality timber for the merchant, fishing and military ships. 'There were areas where the forests stretched for hundreds of kilometres without interruption that were changing with land use, varying by country and time', says Martin Claussen, a meteorologist and climate change expert at the University of Hamburg. Mankind has changed the structure of vegetation and, with that change, the ability to sequester carbon from the atmosphere. The method used by these researchers is highly sensitive, studying the changes in land use and changes in the vegetation cover of Europe, India and China until 1700, and coupling America and other regions of the planet from that time. German specialists have calculated that between 850 and 1850 the increase in carbon dioxide could have reached 5–6 ppm due to the change in the vegetative cover, 'but the most important thing is to note that while in the pre-industrial era the biosphere was capable of capturing up to 48% of emissions, by the middle of 1900 this percentage had decreased to 37%', adds Pongratz. In the final stretch of the industrial period, the biosphere is only able to capture 24–34% of the carbon dioxide generated.

The model also notes surprising facts. After the great plague of 1347, in which up to a third of the Western European population died, the tree cover may have grown by up to 180 thousand trees per square kilometre, as reflected in the fall in carbon dioxide seen in geological cores. Other factors such as volcanic eruptions, invasions and major conflicts, and so on, are collected to construct an approximate model of what may have been the beginning of the planet's transformation as early as the early Middle Ages. That is why scientists like Steven Running of the Numerical Terradynamic Simulation Group at the University of Montana in the United States believe that more attention should be paid not only to industrial processes but also other factors that are sometimes overlooked in climate change models. The carbon cycle on land provides a sink for about 25% of human emissions. 'The models oscillate according to the parameters used,' Running adds, 'so it's essential to make them more and more complex in order to refine the predictions.' For example, the Arctic models initially predicted that, due to longer frost-free periods due to longer spring and summer periods, more carbon dioxide would be captured as the vegetation cover and primary production increased. But these models failed because, at the beginning, the breathing of matter by fungi and bacteria and pure and simple degradation or rot of the vegetation cover were not considered. 'Fire can suddenly change energy flows and balances', Running explains. Apart from the pulse of carbon released first in the form of burnt wood and then decaying wood, the albedo can change from 15 to 20% in the area to just 4%. In a boreal forest, this albedo may take decades to return, due to the slow growth of the northernmost gymnosperms. Other large-scale catastrophes have to be considered for these predictive models. Here, carbon sinks and emitters change not only by burning oil, but also by land use, fires, diseases and parasites.

In this sense, there is a carbon sink that we completely ignore and that could be the key to understanding the capacity of nature to capture this carbon dioxide: animal forests. Sponges, corals, gorgonians…. all those animals that live at the bottom of our unexplored oceans could be (in fact, are) an enormous carbon sink that we don't even consider to be part of the equation. Their extent is immense, the role unknown. But their long-lived structures sequester carbon, and we find them a few metres deep (tropical coral reefs) and deeper (deep corals), covering thousands and thousands of square kilometres. The most worrying thing is that a structure stripped of its complexity—that is, a seabed without sponges or corals, a surface without trees or long-lived forests—may be accelerating the warming process, as it is then unable to immobilize the enormous amounts of carbon that we emit or that arise from the melting of the ice in places that have so far been stable due to the cold conditions.

The more we know, the closer we approach realistic numbers that leave no doubt about the changes that the planet has undergone and that are accelerating, and the little room for manoeuvre that we have with regard to their solution.

The Worst-Case Scenario

What would really be the worst-case scenario? Marty Weitzman of Harvard University has projected it: to reach 1,000 ppm carbon dioxide (in 2016, we had already reached 410 ppm). This possibility is included in the models, but it is at the upper end of the curve. Although the possibility is quite remote, economists, oceanographers, physicists, sociologists and ecologists are also beginning to work with this scenario. As we have mentioned before, the scenarios speak of a range of 1.5 to 4.5 °C. At 1 °C more, 25% of the multi-annual ice cover of the Antarctic ice sheet would disappear. At 3°C more, 90% would disappear. What if the temperature were to rise to 6.4 °C? That is, what if we were to reach 900–1,000 ppm carbon dioxide? The models explain that the probability of reaching these levels is between 5 and 17%—not inconsiderable. Some models indicate even higher percentages.

What is interesting is to see that, with all their failures and all their shortcomings, the models developed from 2001 to 2013 by the IPCC, with all their 'yes but', have confirmed and made this probability more reliable. What will we lose? To begin with, all the high mountain glaciers, without exception. The entire Arctic sea ice cover would also be lost. Perhaps the worst thing is not the loss of that sea ice but that, once the great Greenland glacier and the great glaciers of West Antarctica enter a sustained melting process, there will be no turning back, in the short term.

Most endangered species would be lost and those sensitive to climate change would be lost due to their inability to adapt. Surface coral communities would disappear and deeper coral communities would be severely damaged in various areas of the planet. Up to half of species could disappear, especially those with more specialized physiological and motor needs. All cities located in mega-deltas of Asia, America and elsewhere would also disappear, as the sea level will rise by 10 m. Large monuments and infrastructure would also be below sea level (as we will see below), and there would be hundreds of millions of environmental refugees. Many experts say that this will never happen, that it is an outrage, but there are experts working on it, and seriously. It involves a great deal of speculation, I am aware, because in reality we are still sleepwalking on this subject and still have too many loose ends to tie up, too many pieces of the puzzle, to see the whole picture, yet it is clear that the most disadvantaged countries would be those that are now poor and also those in full development, those that would have the least capacity to react. There has also been speculation about the future legal framework, about sanctions, about processes for complaints from poor countries to rich countries. In the last ten years I have seen how the pace is accelerating, and now I personally do not believe that it is impossible.

The experts have put 450 ppm of carbon dioxide as a magic number that, for various reasons that we will explain in more detail, should not be exceeded (I repeat that we have already reached 410 ppm, in case this has not been made clear). Despite the fact that climate models fail because they are still imperfect, the probabilities gap is narrowing more and more, and they all come to negative conclusions to a greater or lesser degree. According to some experts such as James Harsen, even 450 ppm carbon dioxide is excessive: 'Not only must we never reach this number, but we must reverse the process and go back down to the part where we have the concentration of carbon dioxide in the atmosphere.' Once the carbon dioxide has been released, what is clear is that it takes a long time to recapture it. Only 20% is captured by 'natural sinks'; that is, by biogeochemical organisms and processes that immobilize it without releasing it again in the very short term. In addition, once the seawater is warmed, the inertia until it cools down again, especially at various points, can be very high. 'It will be a long time before we return to pre-industrial levels', says Malte Meinshausen of the Potsdam Institute for Climate Research in Germany. 'We have less and less room for manoeuvre, and we are still discussing whether or not the origin is anthropogenic.'

Deep down, we're not discovering anything that new. As early as 1979, global warming of between 1.5 and 4.5 °C was predicted. However, since 1750, the United States alone has been responsible for 28% of carbon emissions, and China 8.4%, mainly in the last two to three decades. How can we tell the Chinese (or Indians or other developing countries) that they cannot emit more because a series of serious global transformations is taking place partly because of these emissions? In what moral sense do we tell all those countries (they come out as 'others' in the carbon dioxide-emitting stakes) that they are emitting 13.2%, and that they can't grow and have the same comforts as us? What is clear to me is that, whatever the future framework, the next two or three generations (my children, grandchildren and great-grandchildren) are the ones who are going to pay the price for this and other human nonsense that we have accrued, thanks to our total lack of vision.

17

Arctic and Antarctic: Something Moves on the Ice

The three times that I have visited the Antarctic continent, I have been surprised by its beauty and at the same time by its harsh climate. In 2000 and 2003, people were already beginning to talk about climate change, but when they arrived there and saw the huge icebergs surrounding their icebreaker they couldn't imagine that it would be one of the systems that would suffer the most from a future rise in temperatures due to greenhouse gases. But so is, to a greater extent, the Arctic, which has already undergone an exponential change in its loss of ice cover and dramatic change in its ecosystems. In 2011, I returned to visit the Larsen areas A, B and C with the oceanographic icebreaker *Polarstern*, this time in a third approach to an area that is experiencing as a major player the effects of terrestrial climate change (see below).

Overall, the IPCC 2013 forecasts a reduction in land and sea ice, with only the former being in fact responsible for the potential rise in sea level. Of course, the ice is receding, and very much so. Not everywhere, not in all regions of the world equally, but the signs of regression are too clear to ignore, especially in the Arctic, where the temperature has already risen by more than 2 °C, on average, from pre-industrial levels. The coordinated monitoring system in various parts of the world (from the one glacier-foot station to satellite imagery) leaves no room for doubt: although it is still being discussed (in a totally useless and absurd way) what part of the percentage of 'blame' is human or due to a natural cycle, what is certain is that in most key places on the planet ice is sometimes disappearing faster than recent predictive models have established.

More than 20% of the sea ice in the Arctic has disappeared in the last thirty years, and that thaw is accelerating. The natural fluctuation in the extent of

© Springer Nature Switzerland AG 2019
S. Rossi, *Oceans in Decline*,
https://doi.org/10.1007/978-3-030-02514-4_17

sea ice cover in this part of the world is from about 14 million square kilometres in the middle of winter to less than 5–6 million square kilometres in the summer. The fact that the temperature in Siberia and Alaska has risen by 2–3 °C since 1950 is largely responsible for the acceleration of melting in the northernmost part of the world. By 2100, it could have risen to 7–9 °C, and the waters that in 2000 were thawed for only about thirty days a year would go on to being so for more than 120–130 days without forming the well-known Arctic ice floes. On the other hand, one of the most worrying things is the fact that the 'old' ice (multi-annual, more than seven to nine years old) has practically disappeared, giving way to a young and very fragile ice in an accelerating change in this area. On the north-east coast of Canada, for example, frost cover decreased between 1969 and 2002 to 60% in certain areas. We are in the midst of a non-linear phase of accelerating change. What does this mean? The most daring predictive models have failed, and we have entered into an exponential change that we lack the ability to control or even predict: we simply know nothing of what is going to happen, because even our worst forecasts have fallen short (Fig. 17.1).

The changes are uneven, not homogeneous, but they are there. In the other hemisphere we have Antarctica, a thermal regulator of the planet and a unique area. Due to the annual fluctuation of the ice floes, which vary from almost 17 to 18 million square kilometres in winter (September) to just 2.5 million square kilometres at the end of summer (March), life is subject to a high degree of seasonality, as in the Arctic. Here, the greatest losses of ice in

Fig. 17.1 The poles, as regulators of the Earth's climate, are the key to understanding climate change. The climate response to changes at North Pole and South Pole is uneven, being faster in the Arctic than the Antarctic. *Source* Sergio Rossi

both glaciers and sea ice occur in the area of the Peninsula and the West Coast off the coast of Oceania. Of the 244 glaciers that will directly reach the sea in this area, 212 (87%) are in clear regression: they do not accumulate water in the form of snow and ice, they lose it. The rest, 32, gain icy mass, but at usually very low rates. The sea ice surface surrounding the Antarctic Peninsula does not actually make up more than 1% of the total ice floating around the white continent, and the land mass does not exceed 2%. But this area of the world is one of the most productive in the world, where an enormous amount of biomass in the form of algae, krill and other organisms is concentrated. The most drastic changes due to the rise in temperature are taking place in precisely the most dynamic, most vulnerable and generally most valuable areas.

As we have been repeating throughout the book, there have always been changes on planet Earth. Gaia is a dynamic system, and this system depends on several factors (volcanic eruptions, alterations in the inclination of the Earth, tectonic plate drift, transformations in the intensity and orientation of sea currents, etc.). These changes affect the climate over millennia, hundreds of thousands or even millions of years. And those changes have affected the ice cover of the planet. Changes in sea level, related to the amount of water 'captured' or 'released' by the frozen masses, have amounted to more than a hundred metres in the not-so-distant past. These data are collected by geo-logical, chemical and biological approaches, both in the marine sediments and in the deep ice cores extracted from the heart of Antarctica. There are several possible approaches, from using the amount of methane trapped in the form of bubbles in the ice to the abundance of a particular organic chemicals, such as alkenes, in the marine sediment. Like an open book, these can tell us the changes that have taken place on the planet: all we need to know now is how to read it.

The reading of practically all of these indicators tells us that the change that we are seeing now is unlike the patterns that we could consider to be natural: the increase in temperature, carbon dioxide, methane, and so on, is more accelerated than that in other times, as shown by these witnesses of the history of the planet. It is virtually impossible to argue that much of the change is not of anthropogenic origin: there is too much evidence for it. It is global warming that is transforming the cryosphere (or frozen zones of the Earth's sphere).

Until now, places covered by ice and snow have not interested us much, being remote and hostile, but it seems that everyone needs to know more about the potential changes in their extent and dynamics. Rainer Zahn is an oceanographer with a special interest in marine climatology. He works at the

Institute of Environmental Science and Technology of the Autonomous University of Barcelona and, in his more than forty years of experience, has studied future scenarios that try to explain what will happen if the ice masses of the immense glaciers melt: 'In the case of Greenland, if the huge mass of ice were to become liquid, the Gulf Stream in the North Atlantic would be affected: a slight sinking of this current would cause a slowdown, a slight cooling but, at the same time, less evaporation, as the surface water would be cooler.'

This will not happen tomorrow, as it is a slow process in which a 30% deceleration is estimated to take at least a century. Maybe less... But the ice doesn't behave homogeneously. The possible melting of Greenland depends on many factors combined, as this huge island has different areas that create its own climate, and this allows for a certain stability. In fact, the process of recession observed in the Arctic is especially clear in the floating ice or ice floe, but the process in Antarctica is much more complex. In certain areas, instead of regressing there is growth in frost masses: 'while Greenland and the Arctic are exposed to the influences of currents such as the Gulf and the energy balances of continental masses, Antarctica is isolated by its own current, which makes it a little more independent of the global warming process', adds Zahn. But Antarctica, being its own entity, also has differences in the warming process. Thus, the area of the Peninsula and its surroundings, having a more developed continental shelf and being exposed to lower latitudes, suffers more from the processes of water warming.

What would happen if the Gulf Stream, which flows from the Gulf of Mexico and ends up in Northern Europe, going along the east coast of the United States, were to slow down? Palaeoclimatic data indicate that the components would simply reverse the change; that is, the planet would cool down abruptly. We would once again experience glaciation. To explain this, we have to imagine the following: once the continental ice of Greenland has been cleared, the body of water created would be fresher than that coming from the south, from the Gulf Stream. Therefore, despite being warmer, the Gulf Stream water could not rise, as the cold-water body from the north would prevent it. If the body of fresh and cold water were to settle north of the Gulf Stream, preventing it from rising, the heat transported from the Gulf of Mexico would not reach places like Iceland, Great Britain and Norway. The winters would be prolonged, and the cold would increase. In fact, as we have said before, it is not a new phenomenon. The problem is that it would impact us directly and possibly very quickly.

At the other side of the world, it was the changes in temperature in recent decades that influenced the calving of icebergs, such as from the Larsen's shelves, in the southern part of the Antarctic Peninsula. The large agglomeration of drifting ice produced by these events (the maximum extent of which would occupy a surface area the size of Scotland) may already be transforming ecosystems covered by a permanent layer of frozen water (see below an example of changes in Antarctic ice).

The icebergs released by these collapses are a source of disturbance for marine fauna, especially those living in the benthos; that is, on the bottom. The icebergs created literally plough through the seabed. Their depth can be as much as 600 m and the largest can be several kilometres in length. 'They are, along with fires or hurricanes, one of the biggest disruptors of terrestrial ecosystems, in this case marine ecosystems', says Julian Gutt of the Alfred Wegener Institute. In Antarctica, there are more than 400,000 of them, and their lifespan is usually between two and ten years before they melt away completely. Bottom fauna are accustomed to these disturbances, but more frequent 'ploughing' due to more icebergs could cause interference in their operation in certain areas. 'The icebergs are destroying the bottoms, allowing a large part of the organisms to cease to exist because of their direct action', writes David Barnes of the British Antarctic Survey. As carbon sinks, this accelerates change because there are fewer and fewer places where these organisms (sponges, corals, gorgonians) can proliferate without being scraped away by icebergs. Let us remember that, like trees on land, these marine forests at the bottom of the sea are responsible for retaining much of the carbon dioxide that we emit remaining in their structures for hundreds and thousands of years.

In the Arctic, it is not the icebergs that can alter the functioning of the marine ecosystem, as there are far fewer than in Antarctica, but the ice cover that we talked about earlier. 'Different programmes aim to improve the models of operation of the coastal Arctic Ocean in a context of global change', explains Carles Pedrós-Alió, from the CSIC's Institute of Marine Sciences. Indeed, much more important in terms of changing ecosystem dynamics is how ice regression affects microbes more than it does polar bears or seals. Although our ever-present anthropocentrism tends to look more at drowned polar bears and caribou that lack a habitat to adapt to, the truth is that it is the microscopic algae in the sea and the changes in permafrost (frozen soil in places like Siberia or Canada) that attract the scientists' attention. 'The importance of microscopic algae, bacteria and viruses is the same as in any other marine ecosystem', says Pedrós-Alió. These organisms make up a very significant part of biomass, contribute the most to respiration, enable the cycling of elements, remineralize organic matter and produce particulate matter from dissolved matter. The decrease in the ice cover could change the energy flows. Ice, especially that which is covered with snow, has a higher albedo (i.e. the ability to 'bounce' sunlight): snowy ice has an albedo of up to 0.9, while water alone does not go above 0.05. Less ice cover means more warming of the water, and therefore of the whole system.

But in ice, the algae become concentrated, creating their own ecosystem. Many organisms that depend on this food source, such as crustaceans, would be affected if there were changes in algae production. In this case, the tiny organisms in the water column draw not only energy from these algae but also their protection against the sun's UV rays, so changing the dynamics in these polar ecosystems would check them. The sea may become more dependent on a planktonic food web, with species similar to those found today in seas at lower latitudes. Water and air circulation patterns are also likely to change, with consequent changes in climate and energy flows in various affected ecosystems.

On land, changes in the cryosphere are also noticeable. Permafrost (the permanently frozen layer of the northernmost and southernmost latitudes of the planet) covers very large areas, especially in Siberia and Canada. In many places, for some time an increase in temperatures and melting of the surface areas have already been detected. 'This is already impacting on the acceleration of organic matter degradation processes, a change in carbon cycles in these soils through bacterial action, which would circulate, among other things, more methane and carbon dioxide per breath, further contributing to the rise in temperatures', says Richard Spinard of NOAA. It is therefore the changes in the 'small' that are attracting a great deal of the scientists' attention, because that is the basis on which bears, seals, caribou or the very human beings who inhabit that area of the planet are sustained.

At both poles there are currents produced by the tides that cause a periodic movement of the material deposited on the bottom. It is a continuous movement of water that enables organisms to continue to feed, even in the darkest times when the sun is completely absent and surface ice covers huge areas, blocking the algal production system. It has been proven that even at the height of winter polyps of many species are still open, the arms of many organisms are still trapping food and the cilia of many animals are still active. We must add, moreover, that some of these inhabitants of the dark depths of Antarctica can feed on tiny particles, much smaller than in other parts of the world. This means that even when there are only small algae in circulation, their adaptation to the system allows them to survive when the reserves accumulated in spring and summer are exhausted.

This is an adaptation strategy that some authors suggest could become unbalanced in some areas near the planet's South Pole if water temperatures continue to rise. Why? Because an increase in temperature, even if it were only a few degrees, would favour the growth of the bacteria that live on the surface and at the bottom, promoting the consumption of food that would normally reach the mouths of the suspension feeders near the bottom of the sea and depriving them of the oxygen necessary to their survival.

Within this environmental framework we cannot forget krill. Researchers consider it to be the key element between primary production and the rest of the organisms (fish, cephalopods, penguins, seals, whales, etc.) in the Southern Ocean that bathes the Antarctic coasts, and its exploitation is expected to increase more than twenty-fold over the next decade (from some 120 thousand tonnes to more than 2 million tonnes per year). The state of health of their populations is not bad, far exceeding 6 million tonnes, but for a couple of decades now there have been symptoms of change related, once again, to the dynamics of ice. 'Krill has changed its behaviour and distribution according to the dynamics of ice,' explains Dr Croxall of the British Antarctic Survey of the United Kingdom, 'and there are areas of the Peninsula where seasonal ice is scarcer, and if there is no ice, there is no krill.' Indeed, krill are closely linked to the fluctuation of the Antarctic ice floes because they take advantage of both the food and the complexity provided by the submerged area of the ice, which is very rich in nutrients, minerals, phytoplankton and microzooplankton. Bettina Meyer of the German Alfred Wegener Institute and her research team have followed the schools of krill in winter, finding that at this time of year they experience a deep lethargy and low activity, even though the concentration of food around them is very high: 'they breathe less, move less and eat less', explains Meyer, but they are always associated with ice.

It seems that one of the key factors is the currents and the recruitment of new individuals, closely linked to the presence of this concentrated 'manna' that falls from the ice. If there are years of little ice, there are years of little krill. As we have seen, along with the southernmost parts of the Arctic, the Antarctic Peninsula is one of the fastest warming areas on the planet, and one of the places with the most krill. Penguins (among others) are already being affected by these downward changes in their preferred food. 'There are changes in both the abundance and location of krill "spots"', explains Dr Cresswell of the Center of Stock Assessment Research at the University of California. 'The fact that mothers have to swim more to get krill can reduce the condition of penguin chicks by up to 20% as they watch their parents bring less food to feed themselves.' If they are in poorer condition, there is more mortality.

The problem is that the frequency of years without krill rises all the time, therefore the recovery capacity of some animal populations, such as penguins, diminishes. 'Climate change can affect the lives of these birds by changing the weather, oceanography and the very dynamics of their habitats', says Jaume Forcada of the British Antarctic Survey in Britain. 'Penguins are more likely to end up in a dispersal movement than to adapt to new and rapidly changing conditions.' The most affected will undoubtedly be those at the edge of the mass of 'young' sea ice that forms and melts with the seasons.

But humans are already looking at the 'positive' part of the decline in sea ice in the Arctic and certain parts of Antarctica. Despite the fact that there are still and will continue to be many technical difficulties in the exploitation of mineral and energy resources in such parts, from the Russians to the North Americans, including Norwegians, British and Canadians, various nationalities have already made symbolic (and not so symbolic) gestures of presence in these areas, to which some scientists and technicians (not all of whom agree) attribute up to 25% of the world's oil reserves. Also, in Antarctica, despite much clearer legislation than in the Arctic, movements are being detected. The British Foreign Office has already formally requested sovereignty of more than a million square kilometres of the frozen continent, which has particularly disturbed the Chilean and Argentine governments (whose areas of influence partially coincide with those of the United Kingdom).

On the other hand, shipping companies see the advantages of the thaw in a much shorter time frame. Since 2009, for the first time the Northwest route might be crossed without using an icebreaker for a short period of time, and a similar situation is expected shortly on the Siberian route. 'The point of no return for this situation, on a human timescale, will be when the dynamics of melting change the global cycle of ice itself, in which case it will take centuries or millennia to return to a situation similar to that of a few decades ago', says Dr Jacqueline Richter, NOAA's director. In the southern hemisphere, there are now vast areas around the peninsula that were previously inaccessible to boats and are now navigable, so fishing vessels can go where they had been prevented by the ice. So, in the case of krill exploitation, for example, 'although there are many unknowns to be resolved, we have to be very careful about increasing catch quotas now that technology allows us to fish krill more efficiently and treat it for marketing more effectively', explains So Kawaguchi of Australia's Australian Antarctic Division. If the ice is also absent in winter, in certain areas, the problem will become worse, because people will be able to access many places that are now unthinkable.

As a general conclusion, we can say that the major disturbances to influence polar ecosystems will affect several facets at the same time: (1) many animals such as the polar bear, seals, birds or whales would be driven to away and forced to change their habits for subsistence, due to the lack of platforms formed of ice floes (particularly the more specialized ones, those adapted to the cold); (2) the terrestrial forest mass (more than doubled in the last two decades in certain areas like the boreal forest) would become drier, increasing the risk of large and uncontrolled fires, which would influence the biogeo-chemical cycles in the surrounding seas also; (3) the Gulf Stream could be altered by the increase in water temperature, in a slowing down process, altering the general climate of the entire Atlantic area of influence, reversing the process and leading to a new glaciation; (4) permafrost would undergo important changes that would affect drainage and the continent's contribu-tion to the areas closest to the coast, as well as the release of more greenhouse gases; (5) many human cultures that base their way of life on coexistence with the ice (Inuits, Lapps, etc.) may disappear or be forced to change their ways.

This is a very complex panorama that is already affecting the North Pole more than the South Pole in a rather uneven way. The models will have to be more refined, more precise, involve more than just physical variables in a puzzle that surprises us by how quickly everything is developing. But the acceleration that has been taking place over the past three years tells us clearly that things are at the point of no return.

Cryosphere

An ancient legend from the Scandinavian countries proposes an end of the world very different from that offered by many southern cultures. Instead of being consumed in the eternal fire, true hell will be a succession of endless winters that will freeze even the seas. We can affirm that the importance of ice in our lives has been present since ancient times. But it is now time to understand how their formation and disintegration at the planetary level works: in certain areas, ice is disappearing even faster than recent climate models had predicted.

It is a mistake to look at ice as a simple accumulation of solidified water. The cryosphere (those parts of the planet where ice is the dominant element) is a large reservoir of water in solid form, being an important component of the global energy balance and its thermal balance. Of these frozen freshwater bodies, the largest is in Antarctica. The only permanent ice on the white continent is continental ice, which can be more than 3 km thick at certain points. The rest is the ice floe or sea ice that fluctuates with the seasons. And the fluctuations of this Antarctic ice are basic to the functioning of the planet:

its decrease or increase influences sea currents, atmospheric temperatures and the balance of gases in the air.

Ice, as the crystal of life that it is, has undergone a turbulent evolution throughout the history of our planet. In fact, in the present era we find a much smaller mass of ice than in the last ice age, when immense frozen masses in the north formed huge extensions that put life to the test, displacing it and forcing it to adapt to extreme weather.

Many are the men and women who have had to face the ice as an enemy. One of them was the explorer John Franklin, a tireless expedition leader who actively sought out the famous North-West Passage in the northernmost part of the American continent. But the journey cost the lives of him and his 129 men, crew members of the *Erebus* and *Terror* ships. They were stranded at the mouth of the Great Fish River. Rescue explorations did not find the ships that had sailed in 1845 until well into 1854. This gives us an idea of how difficult it must have been at that time to deal with the frozen surfaces, the world of the cryosphere. The case of the Irishman Shackleton, who sought the glory of reaching the South Pole with the vessel *Endurance*, was an historical echo, being trapped with all his crew in the Weddell Sea, at 74° South. However, thanks to his tenacity and good luck he managed to save them and take them home after the nightmare of an entirely frozen landscape.

Although it is one of the most hostile locations, the planetary dynamics cannot be understood without the cryosphere. The boreal is undergoing great changes, just like the southern ice, but without a doubt the ice that matters most to us directly, palpably, is that which we find closest to us: that of the high mountain glaciers. The problem of ice regression in various parts of the planet is that they are closer than they seem. Continental glaciers (in the Pyrenees, Alps, Andes, Rocky Mountains and Himalayas) are among the most affected. Relics of the last glaciation are disappearing in many places due to low snowfall and rising temperatures. In a national park like that of the Montana Glacier of the United States, 123 of the 150 glaciers monitored have disappeared or have disappeared over the past century. And water that comes indirectly from snow and directly from glaciers is the basis of millions of people's livelihoods. Suffice it to say that a river like the Ganges takes almost 70% of the water that comes directly from the Himalayas' icy cover. For populations living in this type of watershed, and for all those living in its estuaries and adjoining coasts, the disappearance of ice can mean a change in the management of water for irrigation, energy, nutrient or sediment inputs, industry and domestic use, especially in poor rural areas where the possibility of obtaining water from elsewhere else is difficult.

The changes in the cryosphere form a complex puzzle that will transform our way of life in the medium or long term, forcing us to adapt to another species (the human species) in a world that is reluctant to change its customs. I wonder if there is really still time to reverse the processes that have been set in motion. But what really keeps me awake at night is this question: What if a new glaciation really begins in twenty or thirty years? Will we be able to adapt?

What Happened at Larsen?

In January 1995, 4,200 km^2 of ice collapsed on the Larsen shelf in the eastern Antarctic Peninsula. Within a few days, the fragmentation of the ice was total, and the satellite images are witness to the disappearance of an area as it had been known until then. In the summer of 2002, a new collapse fragmented and dispersed the ice, leaving a free zone of more than 12,500 km^2. Nothing will ever be the same again in this part of the world, where in less than a decade the 'permanent' ice vanished from an area similar in size to that of the island of Menorca or a third of the area of Catalonia. The collapse is surprising in its magnitude, but not its origins. For decades the air temperature in this area had been rising and was slightly above 0 °C. 'The average temperature in Larsen was about +0.2 °C in 1992–1993,' says Helmutt Rott of the University of Innsbruck (Austria), 'but in 1994–1995 temperatures increased, on average, to +0.6 °C' (Rott et al. 1999). Insignificant, isn't it? In the balance between what is gained and what is lost, one of the main factors in these huge frozen areas is, of course, temperature. At around 0 °C, the balance in summer is positive, even if it is a close thing between profit and loss. From there, upwards, the water lost by the melted ice is higher than that obtained through the contributions from the snow. 'The fact that temperatures were consistently higher than necessary to maintain that balance,' explains Rott, 'was enough to create the collapse.' On the surface, and especially in the interior and the basal part of the immense frozen structures, water flowed, creating the ideal conditions for the ice to crack. The dynamic became stronger as you approached the edge, the part in contact with the sea, because there the effect accumulates of the hundreds of 'rivers' that run, often invisibly, through the interior of the structure. The mass balance takes into account what is eroded at the bottom and what accumulates in liquid form. If the balance is negative, the problems begin. Ice accumulation in the form of frozen snow decreased from 180 mm per year to less than 70 mm per year. Considering that erosion was already around 200–250 mm per year, it is understood that the process was hastened (Fig. 17.2).

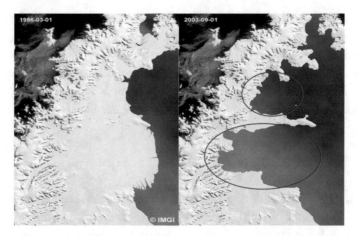

Fig. 17.2 Iceshelves Larsen A and B in 1986 (left) and 2003 (right). Note the vanished ice cover (circled in red), similar in size to the island of Menorca. Now Larsen C has almost collapsed, the future extent of ice cover in this area is uncertain. *Source* IMGI http://imgi.uibk.ac.at/

Added to all this is the effect of the tides, which erode what may at first glance seem undisturbed. These incredible masses of ice are much more exposed to phenomena, both atmospheric and oceanic, than their apparent unassailability reveals to the human eye. Larsen is part of the 11% of the total area of the white continent occupied by the permanent ice shelf, which, unlike ice floes, does influence sea level if it cracks and drifts away in the form of icebergs. The north-west wind blew hard in both 1995 and 2002, adding physical stress to the entire structure, leading to its eventual collapse. These ice masses can withstand specific high-temperature situations perfectly, but if conditions of mild temperatures, high humidity and hot catabolic winds are prolonged, the fragmentation is accelerated. Thousands of square kilometres of ice vanished, in some cases, in less than 33 days. However, the process of erosion had been going on for a long time, and in 1990 the process of 'thinning' the ice shelf had already been consolidated due to a gradual increase in temperatures over the past century.

The Americans were the first to arrive in the troubled area shortly after the 2002 collapse. Led by Professor Eugene Domack of Hamilton College in the United States, various scientific groups managed to penetrate the area flooded by fragmented ice, which had made navigation complex. They surveyed with underwater cameras, collected sediment cores (cylinders several metres long), analysed the water in the transformed area and reported on the situation in an area where sunlight and the dynamics of microscopic algae had not reached in the last 10,000–11,000 years. 'During the Holocene epoch (our own), the collapse of Larsen B is unprecedented', Domack explains: 'Both carbon 14 and the absence of diatoms and the values of oxygen isotopes in foraminifera structures demonstrate this.' In this area they found very little life and an abnormally high grain in the sediments, due to the influence of glaciers and the absence of primary production (microscopic algae). Diatoms in the sediment hundreds of metres deep in that first expedition were almost non-existent, as the system simply had not yet opened up to life as in other areas, where the ice came and went with the summer and winter. 'With the photographs we were able to observe some fauna,' the scientist continues, 'but in a very scattered way and only opportunistic organisms.' The satellite images also indicate that the microscopic algae produced hardly any blooms, unlike in adjacent areas of the same Peninsula.

After the expedition by Domack and his collaborators, scientific concern increased. But even though a vast sea of unknowns was opening up, only a few ships, which could be counted on the fingers of one hand, could access Larsen. One of them was the German oceanographic vessel *Polarstern*, which, under the leadership of Professor Julian Gutt of the Alfred Wegener Institute, managed to visit Larsen-B in the summer of 2006/2007. What kind of life was developing, what were its adaptations, what were its effects, what were the effects of the disintegration of the ice, the underwater topography and the currents influencing the conquest of the new open borders for organisms that had previously lived outside this area? How were the new open fronts connected, and how did warm-blooded animals adapt to this new situation? There were many questions and little time to collect data: just two months, of which only a few weeks were in the Larsen area.

At the beginning of 2007, the *Polarstern* crew members were able to feel the increase in temperatures in the area themselves. 'On 20 January we reached +3 °C, with a relative humidity of 50% (very high for that part of the planet)', explains Gutt in his campaign reports: 'The catabolic winds were warm and exceeded 30 knots.' Scientists began their data collection work. Elisabet Sañé and Enrique Isla, of the Institute of Marine Sciences (CSIC), found sediments that had only a thin film of chlorophyll and lipids on their surface. Over a large area little food was available, and primary production, as Domack had discovered, struggled for nutrients in the midst of areas where ice and water intermingled, not allowing the biosystem to run at full capacity. 'In the first centimetre of the sediment we found some organic matter deposited, but at a depth of 3 cm, nothing', explains Sañé. The pigments that they noted came with some regularity now from the surface of the sea, where the light was from. Nematodes (small worms that live in the mud) increased their presence as food reached them. 'In areas where the ice has disappeared only a few years ago,' says Armin Rose of the Alfred Wegener Institute, 'there is little variety and few individuals, but progress is unstoppable.'

The first to arrive were, as always, the opportunists, those who grow fast but have little capacity to compete. 'The colonization process will be a long and complex one, but it is clear that, as the ice disappears, new opportunities for many organisms are gradually opening up.' Billions of tonnes of ice being removed must have an effect on the surface flora and benthic fauna that depend on the descent of food from above. 'It should be borne in mind that not only does the production of microscopic algae increase, but also the disturbance of the icebergs and the orientation of the main currents', adds Isla, from the Institute of Marine Sciences of the CSIC, Spain. The spectacular changes have left scientists speechless, as they see how a direct effect of climate change is providing new spaces for many organisms to inhabit. In the trawl fisheries and the videos made by the oceanographic vessel, you can see fauna still in transition. 'The sponges in this area are largely considered to belong to abyssal areas,' Gutt writes in an extensive reference article, 'that is, organisms accustomed to an oligotrophic dynamic, practically devoid of food.' These sponges were mixed with other abyssal and pioneering organisms in a sometimes shocking picture of two worlds that collide, where one must disappear. 'To understand the changes that are going to take place,' continues Gutt, 'you have to understand how long they were isolated by the ice that covered them, how far away they are from the populations on the edge of the frozen margin and how much primary production (how much algae) is needed to feed the system.' The changes are faster than in other areas, perhaps because the door had been slammed open, so the ice has collapsed extremely quickly.

Only a few very restricted areas seem to be able to withstand the onslaught of newcomers. Whales, penguins and seals have also begun to arrive, attracted by a functioning system where fish and krill begin to abound. 'For them, this direct and palpable effect of climate change is beneficial: they are moving from an almost desert-like, nutrient-poor, light-poor, gently flowing system to an area of much faster dynamics and increasing wealth', says Gutt. On the edge of Larsen-B, hundreds of taxa have been waiting for the moment to expand their borders. Not everyone will arrive at the same time or have the same opportunities. Mobile organisms such as fish and crustaceans will be the first, and sponges and corals will arrive much later because they have a far more moderate dispersion rate. 'The increased production will create the conditions for substituting more "Antarctic" wildlife for more "Antarctic" life forms than we have in the abysses,' concludes Gutt, 'but the increase in ice in the area and the influence of nearby glaciers and the large amount of mud they carry with them can make this difficult if the collapsed area stretches too far or the ice melts too quickly.' Life will create its own path of conquest in an area that has gone from total darkness to exposure to a whirlwind of organic matter, ready to inject new species into the Larsen area.

Climate Change in My Direct Experience: A Unique Expedition (R/V *Polarstern* ANT XXVII-3)

The end of the world does exist. It is far away, and even with our modern means it is almost inaccessible. Likewise, in Antarctica, getting to Larsen-C (beyond Larsen-A and -B) by boat is almost impossible. In fact, when we reached 66° 12.73' South and 60° 17.00' West, those of us aboard the oceanographic vessel *Polarstern* realized that no one had ever seen this landscape before. The approximately a hundred pairs of eyes watching during the first days of March 2011 were aware that we were living a unique moment. During those days, I spent long intervals outside (I would be lying if I had said hours, as it was too cold) to see the ice, sea and land that made up one of the most beautiful landscapes that I have ever seen. That goal achieved was unexpected. I hadn't planned on going that far. The *Polarstern* had been content with returning to Larsen-A and -B, areas explored in 2006/2007.

The impenetrability of the Larsen area has to be taken into account. 'Since 2002, Americans have not been able to visit this very interesting and inaccessible place again', says Rainer Knust, the campaign manager of that ANT XXVII-3 expedition: 'It's an achievement.' It was, the more so because of the amount of work that could be done, the new findings and the possibility of continuing with research that had already begun to bear fruit, as we have noted above. The tacit race between the United States and Germany was being won by the Teutons and still is! It was not in vain that they had insisted on having data, samples and analyses in their hands that the Americans would not be able to repeat. But getting that far down, getting that far away, came at a price. The wind began to push the ice furiously, which prevented us from reaching the innermost point of the inner part of Larsen-B. We had all wanted to reach this place to check the evolution of the physical, chemical and biological conditions of the area after almost five years of absence. The campaign manager brought us together and explained the problem to us in a somewhat hasty manner. Field meteorologist Michael Knobelsdorf had painted an almost apocalyptic picture for him: 'If we don't get close to the tongue of ice and earth that divides the area in two, we could be trapped for weeks by the ice that is moving due to strong southern winds', the meteorologist had explained. Knust, considerably quieter than his colleague, told us that perhaps it was too risky to journey to that station in Larsen-B. A little dismayed, we made it our priority not to become stuck in the middle of the ice, which would have been a handicap for the campaign and a worrying factor for many.

Many of us wondered how things had changed so quickly. Just a week and a half ago, we had been enjoying some abnormally good weather in Larsen-A. In our shirtsleeves, we had watched climate change come alive: temperatures of up to 8–10 °C were melting the ice, forming spectacular waterfalls in the areas where the ice reached the edge of the sea in cliffs tens of metres high. The splendid sunshine accompanied that strange, unreal feeling of being in the right place but in completely inappropriate conditions. To maintain the equilibrium of the ice masses, the temperature must not far exceed 0 °C, as we have already mentioned. Our arriving when it was 12 °C was something that even those of us who are not used to Antarctica (there were people far more experienced than me, I can assure you) perceived as anomalous. I was thinking for a long time not so much about the consequences (as I have already said, nature progresses not so much because of very specific changes but because of long trends) as about all those who deny the obvious or, in another equally stupid or curious strategy, minimize it. The wonderful spectacle became mixed with a disturbing sensation, from the point of view of the perception of change, of collapse. The winds blew from the north, dragging a tongue of clouds down the mountains and forming a magical landscape.

But the wind was warmer than it should have been. It spurred us on. For all of us who understood what we were experiencing live, being there, seeing what we saw and feeling, it encouraged us to work even harder, to collect more samples and to try to understand as quickly as possible why the Larsen area is changing and what it could mean for the future of this and similar areas of the planet. We observed the changes, saw the progression of the underwater organisms and took more sediment samples in order to make the relevant analyses of the matter that is available to the organisms. We studied the successional states of the organisms and were able to fish, for the first time, animals from the bottom of the Larsen-C ice shelf, where the diversity and abundance of suspension feeders (sponges, corals, gorgonians, etc.), crustaceans and fish were at a minimum. In this remote and still unsettled area, seals and penguins had not arrived, let alone whales. More territory to conquer, new horizons where the new settlers, when environmental conditions allow it, will begin to transform the cleared bottoms of the ice that made the area inhospitable to most species.

There Are Only Four Polar Bears

What about the polar bears, which are the standard for every climate change study? Over the past few years, I've wondered, 'Is it so bad?' More importantly, is something wrong with them? This debate, which is by no means trivial, although some ignore it or turn it into a quasi-inevitable 'side effect', has become a bitter debate among renowned scientists in Canada, the United States and Norway. The worst fate that could happen to this wandering bear has been politicized, no doubt, sometimes filling with contained rage the articles by prestigious scientists. As mentioned above, those most affected by climate change are those organisms with a high degree of specialization and little scope for adaptation to the rapid change that lies ahead. Undoubtedly, the polar bear, as we will explain now, is one of these animals.

Polar bears (*Ursus maritimus*) rely heavily on the dynamics of the ice floe to survive. Most of their energy must come from hunting seals (especially the *Phoca hispida* species) while the ice floe is still present, so that they can wait for the phocid to come out to breathe in order to hunt it or to take it from under the icy layer where it may be hidden, especially newborns. Most of this consumption occurs in March, the month when most changes in the ice dynamics are taking place in the Arctic. It is the beginning of spring, which is moving forward dramatically in certain areas. 'Most of the lipids they accumulate for the rest of the year come from these critical weeks', explains Dr Rockwell of the American Museum of Natural History.

Less ice, less food, and more energy consumption during long journeys. Bears are not afraid to swim, but their efforts may be in vain because the prey available is scarce, especially in certain areas. But there is more: bears need to reproduce, like any other species, and their dispersal or change in migratory habits could isolate them to the point of endangering the stability of their populations. 'There is a relationship between the survival of adult females and the extent of ice', says Eric Regehr of the US Geological Survey in Alaska, USA: 'In 2001–2003, a period when there were 101 days without sea ice, the survival of females was 96–99%, depending on the area; in 2004–2005, when the days without ice reached 135, the number of female survivors ranged from 73 to 79%.' The cubs suffered similar mortality in periods of low ice cover in the areas studied. The effects of more days without ice could be considerable.

'If we continue the current trend and sea ice becomes increasingly scarce, more than a third of the population could become extinct in the next fifty or one hundred years... perhaps less', concludes Regehr.

'Bears are threatened by humans in a number of ways', Pertoldi says. 'Like all major predators in the Arctic, being at the top of the food chain makes them vulnerable to toxic substances such as organochlorine contaminants.' The synergy between the retreat of the ice and the palpable effects on the immune system, hormonal or blood circulation (among others) affected by these toxic substances made the bears of a century ago smaller and weaker than those of the previous four centuries. We must never forget the sum of the factors that can affect a species or an ecosystem, and in the case of bears it seems that everything comes together to make their lives a little more difficult each time.

Future? Adaptations? It is clear that not all polar bear populations are affected in the same way. George Durner of the US Geological Survey in Alaska and his collaborators have applied survival models by validating them with data from existing time series: 'Barents and Chukchi Sea bears in the southern Arctic are the most affected areas; populations in northern Greenland will be the least affected.' The model takes into account something key: the approximately a million square kilometres of ice that was available for bears in summers around 1985 in the areas where they roamed will have been reduced to about 320,000 km^2 in 2090. A reduction of no less than 68%, while in winters it will be 17% less. But changes in polar bear habits are being observed in some areas. By waking up early and getting moving early in the spring, the younger ones are taking full advantage of clutches of bird eggs, such as of the Arctic goose, from which they get more than just the snack that they manage to grab in areas like Hudson's Bay. 'Bear survival may depend (at least partially) on overlap with the clutches of these and other birds, as well as on the search for new food sources', explains Rockwell. 'We have been able to find out in certain areas that up to 90% of the clutches are affected by the incursion of bears', adds Jouke Prop of the Dutch Arctic Center. What is clear is that the changes are rapid and polar bear populations can be damaged in some cases in irreversible ways. The discourse on whether or not to stand behind a species may be correct or otherwise, but it seems to me that all the evidence indicates its decline in this fragile habitat, which is being now exposed to the changing forces of global warming. Although it is a banner, perhaps the problems of this species will help us to change the viewpoint of even the most reactionary on the overall problem of global change.

18

Water Warming

At the end of July 2003, I had to make a working dive in Cap de Creus. Here the waters are some of the coldest in the Spanish Mediterranean, and even in the middle of summer they warm up only moderately. I started to descend in really warm water, but what surprised me the most was to reach the bottom (at about forty metres) and notice that the temperature was still high. I had never felt such warmth at that depth. Generally, after the so-called summer thermocline (sudden change in temperature from hot to cold as the temperature drops), it is typically about 16–17, and 21–23 °C in the shallowest parts. This was 40 m down, the worst years of water warming were yet to come and still it was over 21 °C. It wasn't normal. In the same year, several massive mortalities of organisms were recorded, especially in the most eastern part (France and Italy) of that part of the Western Mediterranean, due to overheating of the waters.

I realize that we are just beginning to understand the potential effects of climate change. One of these stems from the warming of seawater itself, which I have just mentioned. Just as the air heats up, water has to be warmed up, but it happens in a different way. Over the last century, the oceans have warmed at their shallowest point by about 0.67 °C. But this warming is far from homogeneous. Igor Belkin and his colleagues at the University of Rhode Island in the United States have conducted a study that reflects trends in the rise or fall of surface water temperatures in various parts of the world. 'There was a rapid warming of surface water temperature between 1982 and 2006', Belkin explains. Over the past three decades, water warming has been two to four times higher than average in these places, some of which are even higher in areas close to the coast. 'On Long Island the temperature has risen 0.03 °C every year from 1940 to 2012, about 2.2 °C in more than seventy years', says Edward Rice of Queen College in the United States.

The North Sea, the Black Sea, the China Sea and the Sea of Japan have all warmed up the most, with certain areas of the Baltic taking the lead, with increases up to seven times higher than average at the height of the summer. 'It is estimated that by 2100 the waters of the Baltic Sea may increase by 2–4 °C,' writes Agneta Andersson of Umea University in Sweden, 'and this could reduce the ice cover by up to 80% in winter, increasing rainfall by up to 30%.' Other ocean masses, such as the Indian Ocean and the Australian Pacific, are warming at a much slower rate, and some have even had a noticeable drop, such as in the California upwelling areas and the Humboldt Current.

This increase that has taken place so far is insignificant, if we look at it on a global level, isn't it? What harm can a few tenths of a degree rise in water temperature do? A little over a decade ago there were people, like Michael Crichton, who didn't have these numbers. We now know that temperatures are likely to rise by just over 3 °C on average in the ocean. It is no wonder. We have to remember that 80% of the incoming energy on the planet in the form of heat is absorbed and put back into circulation (in part) by our seas. Apart from the fact that the sea currents themselves are affected by an increase in temperature (and with them, let us not forget, the atmospheric circulation itself and the terrestrial climate, which is no small thing), other factors are directly impacted by the increase in temperatures. Elvira Poloczanska of the Australian CSIRO confirms that more than 80% of the work conducted so far on the effects of climate change on marine species confirms the effects on organisms, 'especially in terms of distribution, and their effects may be greater than on terrestrial ecosystems'.

Let's start at the beginning with a concrete example. After more than a decade of satellite data analysis with increasingly accurate instrumentation, Mike Behrenfeld and his colleagues in the Department of Botany at the University of Oregon have concluded that the production of microscopic algae in the world's oceans has dropped significantly. 'As the temperature of the water bodies increases, satellites seem to pick up a decrease in the concentration of chlorophyll on the sea surface,' says Behrenfeld, 'and we all know how important this link in the food chain is for the entire ocean.'

Microscopic algae are in fact the basis of this chain, the basic product on which small crustaceans, fish larvae, bivalve molluscs and many other organisms feed, as we have seen in various parts of this book. Changes in this link affect everything from sardines to large sharks. The changes on a planetary scale could not be observed until we had used satellites and had interpreted the signals correctly. Calibration of these sophisticated devices has been essential to the correct interpretation of the data. It should be borne in mind that the data provided by oceanographic vessels, although essential for understanding how the natural system works, are fragmented and dispersed in an ocean that can only be covered by satellites. 'We estimate a 3% decrease in oxygen in water and a slightly more than 8% decrease in primary production', predicts Stelly Lefort of the French LSCE: 'At high and medium latitudes the size of phytoplankton cells will decrease, decreasing productivity.' They are models based on the consequences on metabolism, the capacity to accumulate reserves and the availability of nutrients for cells in the near future. Microscopic algae, with a small photosynthetic biomass (less than 1% worldwide) are responsible for almost 50% of primary production worldwide, being a true lung for everyone, producing of no less than 40% of the oxygen that you are currently breathing. Any change that affects them will affect the rest of the organisms (Fig. 18.1).

The decrease in the concentration of algae on the surface in future may be due to a change in ocean circulation. It is in the first hundred metres of depth where microscopic algae proliferate, thanks to sunlight, and it is here that the

Fig. 18.1 Phytoplankton cells in the Sothern Ocean. These small algae are already affected in several ways by climate change, especially in areas where water stratification will become more pronounced due to the rise in sea surface temperatures. *Source* Sergio Rossi

changes in the amount of nutrients and turbulence (movement) of water would be greatest. A decrease due to further stratification could be the cause of such large-scale changes. 'The 0.2 °C rise in temperature per decade may be reinforcing this stratification phenomenon', adds Professor Behrenfeld of Oregon.

But why? Why? When water is heated at the surface, it forms a layer in the first few metres where currents and nutrients from the depths decrease. The so-called thermocline is formed, a physical barrier that, since it becomes more and more pronounced as the temperature increases, prevents the exchange of nutrients, particles, and other elements between the surface and the depths. The areas of the planet that could be most affected are the tropical and subtropical areas, where the temperature increase would be more dramatic and the strength of upwellings from the bottom of the sea would be weaker.

Everything becomes even more complicated when we look at the detail in specific areas of the planet. 'The long time series, decades of information collected each week, have proven to be key to understanding what is happening with phytoplankton off our coasts', says Victor Smetacek, research professor at the Alfred Wegener Institute in Germany. 'Despite the enormous variability that exists year after year, the long series have allowed us to observe changes that are explained not only by climate change but also by direct human interference with the ecosystem.' In Heligoland, an island in the North Sea off the German coast where five years of accurate sampling of plankton and other variables has been carried out, the 1.5 °C increase in water temperature has led plankton communities to move the production period earlier, and is the undoing of all the other links in the food chain. Other photosynthetic organisms, however, may be more favoured. This is the case with cyanobacteria. Due to their size, nutrient requirements and adaptation to stratification conditions, it can be advantageous. 'These organisms, which can be very harmful, will increase their presence in coastal areas,' says Hans Paerl of the Institute of Marine Sciences at the University of North Carolina-Chapel Hill in the United States. 'The problem is that they are impossible to control and affect fisheries, water quality and many other factors.'

However, in certain areas off the California coast, it has been the disappearance of mussels and other plankton filterers that has caused alterations to the production of microscopic algae. 'It's not just climate change that's changing the ecosystem,' says James Cloern of the California Geological Survey, 'it's coastal disturbance, overfishing, or a large amount of nutrients and pollutants that seem to be changing the course of microscopic algae populations in many places.' In addition, we know that there are several internal clocks for plankton to respond to various physical and chemical components, and we still have little understanding of how they work. They are in part the key to understanding the response of so many species to changes in the sea. All factors, from changes in the behaviour of predators to changes in the seafloor and changes in nutrient levels, influence the life cycles of such a basic species as phytoplankton throughout the world. The answer is therefore extremely complex, and depends on the species and its adaptability to the environment. Yet, on the coast (where, as we have insisted, the greatest productivity of the planet is concentrated), factors such as the morphology of the seabed and the hydrology of the area are also involved.

There are several working groups focusing on specific areas of the planet, trying to see what happens with these changes at the level of primary production. For example, the masses of the North Atlantic are changing in intensity and direction, changes directly associated with the changes in strength and intensity of the Subpolar Rotation. This causes the displacement of phytoplankton species and the production of existing ones. 'From 1955 to 1995 there has been a decline in phytoplankton production around the Faroe Islands,' says Hjlámar Hátún of the Faroese Fisheries Laboratory, 'so consumers of microscopic algae have also been affected.' But it is still too early to make global models in this and other areas, because of the lack of data. The Antarctic continental shelf is undergoing an accelerated process of change that also affects primary production. 'We are moving in and around the Antarctic Peninsula from a cold, dry polar climate to a more temperate, humid one', explains Martin Montes-Hugo of the Institute of Marine and Coastal Sciences at Rutgers University in the United States. Satellite data combined with data from various oceanographic surveys indicate a 12% drop in chlorophyll concentration north of the 63rd Parallel of this area. Most importantly, south of this parallel, production is increasing. As discussed in the previous chapter, ice cover, also cloud formation (incident light) and wind direction and intensity, promote or impair the ability of phytoplankton to produce biomass.

In this area the ice cover arrives up to fifty days later in autumn and goes thirty days earlier in spring. The air temperature in this area of the planet has risen almost 4.8 times more than the planet's average, with a consequent impact on water temperature. Here, too, there would be an inverse relationship between the mixing layer and phytoplankton production: the thicker and more stable the stratified area, the lower the production, as nutrients do not reach them in sufficient quantity since they are consumed by the algae.

There are also changes in the type of phytoplankton, which have repercussions on biogeochemical cycles. 'We have seen that, during this period, cells larger than five microns are the ones that have been dominating so far,' adds Montes-Hugo, 'but this trend may be changing, with small cells becoming increasingly abundant in the northernmost area, where smaller cells would be more efficient.' But, at the same time, they would provide less food for the next links in the food chain: changes in productivity coincide with changes in zooplankton and penguin abundance and distribution, to give two examples.

Anything that affects primary production will have an impact on the amount of resources that we can extract directly from fishing in our oceans. As part of the future management of fisheries, everything that has to do with climate change will have to be taken into account. 'When we talk about fisheries management and renewable resources,' says Eileen Hofmann of the Center for Coastal Physical Oceanography at Virginia's Old Dominion University in the United States, 'we will have to consider how the temperature change will affect the most sensitive stages of its population, namely eggs, larvae and juveniles.' A lack of proper food or a change in physiology can affect them so much that they can displace species, impair their health or eliminate them from certain places. 'They have to be considered in detail and their impact on spawning grounds,' continues Hoffmann, 'as well as the abundance of larvae year after year in the face of temperature changes.'

We have several examples where predictions of sea production are based on changes in food availability. 'Small crustaceans such as copepods respond poorly to temperature increases', confirms Jessica Garzke of the GEOMAR Institute in Kiel, Germany: 'In warm waters, even with the same amount of food, the size decreases.' That means less food for fish larvae, cephalopod juveniles and other organisms that depend on zooplankton. 'Even with food, in areas with a high concentration of nutrients,' adds Edward Rice of Queen College in the United States, 'we are seeing gradual changes in the small crustaceans on Long Island: smaller, less desirable.' These changes influence fishing. In the Pacific, conclusions are drawn from the models of temperature increase and production decline: 'for aquaculture and fisheries in the reef systems themselves, it is going to be a problem. Stocks are likely to decline by up to 20% by 2050', says Johann Bell of the Fisheries, Aquaculture and Marine Ecosystem Division of New Caledonia. This will involve new management models, because a large part of the gross domestic product of these Pacific islands depends, as elsewhere, on fishing: 'And not only,' Bell concludes, 'because 50% of the protein ingested in these places comes from fish.'

Sometimes the effects are not direct but indirect: changes in local precip-itation, species dynamics, and so on. On the east coast of the United States overlooking the Gulf of Mexico, the oyster (*Crassostrea gigas*) has been heavily exploited. Apart from over-exploitation, environmental factors can greatly affect the viability of the bivalve populations. For example, both adults and new recruits are very sensitive to changes in water salinity. Remember that salinity is the concentration of mineral salts in water. It can range from almost zero per thousand (freshwater) to more than forty per thousand (in the Red Sea, for example). When rivers bring down plenty of water from their upper courses, oysters (and other organisms) can be severely affected. At a salinity below ten per thousand, oysters experience physiological stress, and at less than five per thousand there is massive mortality. Well, it has been proven that the water regime in Galveston Bay, Texas, has increased the amount of freshwater brought to the area in recent years due to a changing rainfall regime, causing oyster production to fall drastically. If they continue to have a low salinity regime in critical times such as regeneration, oysters from this area may no longer be an exploitable option. The same thing will happen in the Baltic, where the 30% increase in rainfall and the increase in water temper-ature that we were talking about at the beginning will fill the sea with cyanobacteria, drastically reduce fishing yields and create more hypoxia and anoxia than is already the case in this sea that is so heavily impacted by mankind.

It is very difficult, given the deep interference of fishing in the dynamics of species, to discern between the effects of climate change and those of exploitation. In the Bay of Biscay, the increase in water temperature over the past three decades is apparently already affecting pelagic fish stocks. 'We know that 75% of the availability of fish stocks on the planet depends on changes in climate', explains Alistair McIlgorm of the University of New England in Australia. 'Fish have been shown to be very sensitive to changes in water temperature both directly and indirectly (changes in food availability). The problem is that public opinion wants nothing to do with long-term changes or problems,' McIlgorm says, 'just look at what's coming to us here and now.' Where resource management is not good, changes resulting from rising water temperatures, changes in currents or problems with acidification (see below) will be more dramatic. Therefore, to the problems of fishing we must add those of a changing climate. This is a bleak picture that will affect the various areas of the planet in a markedly uneven way but that, in general, will bring about a change for which we are not well prepared.

Tropicalization of the Mediterranean

'Species from the Red Sea are being detected and are now reaching the Sicilian Channel.' Ernesto Azzurro of ISPRA in Italy has been one of the main observers of a recent phenomenon with implications that are still little known: the tropicalization of the Mediterranean. Some fish species such as trumpetfish have already been detected in the island archipelago between Tunisia and Sicily, and many species are rapidly moving from the Suez Canal, and their reproductive capacity is proven. Others, such as the lion fish, have been confirmed as expanding their presence and are now reliably reported to have been found in Chopre waters. If the species reaches a certain point and manage to reproduce, it can be said that it has successfully penetrated the ecosystem. Already in 1869 when, thanks to the efforts of the engineer Ferdinand de Lesseps, the Suez Canal had opened, they had begun to spread along the Mediterranean coasts of the Sinai Peninsula. But it was not until the last few decades that they began to be detected far beyond the Mediterranean coasts of the Middle East. This is partly related to the phenomenon of invasive species discussed in a previous chapter, but for these tropical species there was a physical temperature barrier that now seems to be dissipating. The Nile, with its enormous freshwater supply, was another major barrier to the movement of species. But the construction of the Aswan Dam and the use of water for irrigation in the world's longest river basin has brought about another physical challenge for newcomers to face: tropical species are becoming more and more comfortable.

More than sixty species of tropical origin have already been detected in the Mediterranean area, far from the Suez Canal. Here, the waters are very deep and the movement of certain species is difficult. However, according to research professor Carlo Nike Bianchi of the University of Genoa, there is a second barrier that seems to be weakening. 'The temperature in February in the Sicilian Strait is about 15 °C, and many species do not move and colonize the Western Mediterranean because at the same time of year temperatures are lower, 14 and 13 °C in the northernmost part'. The species encounter a physical temperature barrier that prevents their movement and, above all, their reproduction. 'The Mediterranean ecosystems are changing very rapidly,' says Professor Ferdinando Boero of the University of Salento in Italy, 'and the rapid increase in the temperature of their waters is one of the key factors in understanding this drastic change'. We should focus more on observation and identify how the mechanisms of expansion and invasion work, not just on monitoring them, if we want to understand where the evolution of our seas is heading.

Anna Sabatés of the Institute of Marine Sciences (CSIC) demonstrated almost a decade ago that an important commercial species was moving from the south to the north of the Spanish Mediterranean coast. It is the round sardinella (*Sardinella aurita*), similar to the sardine but with a slightly coarser taste. 'What we have been able to verify is that from 1950 to 2003 there is a positive relationship between the increase in temperatures in our waters and the northward displacement of the sardines', comments Sabatés. 'The catches a few decades ago were almost non-existent, and now they are very abundant.' The water temperature in the Western Mediterranean has increased by 1.1 °C on the surface and up to 0.7 °C in deeper waters. 'This has favoured not only the displacement of adults, but also the success of larvae and juveniles that in 1980 were not detected and that in the early 2000s could be found even in the area of La Escala in Girona, in northern Spain', concludes Sabatés.

It is not the only place where this type of displacement occurs. On the southwestern coasts of Australia, temperate water spermatophytes are moving southward under pressure from tropical plant species and also from herbivores. 'There may soon be a major change in food chains in this type of ecosystem that is key to coastal productivity and biodiversity', says Glenn Hyndes of Edith Cowan University in Australia. In places like Australia, these plants may tend to move gradually south, but in the Mediterranean there is no exit to the north. Many species may be forced to colonize only those places where conditions of temperature, nutrients, light and other factors are favourable.

'The Mediterranean has always been a breeding ground for invasions', says Professor Bianchi: 'As in other interglacial times, the waters of the Atlantic bring species from warmer, subtropical waters. The problem is that there are now two more sources of entry, completely artificial: lessepsian species and those introduced by various human means. And the changes we detect are very rapid.' The ecosystem matrix, however, does not seem to be changing. The Mediterranean, especially the Western Mediterranean, has no coral reefs as such and the species known as bioengineers (algae, corals, sponges, gorgonians, etc.) that create structures and shelter for many species remain the same. However, the profound change may come from somewhere else. Ferdinando Boero insists that the changes in the formation of cold waters in the north could transform food chains and diversity, especially in deep areas. 'It would be a catastrophe that would undoubtedly affect the productive system, so we cannot look at isolated cells, we must see the problem as a whole connecting points to be able to respond to fishermen and other sectors of economic interest that do not understand what is happening.' We do not really know to what extent these changes and introductions will have an impact on the ecosystems of the Mediterranean, but over the next few decades we will see a tropicalization of the waters of the southern Mediterranean, and the stabilization of rising temperatures will provide the ideal environment for species completely alien to our underwater landscape. It will be a silent transformation that may change the appearance of one of the most iconic seas on the planet.

Coasts that Make Water

On our planet, sea level has risen and fallen continually. At the height of the ice age, the ice masses had been able to retain a considerable amount of water, leaving the sea level more than 120 m below the current level. Caves, tunnels and many coastal formations such as underwater canyons are due to periods in which the waters regressed considerably, allowing rain to make its mark and eroding calcareous or sedimentary formations. Now, in the middle of the interglacial period, the sea level continues to rise, although it seems at a faster and faster rate, as some of the water that had been retained by glaciers now reaches the sea without renewing the glaciers by falling as new snow in extremely low temperatures.

Continental glaciers in most parts of the world are shrinking, especially those most isolated or in a tropical or subtropical climate. The water that is not retained in the form of ice is found in the form of vapour or liquid. It is the circulation of the water from these glaciers that is causing the rise in sea level, not the water from sea ice, the formation of which (as the word goes) is from the freezing of the water in the Arctic and Antarctic oceans. According to the IPCC 2013 models, the sea level rise could be between 0.3 and 0.5 m in about ninety years, with an estimated maximum of one and minimum of 0.1 m. This is a benevolent scenario in which there is no positive feedback, so that the large glaciers, those of the central Antarctic continent or Greenland, do not begin to melt. 'A positive forecast speaks of an increase of 19–30 mm from the glaciers of this immense island alone in the next two hundred years, but with the way that we are going (having increased the atmospheric temperature by 4.5 °C), the contribution may be more than 50 mm,' writes

Faezeh Nick of the Free University of Brussels in an article. This is the case of Greenland, which already shows symptoms of severe melting. In the almost impossible event that the entire Antarctic continent were to melt, sea level rise would be several tens of metres. So, in the cases we're talking about, a 0.5–1 m increase, what could happen?

Let's take a concrete example. New Orleans residents would see their city and the surrounding Mississippi delta flooded by the Gulf of Mexico in 2100. These are the conclusions of a recently published study, in which, like others, the model points not only to a rise in sea level due to climate change but, above all, to the loss of sedimentation by the Mississippi River. 'Because of the river's dam system, more than 50% of the sedimentation that kept the delta alive and its extensive wetland system has been lost', says Michael Blum of the Department of Geology and Geophysics at the University of Louisiana: 'There is a clear shortage of material, that is already taking its toll on the delta system.'

By 2100, according to Blum's calculations, between 10,000 and 13,500 km^2 could be lost, leaving the city of New Orleans almost isolated on a narrow isthmus. Forecasts have been made based on the increase in water levels of a minimum of 3 mm per year and a maximum of 8 mm per year proposed by the IPCC in 2013, as well as estimates of material loss due to lack of sedimentation of between 1 and 4 mm per year. 'Between 18 and 24 billion tonnes of sediment would be needed to counteract this situation', adds Harry Roberts, a collaborator with the Institute of Coastal Studies at the same university. In any case, both scientists see flooding as inevitable, as the Mississippi delta's management plans do not include a sea level rise that is up to three times faster than the current rate. We have to think that the delta was already slowly regressing due to a slow but unstoppable rise in sea level for more than seven thousand years. 'We are still in a transgressive phase, i.e. of the conquest of the mainland by the sea,' confirms Blum, 'but nobody expected such a rapid rise in the water level.'

We have to remember that the rise in sea level will not be homogeneous. It never has been. Currents, topography or geographical location are factors that will influence the rise in sea level either more or less. 'In Cape Hatteras, for example, in the United States,' says Asbury Sallenger of the Geological Survey, 'the rise in sea level between 1950 and 2009 was four times the world average.' It is necessary to plan according to the area, and what is clear is that this change is already affecting the whole planet, in some cases in a dramatic way. By 2100, Louisiana estimates a sea level rise of between 0.2 and 0.6 m, 'which means that areas below one metre in height will turn into marshes or shallow beaches', concludes Roberts. According to the authors, similar studies have come to a similar conclusion; that is, that we must take action if we do not want to see cities like New Orleans in a critical situation.

The alarm comes not only from the flooding of vast areas of the planet, but also from increasingly adverse weather conditions (floods, storms) that would damage islands and coastal areas whose emerged parts are no more than 2 m above sea level and which would be particularly vulnerable to these changes. For example, in tropical environments the flooding of large areas could increase the risk of more violent and prolonged heavy rains and monsoons. Places as diverse as Bangladesh, the Ebro Delta or entire Indian or Pacific archipelagos could feel the consequences of a slight rise in sea level also in land management. Beaches and coral reefs are expected to experience more severe erosion as sea levels rise. In Tonga, the elevation of a single metre would cause a 14% loss of land in the Majuro Archipelago (Marshall Islands), and up to 80% of this land would be lost.

The strategy is clear in this and other cases: understanding the real extent of losses, adapting to new climate trends, including new modalities and clauses in which this and other climate change-related phenomena are put at risk. I am honestly surprised that there are still people who, having accepted the fact that climate change is a reality, are now engaged in discussing whether or not it is anthropogenic in origin.

Storms

From time to time the easterly storms in the Mediterranean are stronger, more intense than normal. On 26 December 2008, a strong storm devastated the Costa Brava, eradicating many organisms down to 20 m deep, leaving fish stranded on the beaches, thrown up by waves several metres high. The experience of the people who experienced it was simply an attack of unparalleled fury by the sea.

The dynamics of marine communities must regularly withstand these natural 'disasters' of particular intensity: depending on their virulence, they will affect organisms to a greater or a lesser extent, especially those at the bottom, which sometimes recover or leave space for new generations. The problem may arise when these storms become more frequent than the system is able to withstand, not giving communities the chance to restructure. Some organisms such as photophile algae are annual, but others, such as gorgonians, have a longer lifespan. If the frequency is increased, the resilience would be less in the shallower areas, which are always under more stress than the deep ones.

Is this happening? Are easterly storms now more frequent in the Mediterranean, and hurricanes in tropical areas? Studies are far from conclusive, but it appears that warmer water could lead to more violent or frequent storms. Let's see two opinions from very different places, with different ocean dynamics. 'In north-west Europe, off the coast of Great Britain, Ireland or Norway, storms have increased in frequency since 1881,' explains Hans Alexandersson of the Swedish Meteorological and Hydrological Institute, 'but they seem to be weaker than in the past. On the other hand, Kerry Emanuel, who focuses on tropical storms and hurricanes in the Caribbean, says that 'the time series since 1970 do not seem to show them to be more frequent, but they are longer and more intense'. These are not contradictory data. They are simply different places on the planet. 'We are beginning to quantify the increased intensity of cyclones and hurricanes', says Robert Mendelsohn of the Yale School of Forestry and Environmental Studies in the United States. His working group found that the areas most affected by this change in intensity (the proportion of hurricanes of force 4–5 has increased by 25–30% for every degree Celsius that the water temperature has increased) are eastern North America, East Asia and the Caribbean.

In any case, the authors, like most of the other experts in this field, to explain the trends insistently allege links to the increase in the surface temperature of the seas and oceans. Returning to the Mediterranean, specialists do not yet have conclusive data. 'On the coast of Israel, the westerly storms seem to have increased sharply in recent years,' says Hadas Saaroni of Tel Aviv University, 'but we lack the data to be conclusive on the subject.' This is one more effect to take into account, because storms have many effects on biota, infrastructure and tourism. At present, damage from cyclones, hurricanes and major storms is estimated at some $26 billion (0.04% of Israel's gross domestic product). In the coming decades, this figure could even double. An increase in virulence or frequency would mean an additional twist to the system as we know it.

19

More Acidic Seas

In 1956, Roger Revelle and Hans Guess, two visionary geochemists, warned of a possible increase in carbon dioxide in the atmosphere. We all know that this has happened, yet they saw beyond the rise in temperatures, because this and other gases that are responsible for the so-called greenhouse effect were forming higher and higher concentrations due to industry, transport, agricultural uses, and so on. The increase in this gas could lead, among other things, to acidification of both inland and marine waters. The issue is complex, but it arises in the following way: the carbon dioxide entering the sea combines with water and gives rise to another compound, weak carbonic acid (H_2CO_3), capable of releasing hydrogen ions (H^+) with ease. Having lost this hydrogen, the carbonic acid remains as a bicarbonate ion, HCO_3. When this happens, hydrogen ions remain in the water, acidifying the liquid medium. In other words, the more carbon dioxide in the atmosphere, the more hydrogen ions will be put into circulation. This acidification, measured by pH (or concentration of H^+ ions in water), is already occurring in our oceans. Many marine organisms will be affected. The pH of the waters that make up our seas is slightly alkaline (pH 8–8.3); that is, it exceeds neutral pH (which is quantified as pH 7). The tendency to lower the pH, especially in certain areas of the planet and especially on the surface, may lead to changes in food ratios, material and energy flows and ocean geochemical balances that are still unpredictable. In all the oceans. It is a global process, like that of heating, which is extremely difficult to reverse since acidification is a much faster process than de-acidification, for reasons of chemical balance and where the acid ends up once in circulation.

© Springer Nature Switzerland AG 2019
S. Rossi, *Oceans in Decline*,
https://doi.org/10.1007/978-3-030-02514-4_19

Although there has been substantial progress over the past fifteen years, little is known about this issue still. In fact, we could say that in a way it is in its infancy, and there is a great deal of controversy about what is actually going to happen over the next few decades. There were few scientific articles prior to 2004, not because the problem was not predicted but because nobody took it seriously until just over ten years ago. When I started talking to specialists like Patrizia Ziveri from the Institute of Science and Technology of the Autonomous University of Barcelona I realized that the subject was much more complex and not as obvious as I had thought. In fact, the first time that I heard about acidification was from a professor of marine biology, Joandomenec Ros at the University of Barcelona, in our marine biology class. The class was not talking about the global absorption of carbon dioxide by the sea but about the possibility of injecting our carbon dioxide directly into it, and the scheme was ruled out precisely because the affected areas would become acidic.

Both carbon dioxide and acidification levels have soared since the late 1700s. However, the acidification of the oceans that we are now beginning to understand is not new. We know, for example, that some 55 Ma ago, in the so-called Palaeocene–Eocene Thermal Maximum (PETM), the seas were far more acidic than they are today, and this led to the disappearance of many marine plants and animals from our planet. According to the fossil record, there have been appearances and disappearances of coral formations over the last 200 million years. It should be remembered that corals base their structures on calcium carbonate ($CaCO_3$), a compound that is highly vulnerable to acidity, so slightly alkaline water only is conducive to precipitation. We know, however, that corals come and go. They have not become extinct, despite changes in the acidity of the water, presumably due to their physiological adaptability. However, some specialists, such as Samantha Gibbs of the National Oceanography Centre in Southampton, in the United Kingdom, believe that it was the temperature rather than acidification itself that affected many of the calcareous organisms: 'The coccolithophorides took refuge in high latitudes, where the acidification conditions might have been more acute, but they were wary of excess temperatures', she says. It is possible that in times of transition, when the planet's oceans acidified, some of these plankton or bio-builders were able to reduce their size or rid themselves of their calcareous structures—to become what are known as soft corals, also found in the world today.

It took about 2,000 gigatonnes of carbon dioxide to reach the pH levels that hindered the cycles of certain marine organisms in PETM but, according to recent estimates, we could be producing up to 5,000 gigatonnes of carbon dioxide over the next 400 years. That would be much more carbon dioxide, so the oceans would acidify faster and faster than at that earlier time. Today, estimates up to the year 2007 speak of 560 billion tonnes of accumulated manmade carbon dioxide.

The problem is that the rate of acidification that we are detecting today is far faster than in that distant period. In other words, the pH is dropping at an unprecedented rate. What is known for certain is that, compared to the so-called pre-industrial period (late 1700s), to date the pH has fallen by 0.1 points. At first glance it doesn't seem a major problem that the pH in the oceans will drop by 0.3 or 0.4 points, which is the reduction predicted for a period of 400 years. But let us remember that pH is measured on a logarithmic scale. In other words, every time we lower the pH by 0.1 point what is put into circulation is an immense amount of hydrogen ions. You don't have to go so far. In the year 2100, the pH could have fallen between 0.15 and 0.25 points compared to the pre-industrial period, according to the most conservative models, which are being continuously reviewed. These revisions are leading us increasingly towards faster transition scenarios than we had thought, so these figures could be reached even earlier.

The effects of acidity seem to be concentrated in marine areas where biological production is higher. At the sea surface, at depths of less than 200–300 m, the acidity is much higher and more concentrated than in deeper regions. In other words, acidification would affect the most productive layers of the ocean (the three oceans surveyed have indicated this), especially in temperate and equatorial zones. It should be borne in mind that the precipitation of calcium carbonate is favoured and the absorption of carbon dioxide also increases in those regions where the water temperature is higher. The balance between physical and chemical agents is not yet fully resolved, but there are signs of acidification in these vulnerable areas.

In upwelling areas, such as the Benguela (Namibia–Angola) and Humboldt (Chile–Peru) coasts, the effect could be mitigated by the emergence of more alkaline currents that would counteract the injection of carbon dioxide into the surface layers, but this is far from clear as the acidification process depends on temperature, and depth is also an important factor. The more alkaline, less acidic waters would be kept in the deeper waters, and there the low temperatures (and pressure) would make it difficult for the organisms to calcify. In all oceans, the dissolution of $CaCO_3$ increases as we descend, therefore acidification will be especially acute in the first 250 m, the most productive layer of the sea.

Today, marine organisms have adapted physiologically to slightly alkaline marine waters, but if acidification occurs (and even more if this is sudden) they will undergo drastic environmental changes that many of them will not be able to withstand. There are a huge number of organisms that need calcium carbonate to create their structures, shells and protection. In addition, many planktonic organisms, such as small algae called coccolithophorides or the abundant protozoa, foraminifera, will find it harder to survive. The former are cells capable of photosynthesizing and constitute an important part of primary production in certain places. The latter are protozoans capable of eating algae or debris and represent an important link in the food chain. Both are at the base of highly complex food chains and the fact that they cannot create their calcium carbonate structures (mainly aragonite and calcite) will have consequences on their life cycle.

The coccolithophorides, for example, produce large blooms like other algae. The difference is that, being of a calcareous constitution, these small cells form carpets of a whitish colour that can extend many square kilometres over the surface. 'In a bloom of coccolithophorides the reflection of light is very intense,' describes William Balch of the Scripps Research Institute, 'and at just one and a half metres of depth we cannot see a submarine probe because of their density and the light that they can reflect.' More than 100 thousand cells per litre can form spots on the ocean that are capable of reflecting sunlight, increasing the Earth's albedo. 'Coccolithophorides may be responsible for up to 0.13% of the Earth's albedo', continues Balch. As ocean particles, they can reflect 10–20% of light at normal concentrations, but when they form intense blooms (which can be seen by satellite), the reflected light can reach 90%. Changes in the quantity or formation of these calcareous carpets, or even their partial disappearance from the sea, could increase the absorption of sunlight by the sea, in full synergy with other factors such as decreases in either ice or clouds cover. These are all factors that provide positive feedback to the global warming of the planet. To this end, a further piece of information should be added. Coccolithophorides produce dimethyl sulphides (DMS), a chemical compound that, when released into the atmosphere, is partly responsible for the nucleation that forms clouds above the sea surface. If less DMS were produced, cloud formation could be affected, again influencing the albedo issue, because clouds also participate in the reflection of some sunlight, preventing it from reaching the surface. This appears to be an indicator for organisms such as birds, fish or food mammals that grow at sea. Its reduction in productive areas could mean a decrease in many organisms' ability to spot the signals for locating food, although this point is still much debated. Over the past few years, an effort has been made to understand better the evolution of short- and long-cycle species by proposing long-term experiments. Kai Lohbeck of the German GEOMAR

Institute carried out experiments in which several generations of coccol-ithophorides were analysed: 'After 500 generations under low pH (remember that they are algae that divide every two or three days), *Emiliana huxleyi* reduced its size and responded by adapting its calcification rate to the new conditions.' Several groups insist that the experiments must be long term, with enough time to see adaptations to the new conditions. This is not to say that nothing will happen to them, because in this case the decrease in size represents a factor that will affect all the organisms that feed on this algae: 'These experiments have to be done in a more robust way, crossing the effects of temperature and acidification so that we have an idea of where the evo-lution of species like this or more complex ones will go', adds Lothar Schlüter from the same GEOMAR Institute in Kiel.

Another problem related to acidification is that it affects the extensive coral reefs. As mentioned earlier, corals have appeared and disappeared over the last 200 million years. Lowering the pH decreases the availability of material to build shells, structures or other protection. The organisms adapt to these situations, but in the case of coral the problem may be the disappearance of the reef itself. These biostructures are cradles of biodiversity, refuge and source of food for an immense number of species, as well as an economic source for the humans in several ways—fishing, tourism, medicines, and so on. According to some models, by 2090 surface coral reefs will be severely affected by the acidification process.

While some experts believe that there will be a transition only to less 'calcareous' structures and species, others believe that there will be a signifi-cant loss of diversity due to the survival of few species. However, scientists like Paul Jokiel of the Hawaii Institute of Marine Biology argue that the physi-ological barrier of coral and its symbiosis with microscopic algae may be sufficient for medium- to long-term adaptation. 'The effects of acidification are very difficult to assess because there are other factors that affect the barrier reef, especially temperature increases', writes Rebecca Albright of the Carnegie Institution for Science in California: 'We should experiment with pre-industrial levels of carbon dioxide to see if the reef is calcifying again at higher rates.' Not only symbiotic corals will be affected by acidification. Lorenzo Bramanti of the French CNRS and I designed a long-term experi-ment (about 13 months) in which we saw that red coral, valuable for jew-ellery as we discussed in a previous chapter, lost more than half of its calcification capacity at temperatures of about 14 °C. 'Most of the coral of interest to fishing is below 40–50 m deep, where the temperature is almost constant', says Bramanti. We were able to observe that the increase in diameter was reduced by more than 30%, which would greatly affect this already slow-growing species. But most interestingly, to counteract the effect of acidification, red coral changes its metabolism. This factor is key, because if you have to invest all your energy in survival, you don't waste it on things like growth or reproduction. Núria Viladrich, a doctoral student of mine, in other work has demonstrated that energy investment in reproduction varies according to the type of food and the reproductive strategy. 'The most vulnerable species could be those that need to invest large amounts of reserves in the form of lipids to reproduce', says Viladrich. 'Metabolism is the battle horse that we still need to understand', argues Dr Godbold of the University of Southampton in the United Kingdom. 'Species can adapt, but we need to understand the energy cost of that evolution in fact.' Godbold also stresses the importance of species interaction, factors such as diversity, food chains, competition, and so on.

Unfortunately, as we have said, acidification is yet another factor that will add to those already affecting these complex communities, such as loss of colour (coral bleaching, see Chap. 20), overfishing, tourism and eutrophication of waters. The result will be the local collapse of many of these biogenic structures. If we build a survival model for coral reefs, we will see temperature rise and acidification attack on several fronts at once. On the one hand, acidification decreases the coexistence and roughness of the corals, which increases the intensity of predation, in particular by fish. It would increase the number of macroalgae, also favoured by the reduction of sea urchins (remember that their skeleton is calcareous). In addition, many of the larval stages of corals, urchins, bivalves and other animals that base their survival on the construction of calcareous skeletons would be compromised (among other things because of what we mentioned earlier about the metabolic priorities of species under physiological pressure). In other words, both the recruitment and growth of these corals would be affected by acidification, without taking into account other factors such as water warming or local resource exploitation.

In general, the existence of all those organisms that base their life cycles on the construction of calcareous skeletons, both internal and external, would be compromised—gastropod or bivalve molluscs, gorgonians and certain groups of sponges, among others. This process is called acidosis, and it is suspected that even fish and other higher organisms may suffer from it. Acidosis may depress the immune system, alter metabolic relationships and cause breathing difficulties.

Few experiments have yet been carried out on fish, but most indicate a weakening of structures and significant physiological change. It is not only that water is no longer acidic but the interaction between molecules, the change in the spectrum of waves that penetrate the sea, the acid-base balances of organisms—in short, a complex series of factors that we still have a long way to go to understand. In some cases, it has been seen that the survival of more than 70% of fish larvae may be compromised, but longer-term experiments are needed to understand their true impact. In fact, Sam Dupont of the University of Gothenburg in Sweden found in a four-month experiment that sea urchins' reproductive rates were severely affected, but if we conduct an equivalent acidification experiment lasting longer, sixteen months, the effect is almost imperceptible. 'Many of us insist that longer-term experiments are needed to see the adaptability of species; if you subject them to physiological shock, the response may be momentary stress.'

Acidification and other factors will also lead to a change in the distribution of nutrients across the planet, which will affect the survival of certain primary producers (algae) and thus the circulation of matter and energy in each area. 'It is clear that the increase in acidification and carbon dioxide in the system will have, along with other changes, important repercussions on biogeochemical cycles', says David Hutchins of the Department of Biological Sciences at the University of Southern California. Early experimental attempts to understand what may happen in this regard indicate increased nitrogen fixation, thus decreasing nitrification. This may mean less nitrogen available for certain groups in primary production but not others. 'Cyanobacteria seem to be comfortable with this future scenario of slightly warmer, more acidic waters', concludes Hutchins.

Neither are the effects on the cycles of phosphorus and silica, nutrients essential for life functioning, at all clear, but it appears that they may be less affected. More affected is the availability of certain micronutrients. The availability to living beings of atoms in their ionized forms, such as iron, aluminium, magnesium or chromium, may have altered. 'The decrease in OH^- and CO_3^{2-} can affect the solubility, adsorption, toxicity and speed of oxidation-reduction processes in seawater metals', says Frank Millero of the Rosenstiel School of Marine and Atmospheric Science in Miami, USA. The solubility of metals in water is highly dependent on pH, as is the case with trivalent metals, as they are more soluble in acid solutions than in basic solutions.

What is clear is that by modifying all trophic levels of the system we may face unprecedented uncertainty, with a series of physiological adaptations of which we still know very little, and eliminating some groups may replace them with others. 'Man is doing a large-scale geophysical experiment,' says Scott Doney of the Woods Hole Oceanographic Institution in the United States, 'and he is doing it in a way that has never happened before and will never happen again in the history of the planet.' The upheaval may be tremendous, but not everyone will lose out: there are many groups of algae, such as diatoms or dinoflagellates, whose existence is not based on the formation of carbonate structures. However, the alteration due to the acidification of biogeochemical cycles—that is, carbon and nutrient exchanges in our oceans—is a real unknown. 'While the reefs will be under profound stress from several components of stress at once, kelp forests and other non-calcareous algae may remain the same or even be favoured', concludes Dr Comeau of the University of Western Australia: 'There are species that will resist acidification, others that will be severely damaged; we need to understand the mechanisms and time is not on our side.'

According to Rob Moir, Director of the Ocean River Institute in the United States, 'Acidification can no longer be a neglected issue; it can be as serious or more serious than the rise in temperature or sea level due to climate change.' We still have a long way to go to understand the extent of ocean acidification. It is also the factor related to climate change that worries me the most, along with that of a possible new glaciation. It is yet another problem to add to the already complicated picture that we are facing in which more and more pieces of the disturbing puzzle are beginning to fit.

The Keys to Understanding the Problem

Over the last 200 years, between 25 and 35% of the carbon dioxide emitted into the atmosphere from burning fossil fuels has been absorbed by the world's oceans. This apparently good fact about how to mitigate the effects of this greenhouse gas has a perverse effect that is still little known but has already begun to be quantified: the acidification of our seas.

There are three basic concepts about the acidification of the seas:

1. Carbon dioxide from the consumption of fossil fuels is partly absorbed by the ocean.
2. This carbon dioxide forms carbonic acid in seawater and reduces its pH, which is currently slightly alkaline. It also causes a change in the balance between carbonate and bicarbonate ions.
3. Changes to a more acidic ocean complicate life for species with hard parts made of calcium carbonate, such as corals, bivalve molluscs, echinoderms and some important plankton groups.

A Natural Laboratory on the Move

There is still a long way to go before we understand the phenomenon of ocean acidification and its impact on fauna and flora. Various research groups around the world are working against the clock to see what the future outlook might be if pH levels were to fall to levels significantly below the current levels. However, apart from the ongoing experiments, there are certain places in the world where the whole community is subject to acidification similar to that predicted by the models for the future. In the sea off Ischia, an island off the coast of Naples, natural carbon dioxide wells up to make this site one of the few natural acidification laboratories of the Mediterranean—and of the planet. In one of the shallower areas, scientists can see what might potentially happen to organisms living in the sea under the effects of acidification.

Unlike other places where there upwellings of carbon dioxide, in this case the gas is not accompanied by sulphur or other compounds that are usually present in emissions from the bowels of the Earth, nor does it emerge at high temperatures. Carbon dioxide is released and then it is dissolved in the water, creating a microclimate. 'Corals, calcareous algae, sea urchins and snails are very much affected', explains Jason Hall-Spencer of the University of Plymouth's Marine Biology and Ecology Research Institute. The pH is normal in the nearby area that is unaffected by this type of emergence, ranging between 8.1 and 8.2, as in the rest of the Mediterranean, but close to the emissions it can drop to 7.4–7.5. This is the value that, if global carbon dioxide emissions continue, could become a reality on a large scale within a few decades. Here, for example, snails and sea urchins lose their robustness as their calcareous structures are eroded; they become fragile and vulnerable. Experiments show the vulnerability of many of the organisms that are unable to withstand slightly more acidic conditions than normal—sponges, for example. Up to 30% of species disappear from the acid zone, especially calcareous ones, which avoid it or become damaged. 'Only one sponge, *Crambe crambe*, survives at the lowest pH, with a clear gradient of diversity from more acidic to more alkaline areas', says Claire Goodwin of the National Museums Northern Ireland in the United Kingdom. On another island, Vulcan, similar experiments have been conducted. Vittorio Gavilli of the APEMA Institute in Palermo, Italy, notes that on these islands, where there are also carbon dioxide surges, the conches that survive are the smallest and have a more fragile calcareous structure. But most importantly, as the habitat changes, the entire food chain in this area is affected. Many fish look for food elsewhere, where there is more food diversity.

While the authors acknowledge that these areas are a peculiarity in the midst of a Mediterranean dominated by alkaline waters, they insist that the study demonstrates that slight acidification can radically change the marine landscape across the globe, and the natural laboratory provided by these unique sites is already being used by research groups from around the world.

The Mediterranean: A Special Case

The Mediterranean Sea is an alkaline sea par excellence: of the values in the various oceans on the planet, its pH is above average. Fishing, one of the area's biggest headaches, is at a low level compared to some places, yet the organisms living in the Mediterranean have become adapted to the water's intense alkalinity, which allows them to form shells, bone plates, skeletons and calcareous coatings. Moreover, the level of tourism is high, and some 200 million people live on the coast. This is why, faced with the problem of potential acidification, nobody considered this 'small basin' to be too small to worry about. In 2009, under pressure from a large group of scientists and experts, the European Union created a topic for future programmes on the effects of acidification in the Mediterranean Sea.

The circulation of the Mediterranean is complex, with both surface-water and deep-water flows through the Strait of Gibraltar, to and from the Atlantic. The waters are supersaturated with calcite and aragonite, from the surface to the bottom. However, several working groups have already detected the penetration of carbon dioxide in deep waters in both the Eastern and Western Mediterranean that may be causing the pH to decrease. In fact, contrary to what had been claimed, acidification would be a phenomenon that would have enormous repercussions in the Mediterranean, precisely because of its adaptation to this alkalinity. 'Changes in plankton have been observed', explains Patrizia Ziveri of the Institute of Environmental Science and Technology (Autonomous University of Barcelona): 'Some groups will be more affected than others.' But what we don't know and need to understand is how it might affect the rest of the food chain, like copepods or cladocerans. It is estimated that between 2010 and 2050 acidification in this part of the world could increase by up to 50%. 'If we continue at this rate to the turn of the century, it could have increased by as much as 150%', says Ziveri.

Changes in plant plankton can lead to changes in zooplankton that are yet to be understood. Coccolithophorides, foraminifera and pteropods are the focus of several working groups, and also organisms such as red coral or hexacoral organisms such as *Cladocera caespitosa*, both eco-engineers. How will it affect the young, the most vulnerable? How will it affect their growth, their consistency? These are recurring questions everywhere, but in the Mediterranean, a sea of small dimensions, changes could be faster than elsewhere. Everything can be affected, from nutrient recycling to benthic biodiversity. As we have mentioned before, one of the most iconic communities here is the coralligenous one, a community of calcareous algae of slow growth, and diverse animals (sponges, gorgonians, bryozoans) also with low population expansion potential. With almost all of them based in one way or another on calcium carbonate, how will the change affect this community? Other organisms of commercial interest, such as mussels (the shell is made of calcium carbonate), will also be affected. 'Depending on the species and its calcareous and metabolic needs, we will see profound changes in the survival of mussels, oysters or clams', explains Frédéric Gazeau of the French CNRS.

But there is a lack of data, not only on what will happen from a biological or ecological point of view but above all from an economic and social point of view. In a paper by Luis Rodrígues and my ICTA-UAB group, we found that people were willing to pay to conserve the seas as they were, yet needed to be informed of this and other problems associated with the disappearance of species in the Mediterranean. The economic and social side is perhaps the most complex to foresee, which is why specific projects have been launched in this respect for the world's most historic sea. The final question most of them ask themselves is: how much will this change cost nature and society?

Climate Change 'Mitigation' Experiments

In recent years, and in view of the increase in ocean temperatures and acidity that we have discussed throughout these chapters, several projects have been implemented involving so-called environmental engineering for climate change mitigation, also known as 'geoengineering'. These large-scale experiments are not without controversy, and one of the most recent (the LOHAFEX project) was carried out from the German oceanographic vessel *Polarstern* in the polar waters of the southern hemisphere. German Professor Victor Smetacek of the Alfred Wegener Institute in Germany was one of the main promoters, in collaboration with an Indian research group led by Professor Wajih Naqvi of the National Institute of Oceanography in Goa, India.

Algae, especially the diatoms that are so important to marine production, have elements that limit their growth. Among them, there are so-called micronutrients. Unlike nitrogen or phosphorus, these are scarce, are found in very low concentrations in the sea and are essential for cellular functioning.

One of them is iron, which long ago proved its importance in the production of these algae. In the experiment carried out between February and March (in the middle of the southern summer), soluble iron was added to create a large proliferation of algae that would be consumed by small planktonic crustaceans or fall into the sediment after having absorbed carbon dioxide for growth.

The promoters of the large-scale experiment state that the aim is to promote climate change mitigation through the sequestration of carbon dioxide by microscopic algae. The first results effectively showed a major proliferation of algae and an increase in the concentration of copepods and other small crustaceans. But there are voices within the scientific community that are diametrically opposed to this environmental engineering tactic. Professor John Cullen of Dalhousie University in the United States stresses that little is known about the effects of this type of undertaking on the rest of the ecosystem: 'It is very possible that what falls by gravity will return to the system, breathed by bacteria (which are the main biological producers of carbon dioxide on the planet), so we would not be doing anything.' Several experts have carried out fertilization experiments (in the Pacific, the North Sea and off the coast of Ireland) and found something with a little promise: by supplying iron, one toxic species of diatom (*Pseudo-nitschia variabilis*) appears to prevail over the others, forming extensive red tides whose effects on the ecosystem have been shown to be harmful. A system of immense pipes has been arranged that, submerged in water, would stimulate primary production by bringing nutrients from the deepest area (where less light reaches) several tens of metres away, by means of pumps, so that algae would proliferate unchecked in the photic area (where light is abundant), capturing carbon dioxide from the atmosphere. 'We should not be tempted to experiment without being absolutely certain that we will not harm the environment with our "remedy"', said Ken Caldeira of Stanford University's Department of Global Ecology.

The issue is, of course, very complex and requires precautions to be taken before applying measures when we are unaware of the potential side effects. Let's take two more examples. Why not suck carbon dioxide and methane directly out of the atmosphere to mitigate the greenhouse effect? This is the proposal of several scientists such as Nicola Jones, whose aim is to create in desert areas a farm of huge collectors, where filters would selectively retain the gases. 'They would be immense purifiers of the atmosphere', explains Jones: 'With these farms we could recapture a significant part of the 9 gigatonnes of carbon dioxide generated in a single year.' But critics see this measure as very expensive, from an energy and monetary point of view, at between €50 and €200 per tonne of carbon dioxide. Taking into account an 'average' cost of €100 per tonne and rounding up, that makes about €900 billion.

Another, apparently much cheaper, method would be to create micro-drops, about a thousand times smaller in volume than the droplets produced by normal clouds, to sow into the cloud formations in the central areas of the oceans. 'In this way we would increase the Earth's albedo very considerably', concludes Stephen Salter of the University of Edinburgh: 'Thousands of ships around the world would produce these cloud formations capable of reflecting the incident light.' The project has been sounded in various government agencies, and tests were carried out off the Chilean coast in 2009. 'We don't know what effect these types of formations will have on, for example, ozone', says David Keith of the University of Calgary, who is one of the biggest critics of environmental engineering operations. The particles put into circulation as nuclei of condensation would end up somewhere, and the artificial clouds would have a dynamic that we do not know about. 'This is a very economic measure; for example, reconverted vessels from the fishing industry could be put into operation', adds Salter. About 25% of the surface of the oceans is covered by clouds, and the purpose of this geoengineering project would be to make them brighter. But we do not know what the precipitation patterns of these artifacts would be, either, as the nuclei would be much smaller and their dynamics in the atmosphere would be different from what we know. The most daring talk is about sending rockets full of tiny solid particles into the stratosphere to exert a permanent 'shadow', which would limit the amount of solar radiation arriving on the planet.

'Undoubtedly one of the biggest problems for geoengineering is the lack of clear legislation', concludes Edward Parson of UCLA's Emmett Center for Climate Change and Law in the United States. 'Some scientists advocate that there should be no restrictions, but it is clear that if someone unilaterally takes the decision to do something that affects us all, we could be faced with a general conflict that is difficult to resolve.' We know little about abrupt climate change, so any mitigation effect due to the 'need' for action should have broad consensus.

There are many ideas, much haste and, in my view, little reflection. 'The problem I see is that if this type of prototype worked,' says Cullen, 'people would put the greenhouse gas problem aside. We would jump to alleviate the problem simply with large-scale environmental engineering.' I agree with Cullen. I am simply horrified by this approach. The problem is that we cannot simply wipe the slate clean by creating a series of mitigation systems, moreover ones whose consequences we are unaware of.

20

The Slow Death of the Reef

When I first visited a coral reef, I went down from a beach next to the apartments occupied by a colleague and I near Sant Genís on Reunion Island (near Madagascar). It was the first thing in the morning in mid-December back in 1997, and I was looking for a place to take some photographs as a souvenir of the trip. We had gone there to study the feeding of various species of cnidarians in the reef area and, being a newcomer, as soon as I had the chance I rushed into the sea to see the coral structures that I had heard so much about. I found myself in a scattered and poor environment, nothing like what I had seen in documentaries and photos. This was not surprising, as the reefs on that island are considered immature, due to volcanic action, but I found myself a little disappointed to find nothing but clumps of an undefined species of staghorn coral (*Acropora*) here and there. The few formations I found, moreover, were defended by fierce little fish jealous of their home, and these kept harassing me by biting my fingers and legs when the occasion presented itself, so the underwater excursion ended up being a disaster. A few days later, my colleague and I located a place to carry out our experiments where the fauna was a little more abundant but, despite the polychromatic hues, I was not much encouraged by this spot where, apart from four small fish and a handful of species of coral, anemones and small gorgonians, the most exciting thing I saw was a brightly coloured nudibranch (sea slug) gliding smoothly past.

Years later I wanted to make amends, so in 2003 I went to the north Red Sea to join three dives a day and enjoy the authentic, pristine reef that thousands of tourists explore every day from the barges flocking out of Sharm-el-Sheikh. I had a great time with an excellent crew and super

© Springer Nature Switzerland AG 2019
S. Rossi, *Oceans in Decline*,
https://doi.org/10.1007/978-3-030-02514-4_20

colleagues, and took good photographs, but I still did not see what I had expected. The reefs I saw were alive, but I had been convinced that, apart from coral cathedrals, I would find all kinds of other organisms, day and night. I saw fish, I saw some small elasmobranch, I saw some crustaceans and I even saw a couple of confused turtles, but nothing indicated that it was an oasis of life. I was already working as a marine biologist, specializing, among other things, in the marine animal forests (which we have discussed on other occasions throughout the book), so was aware of the degradation of the coral reef, especially in areas under pressure from tourism, urbanized areas and nearby fishing grounds.

This concern of mine, knowing that what I was seeing was not what it once was, intensified after delving into scientific works such as those by Jeremy Jackson, Nancy Knowlton, Ove Hoegh-Guldberg and Terry Hughes, to single out a few top names. The diversity, frequency and scale of the direct or indirect human impact that we are seeing on reefs are increasing in such a way that their structure and dynamics, as we understand them, are being compromised globally. The increases in carbon dioxide and temperature are the final blows in a phenomenon that began many decades ago—in fact centuries ago, as we have explained (Fig. 20.1).

Coral reefs comprise only 0.2% of marine environments but are considered the most biodiverse type of all, with no less than a third of the world's marine

Fig. 20.1 Scenarios of reef degradation due to a combination of acidification and sea warming. Left image: a coral reef in current conditions (375 ppm of CO_2 and seawater temperatures have increased by 1 °C; in some places, CO_2 levels already exceed 400 ppm); Centre image: scenario of 450–500 ppm of CO_2 and seawater temperatures increased by 2 °C; Right image: CO_2 at levels over 500 ppm and seawater temperatures increased by 3 °C. In this case, coral reefs disappear, as they are doing now. *Source* Hoegh-Guldberg et al. (2007)

species, an estimated minimum of 835 species of corals capable of creating biostructures that house one to nine million different species (we are far from knowing how many there are). In more than a hundred tropical countries, these biostructures protect the vast coastal strips from the erosive effect of waves, typhoons and sea currents. Furthermore, directly or indirectly, they supply the more than 600 million people who live less than 100 km away— plus those who live far away yet rely on them for fish, pharmacological products or leisure, to name three examples.

Our economic dependence on reefs is only increasing, despite the impacts that they have survived throughout our recent history. It is estimated that the financial turnover of these ecosystems is from €125 billion to €275 billion. This without taking into account the value of the ecosystem itself, which, as we will see in the chapter on nature reserves, is gradually being quantified in order to implement market criteria for the management of the various ter- ritories (future benefits according to the type of use of the ecosystem). What surprises me most, however, is that these figures do not take into account two things that are fundamental to the assessment of the damage that the loss or transformation of these biogenic structures would cause. First, there is the loss of quality of life of the people who see their environment deteriorating alarmingly. Second, and more worrying, is reefs' still little-known global role in this dance around the changes produced in carbon capture by mature systems such as this one; that is, the primordial role that a healthy and steady system, with slow growth, must play in the capture of carbon dioxide and therefore in the reduction or slowing down of the effects of greenhouse gases. In other words, the upheavals of this system that captures carbon dioxide may imply one more step towards a real and palpable imbalance in the chemistry of the oceans and therefore, I insist, of the entire planet.

In fact, the degradation seems unstoppable. An estimated 20% of coral reefs have now disappeared and 35% are severely affected. 'The different actors affecting the world's corals, especially the effects of water warming and acidification, can lead to the disappearance of up to 60% of tropical reefs by 2030', explains Terry Hughes of Australia's Centre for Coral Reef Biodiversity. In the past, a genus such as *Acropora* occupied up to 30–50% of the Caribbean reef, but today, in the vast majority of sites, this genus has ceased to exist between zero and 20 m deep, especially in the areas most affected by humans. 'In 1956, the coverage of live corals in the Maldives was up to 75%,' says Carlo Nike Bianchi of the DiSTAV of the University of Genoa, 'but after the 1998 El Niño phenomenon the coverage has been reduced to 10%.' Nike Bianchi notes that 16 years later the coverage, despite other bleaching episodes, had recovered to 60% but that the complexity, the diversity of species, is far from being as it was in earlier decades. 'It erodes the three-dimensional structure, the complexity of the forms, and therefore their functionality.' Another scientific group studying this loss of function is the scientist Lorenzo Álvarez-Filip, from the Institute of Marine and Limnological Sciences of the National Autonomous University of Mexico: 'Under a major thermal stress such as that suffered by reefs, a loss of the "roughness" of the reef is expected', indicating 'what is already affecting and will affect many species associated with these complex structures.' It is not something new per se because, in the 240 million years that they have been triumphing as a group of eco-engineering species, they have undergone collapses and disappearances on several occasions. However, over the last few years, this current degradation has proved to be both general and particularly rapid.

Here's a comparison that gives an idea of the degradation. Coral reefs such as in the Kingman Islands (which we discussed at the beginning of the book), where human pressure is almost non-existent, have as much as a kilo of fish per square metre and a healthy coral cover of up to 50%, while in Jamaica, where the reefs are highly degraded, in many places they do not have as much as 40 g of fish per square metre and just 10% coral cover, and in many cases on the verge of collapse. This is without taking into account the complexity of the system, biodiversity or many other factors that are also comparable and in which Kingman also wins. There are too many variables that affect reefs synergistically. 'When we try to understand what happens to reefs, we cannot isolate effects except in very specific cases', says John Pandolfi of the University of Queensland in Australia. 'Multistress is something that is far from being understood but which we must urgently address in order to provide concrete answers to the problems that choke our coral reefs.'

Let's talk about those variables. The increase in temperature causes coral bleaching and occurs in several species when the maximum summer temperature in surface waters (where most reef corals are concentrated) is exceeded by 1–2 °C. Bleaching actually consists of the expulsion of the coral 'tissue' (corals have no real tissue) from the polyps and the rest of the living structure of the so-called zooxanthellae (*Symbiodinium*), plant cells that live inside in a symbiosis of convenience. At high temperatures, corals lose their symbiotes and become white. These phytoplankton cells seek refuge in the nutrients excreted by the corals in a nitrogen- and phosphorous-poor environment, giving in return products resulting from photosynthesis, such as sugars and other molecules. The algal cells benefit from the protection and also from the light reflected by the calcareous skeletons of the corals, which act as a screen to bounce the sun's rays off the photosynthetic material from several places at once, not just from above. These screens and the complex mechanism of interaction between the two organisms (the animal and the plant) are the key to understanding the tremendous specialization of corals, which can obtain from these small tenants up to 85–90% of the energy that they need. The balance is more fragile than you might think, because both seek to make the most of their relationship and, if they don't find it, the union is broken. If the temperature is too high, what was a blessing becomes quite the opposite, because the photosynthetic apparatus accelerates—it cannot stop—and begins to emit toxic substances that break up the bond.

For corals, the loss of *Symbiodinium* due to a rise in temperature implies a change from being a 'plant' (feeding on much of what the plant cells give it) to an 'animal' (remember that all corals are animals). Without algae, the effort that an animal has to expend to maintain itself is much greater, as it has to capture ciliate, small crustaceans or larvae from other organisms to survive in much greater numbers than they usually do, which implies more energy effort and fewer reserves to accumulate. But it is not only that. The symbiont algae are the key to understanding the calcification of its skeleton. Without them, the mechanism does not work and stops growing as it normally does. Those species that can meet their metabolic demands during the period of energy stress, during the abnormally hot phenomenon that leads to bleaching, will be those that persist and gradually come to dominate the reefs. But although there are corals that are capable of surviving bleaching, the mechanisms by which they are able to resist are still little known. 'We know that there are different kinds of *Symbiodinium*, some of which are more or less "resistant" to high temperatures', writes Stephen Palumbi of Stanford University in the United States: 'By transplanting temperature-resistant species we have seen that they can be the key to understanding who can and cannot withstand future conditions.' However, there is much controversy at this point. Scientists like Peter Edmunds of California State University at Northridge believe that simplifying things cannot explain the phenomenon of winners and losers. 'It is not going to be so simple to define who wins and who loses; we are facing a transitional landscape, and there are factors such as the local environment, ecosystem functioning or specific physiology and competition that need to be taken into account to understand the succession of "underwater landscapes".'

Between 1982 and 1983, there was a phenomenon of bleaching that moderately or severely affected more than 14% of the world's reefs. It coincided with the El Niño phenomenon, which was particularly severe at that time. It showed that reefs are not, by any stretch of the imagination, all alike in their response to this phenomenon. The maturity of the reef, its so-called resilience and the very biogeochemical cycles that govern them are completely different across the planet. 'There is a great deal of genetic variability that will surely allow different responses to the same stress problem', says Dr Wakeford of the Australian Institute of Marine Science. However, all specialists are aware that the theory of adaptation is very inconsistent, because changes are happening too fast. There comes a point of no return on a medium scale (as always, on a human timescale), beyond which the community is unable to recover, to return to its former maturity and complexity. In what we might call 'normal' dynamics, corals have to be concerned above all with their competitors for space, occasional physical disturbances (major storms) and predators. 'In Florida, we have seen in ten years that the corals have left room for gorgonians', says Dr Ruzicka of the Fish and Wildlife Research Institute in this state: 'At shallow depths, where the thermal effects are most acute, the underwater landscape has changed completely.' It's not the only place. Corals today have more and more problems with manmade disturbances, from bleaching, to pollution, to over-exploitation through tourism.

So, what we are looking at today is a transition from coral reefs to dead areas of coral where the dominant species are macroscopic algae, gorgonians and sponges. If the site is particularly nutrient laden, the balance, after the death of the corals from bleaching, disease (see next section) or predation, tilts in favour of these fast-growing plant organisms occupying the space. The biggest problem is that the site has nothing to control the populations of algae because, due to overfishing and the disruption of the food chain, there are no herbivores to eat them. When much of the fish population disappeared from the reefs of Jamaica, there were still sea urchins (*Diadema antillarum*) that kept algae populations at bay across the reef. But during the massive mortalities caused by El Niño in 1982 to 1983, a disease affected these peculiar long-spined urchins, slashing them from frequencies of sometimes ten individuals per square metre to just one every ten. This promoted an explosion in the numbers of these algae, which saw the field clear and grew faster than corals, and then had a great deal of new surface to occupy when the corals ceased living. We have gone from having many species of herbivores that browsed various types of algae to few species (and with low densities, due to overfishing) that eat only one type of alga in particular and ignore the others.

However, it must be borne in mind that, while the problems of overfishing and local pollution can be addressed with some effectiveness, those arising from climate change cannot. It is 'a fish that bites its own tail', because the very survival of fish and other organisms depends on the diversity, roughness, biomass and complexity of the reef: the very disappearance of the reef means that more species disappear, and the reef disappears.... In the coming decades, the bleaching phenomena will become chronic, especially in light of forecasts of increases in temperature in these tropical areas exceeding 3 °C, in the summer maximums. The last one, from 2015 to 2016, was on a large scale. A terrible bleaching had affected, among other areas, the great Australian coral reef where, in certain areas, up to 95% of its scleractinia colonies had been bleached. Only four of the 520 sites observed were intact, and the rest had signs of bleaching to a greater or lesser extent from late 2014, due to El Niño: too frequent, too intense.

As the reef is eroded over and over again, the chances of these building bodies recruiting new individuals are diminished. Two examples: recently it has been shown that local winds influence the mass reproduction of corals. Places with brief calms have an explosive but short period of reproduction, with many species participating in a synchronized release of eggs and sperm. On the other hand, places where the calms are prolonged have a longer period (up to seven months) for the expulsion of eggs, during which the various species reproduce in a much less coordinated way. The wind and the currents associated with it are a problem for new recruits, who may be displaced away from the mother colonies to places less suitable for their establishment. Due to climate change, at the local level the wind regime will alter over the coming decades. The models are not yet precise, but places where there are now prolonged calms could have a higher intensity and frequency of winds in the near future, so that many species could be harmed in one of the most crucial stages of their life cycle, the production of young.

In another specific case, acidification (as we have discussed extensively) would affect not only large corals, but also the small polyps responsible for maintaining these habitats. It further reduces growth, fertility and the capacity to build up reserves. Recent experiments have shown that newly established colonies have their capacity to form calcareous cover reduced by up to 50%, making them much more vulnerable to predation and erosion. The last example is that corals have a generally short dispersion. They do not travel long distances. We find few new additions because it is only in certain places that they are they capable of colonizing the terrain and because the recruitment of new individuals is successful in pulses; that is, that only on certain occasions is there an arrival of a large number of recruits to succeed in colonization. According to a rough estimate, specialists explain that the larvae would have to be able to move more than twenty kilometres to become successfully established, and the new individuals would have to expel larvae another twenty kilometres every year to compensate for the increase in temperatures that will come in the next century. In other words, the corals would never be able to win the race to more suitable areas (in the north and south) by escaping areas too warm for their survival. The rising sea levels will pose an additional problem for corals in the not too distant future, because communities will have to rush to avoid being overtaken by the sea in a short period of time, leaving them too deep to perform their vital functions properly.

These are examples of the countless effects that can affect coral's crucial early stages of existence. We have failed to protect our planet's coral reefs. In many cases, the accumulated stress has exceeded the resilience and recovery capacity of the system. 'At best, the collapse of these communities will be equivalent to the collapse of some 420 thousand years ago, when many dominant species disappeared due to climate change; the difference is that current climate change and reef disturbances can be a hundred to a thousand times faster than in that remote past', explains Ove Hoegh-Guldberg of the Center of Marine Studies in Australia. He insists that, 'despite the adaptive capacity of corals, the current frantic pace of change exceeds their capacity to adapt'. It is not necessary to go so far back: 18,000 years ago, in the midst of the regression of the waters due to glaciation, the two dominant species of coral became completely extinct when up to 90% of their habitat disappeared in a short time.

It is difficult to identify direct and indirect human effects on reefs. Because there is a great complexity of interactions between species in such a diverse system, the degradation trajectories will depend on the area and the dominant species, and the time scales of degradation can also be diverse. 'The healthiest reefs, the really pristine ones like Kingman's, are the ones that can best withstand bleaching or disease', says Jeremy Jackson. Unfortunately, a small number of uncontrolled people can do a great deal of damage and have a huge impact on reef communities. Only large Marine Protected Areas (see below) will give hope to these wonderful and useful biostructures. But we are in a world that is not at all linear, not at all simple. Getting the protection message across to society is difficult, considering that even first-world countries like the United States have less than 5% of tropical protected areas. Knowing what will happen in the future is almost impossible, but we can understand that the degradation of these and other systems in our oceans cannot be left for later: we must act now.

The Sick Reef

Diseases are natural in any community, in any population. Ultimately, they are one more form of natural regulation of populations, either by parasitism or because one or more species takes advantage of the metabolic or immunological weakness of another to proliferate. However, reef diseases have increased exponentially over the last thirty years. During this time, more than 30 diseases of different coral species have been mentioned in publications, although only 18 have been rigorously described and in only five do we know the exact pathogen that induces them. 'There is no species-specific pathogen,' explains David Kline of the Smithsonian Tropical Research Institute in Panama, 'related organisms are usually generic, attacking more than one species at a time when present.' Some diseases can affect more than sixty different species of corals.

Stress in coral may be caused by changes in water bodies due to temperature and, above all, eutrophication. 'Stressed corals release more mucus, which is a perfect breeding ground for bacteria that consume more oxygen, produce harmful secondary metabolites or expose coral polyps that are thus vulnerable to predators', continues Kline. These same bacteria, under normal conditions, are harmless; indeed, they are part of the cycle of regeneration of nutrients and organic matter that the corals themselves take advantage of. But the local change in the physicochemical conditions of water is accelerating a process of decline that seems unstoppable. Also, the microscopic algae that are found in coral's interior and, as we have seen, are a fundamental part of the metabolism of the coral reefs, are fundamental to this balance, but if they find conditions inimical they leave the polyps and coenenchyma (pseudo-tissue) of the corals, promoting the proliferation of disease. The processes of infection and mortality are highly variable, even for the same species.

In early 2000, a rigorous test was conducted in the Caribbean to see the cause and effect of disease in 150–200 locations. 'What was found was that in almost all of them there was a direct or indirect effect of changes in the water bodies due mainly to organic contamination (eutrophication)', said Ernesto Weil of the University of Puerto Rico. There is a relationship between the dissolved organic carbon produced by macroalgae (algae that proliferate in eutrophic areas) and reef diseases. The bad news is that there is positive feedback: the more algae, the more eutrophication and more disease, the more dead coral and more room for the macroalgae to proliferate on the dead structures. It has also been proved that the spread of contagion from one place to another is partly due to human movements. Uncontaminated areas become infected in a fairly short time as maritime traffic (especially from small boats) increases between one place and another, especially when the bodies of water begin to be enriched with nutrients and organic matter from crops, human solid waste, new ports, and so on.

Despite being such a serious problem that it has affected more than 95% of coral populations in some parts of the Caribbean, Indian Ocean or Red Sea, very little is known about either the proliferation or defence mechanisms of tropical corals. Steven Vollmer of Northeastern University of Massachusetts in the United States conducted a pioneering study that discovered that 6% of the population of one of the most important corals in the Caribbean (*Acropora*) is resistant to white band disease. It gives us a little hope, a little respite, through of the possibility of transplanting fragments of these resistant colonies to places devastated by the disease to save the reef. If we expect the corals themselves to expand we will have to wait decades, perhaps centuries, because of their low ability to spread to recolonize space. 'Climate change is accelerating processes', insists Weil, 'and the effects on the underwater landscape are already devastating.' A serious problem that, together with the other problems we have explained, is prompting various groups to propose more than two hundred species for the endangered Red List for the protection of organisms by US entities—a drastic solution, in the midst of the storm.

Mass Mortalities in the Mediterranean

During the last few years, the Mediterranean Sea's topmost layer has not been free from lethal phenomena due to increases in water temperature in the communities that live fixed in the bottom, far below. As records in various locations show, mortalities are not a new phenomenon on our coasts, but have always been observed locally and sporadically. The warming of the waters was followed by the great mortalities of sessile organisms recorded in 1999, 2003 and 2006, when between 25 and 30 species of vital importance to the structure of the community (sponges, gorgonians, hexacorals, bryozoa, bivalves and calcareous algae) were much affected between down to 30 and 35 m deep.

Partial or total mortality in marine sessile organisms may occur under particular circumstances, but in these years the Mediterranean, especially on the Italian (from Liguria to the Gulf of Naples) and French (Côte d'Azur) coastal strips, and to a lesser extent those of Sardinia, Corsica, the Balearic Islands and Costa Brava and Columbretes, was hit by a heatwave that swept entire populations away intermittently along more than 500 km of coastline. The organisms lost much of their living tissue, if not all.

It was a widespread phenomenon that has been well documented. 'The mortalities became evident especially in autumn', explains Cristina Linares of the Ecology Department of the University of Barcelona: 'After years of monitoring in the Port-Cros Nature Reserve, we were able to observe up to 95% of the gorgonian colonies affected to varying degrees, with the "internal skeleton" bare due to lack of tissue in certain observation sites.' The shallower the depth, the greater the damage. At a depth of 40 m, only 5% of the population was affected. The worst thing is the long-term impact, as serious effects on the reproductive capacity of some species could be observed. In the case of the chameleon, or red gorgonian, one of the most important of the coralligenous organisms, the female colonies carrying the eggs that serve to maintain and expand the populations were noted to have a decrease of between 22 and 35% in their fertility, and the males 12%. But colonies that had more than a third of their area affected had a reduction in their ability to produce eggs of up to 75%. 'Population fertility affects population viability', concludes Linares. But when the gorgonians disappear, something more than the colony itself disappears, as they are the 'trees' that comprise the marine forest. 'The disappearance of the gorgonians can lead to the decrease of many organisms associated with the "forest"', writes Massimo Ponti of the University of Bologna: 'The system is simplified, which entails a change in hydrodynamism, biogeochemical cycles or the capacity to recruit.'

Many of the organisms living in coralligenous communities have a low dispersal capacity; that is, their larvae or fertilized eggs do not travel far. New recruits may have problems becoming established, and generally have to compete for space and not be preyed upon. Their success rate is usually not high. 'In the shallower populations of red coral, at a depth of about 25 m, we were able to observe a higher mortality rate among newcomers', says Lorenzo Bramanti of the French CNRS: 'The heatwave had hit the new generations, the newcomers.' In the case of red coral, it had also impacted on adult colonies, affecting between 5 and 80% of the colonies, depending on area. Many of these sessile organisms mature extremely slowly, taking more than a decade to produce colonies that are in top reproductive form, so a high frequency of these phenomena could reduce their viability in the near future.

But what could have been the cause of these mortalities? It is most likely that a persistence of very warm water during the summer would cause too much 'hunger' and force the organisms to breathe more (as the temperature increases, we breathe more and metabolic processes accelerate). When the waters are calm in summer, the animals cannot feed as there are no currents to carry in their food (in the form of particles). Because this warmth lasted weeks (certain areas were up to 29 °C at the surface, between 3 and 4 °C above the summer average), no food came in the water reaching the mouths of these sessile organisms, preventing adequate feeding. Some organisms were also affected by pathogens that would have taken advantage of their weakness and the high temperatures to feast. Perhaps the least observed fact is that a few weeks later, in the autumn, there was still a lack of food, this time due to excessive turbulence and the poor quality of the particles in suspension, potentially finishing the organisms off.

It is important to point out a feature of the Mediterranean coralligenous reefs unlike those in the tropics: most of the organisms, those with the greatest exuberance, are found below a depth of 40 m. 'We have been able to see that the densest, healthiest and longest-lasting populations are between 35 and 90 m deep', explains Andrea Gori of the Institute of Marine Sciences (CSIC). As we have said before, Linares and his colleagues found populations at shallow depths that had been severely disturbed by the heatwave, yet below 40 m the damage was minimal and they recovered immediately. This is because these deeper populations do not experience the immense changes that the superficial ones do, so in a certain sense the latter are more fragile and pioneering in terms of the water warming caused by to the formation of the summer thermocline and the strong easterly storms. Ivone Lilian Pairaud of the French IFREMER draws a map that shows that, in future, places like Liguria or the Tuscany coast may be much affected by the increase in temperature. Unfortunately, surface bodies of water are getting far warmer in the hot summers of current climatic trends, so mortality phenomena will perhaps occur more frequently than can be tolerated by the benthic communities, making them disappear from the depths that we can access easily by scuba diving. The frequency of these mortalities will increase, completely transforming the space occupied by iconic Mediterranean species.

Why is the 350 ppm Carbon Dioxide Figure Important?

I maintain that 350 ppm carbon dioxide is a vital figure. Scientists from various disciplines stress the importance not only of our reefs' temperature, pollution, invasive species, biodegradation, and so on, but also of the perverse and still little understood synergistic effects of our water bodies' unstoppable acidification. It was found that the coral bleaching phenomena of the 1980s coincided with the El Niño phenomenon and also a carbon dioxide concentration of 320 ppm. Between 1982 and 1983, 14% of the world's reefs experienced moderate or severe bleaching. The temperature had risen between one and two degrees above average, so the waters had warmed up more than normal. But the important thing is that in the water an increase had already begun in the concentration of this component (carbon dioxide), so basic for life and at the same time so crucial to the regulation of certain components. After a level of 340 ppm of carbon dioxide has been attained, sporadic but highly damaging events have affected large areas of the planet. Today, at 410 ppm carbon dioxide, coral communities around the world are considered to be in irreversible decline, on a human scale. We have exceeded the level of 350 ppm of carbon dioxide, the concentration considered to be the key to the health of the calcareous structures in the complexes that form the spectacular reef formations. Currently, levels of aragonite (see chapter on acidification) in tropical environments exceed saturation, but we have gone from 4.6 to 4 Ω ppm aragonite in less than three hundred years. At about 540 ppm of carbon dioxide, pH could be reduced by 0.24 points, from 8.2 to 7.9 (at the very most), which would imply a practical inability on the part of structures such as corals, gorgonians or sponges to use aragonite in the same way as at present to form their calcareous endoskeletons.

If the level were to reach 600 ppm carbon dioxide, what would eventually be eroded is the geological structure of the reef itself. The marine biota, as we know it today, would only survive in niches. However, in such a widespread phenomenon of degradation, a microhabitat cannot be considered as a salvation to allow the recovery of structures, in the medium term. Coral reefs are some of the most sensitive systems on the planet to changes in both atmospheric and aquatic chemical balances, demonstrating a great capacity for restructuring and restoring diversity once conditions return to favourable. But we are heading to certain collapse, to the disappearance of a system that is particularly sensitive to carbon balances and flows—that's why they have appeared and disappeared over the past 240 million years of the planet's geological history. It is important to emphasize an inescapable fact: the absorption of heat and carbon dioxide by the oceans is taking place extremely quickly, but going back to the previous conditions is something that is going to happen incredibly slowly. It is a question of physical-chemical exchanges, which do not occur at the same rate when going from A to B as when returning from B to A. If we cause the reefs disappear again (with all that entails), we will add one more problem to our own survival as we understand it today.

Part V

Future Prospects

21

Farms in the Sea

In the wake of the Green Revolution, which sought to promote agriculture scaled to providing food for the planet to end hunger among the less-favoured countries, the Blue Revolution sought to complement the efforts made in the countryside, creating farms in both inland and marine waters to breed species in captivity to alleviate our overstretched fishing resources and the collapse of fishing grounds. The idea, of course, was good, but its implementation has experienced many considerable problems and perversions. We consume per capita on average about 17 kg of fish per year, of which about 9 kg currently comes from direct fishing and no less than the remaining 8 kg from aquaculture. Aquaculture in 2014 supplied almost half of the world's fish. Today, the latest available FAO report (2016) indicates that, of the total 167.2 million tonnes of fish, cephalopods, molluscs, and so on, that are taken from the aquatic environment, no less than 73.8 million tonnes come from some form of aquaculture, with fishing fleets, both industrial and small-scale, responsible for almost 27 million tonnes per year.

Aquaculture has grown from 3.9% of world production in the 1970s to 44% in 2014, from less than a kilo of fish per person per year to more than eight. 'Aquaculture will grow steadily until 2025,' says James Diana of the School of Natural Resources and Environment at the University of Michigan in the United States, 'becoming the main source of food from both inland waters and the oceans'. While the catch of wild fish has stabilized at about 90–100 million tonnes, aquaculture has grown from 12 million tonnes in 1985 to over 50 million tonnes in 2007, an annual growth rate of 9%. By 2020, we will need at least 10 million more tonnes of fish a year, food that cannot come from fishing grounds. It can come only from captive breeding.

© Springer Nature Switzerland AG 2019
S. Rossi, *Oceans in Decline*,
https://doi.org/10.1007/978-3-030-02514-4_21

It is not surprising, therefore, that such a business today moves a huge amount of money, almost €120 billion a year in 2014, directly employing more than 24 million people and expanding its production more than any other food on the planet. Therefore, aquaculture—that is, the 'domestication' of the sea (and of rivers and lakes)—is on the rise.

As we have commented throughout the book, until recently the only option that we had was to continue 'hunting' in the sea; to continue extracting renewable resources through direct action in the aquatic environment, both small scale and industrial. Aquaculture has changed this perception, and today the 'farm of the sea' in its different aspects has to be considered as a clear element for the future in the complex food organization chart needed by the more than 7,000 million people who inhabit our planet, and who could soon exceed 9,000 million.

All types of mariculture—marine aquaculture—will alleviate the pressure on the fishing grounds. It would potentially be a tool to enable us to leave the ecosystems in peace, to restore the battered biodiversity of the oceans and wild stocks of various species. Moreover, this type of large-scale practice can ensure a certain stability for seafood supply, with less reliance on the environmental or biological fluctuations that may affect wild stocks depleted by overfishing, and provide a large number of jobs that can gradually be converted into new jobs: in China alone (the country with the highest fisheries and aquaculture production on the planet) more than 14 million people work on farms in rivers, lakes and the sea. If this is the case, why is it not working as it should? What are the actual problems with aquaculture?

Any such abrupt expansion must have social, economic and environmental repercussions. The first to be taken into account is that there are two types of mariculture. One is based on closed cycles (i.e. starting with reproduction in captivity until the extraction of adult specimens) and one for rearing (i.e. collecting larvae or juveniles of a species, enclosing them and feeding them until they are economically viable). The first one is practised with certain species, but it is extremely complicated to ensure mating in captivity and to prevent high mortality rates in the new generation. Just ask the hundreds of scientists who study everything from fish hormones to feeding young fish. The second type of aquaculture, rearing, brings together many other species, but in reality it depletes the original stock, the wild specimens; in other words, it cannot be called true mariculture. You are not taming the species: you just lock it up, fatten it up and consume it. Of the 24 species at the top of the world fishing list, 14 can be bred in captivity but few have a fully closed cycle. Moreover, some, such as that for tuna, as we have seen, involve a highly questionable type of husbandry, in all its aspects.

A second major issue is the environmental problem; that is, the direct impact of the mariculture facilities on the area surrounding the farms. Fish and shrimp, for example, are squashed into a tight space where they are supplied with food, but this is only partially consumed. The rest goes to the bottom, where it accumulates in the form of organic matter that chokes the benthic communities, especially those with a simple way of life and those that form structures of slow growth and great biodiversity. Added to this, of course, is the excretion of faeces and thus ammonium in the environment, which only increases the problem, besides the contaminants derived from artificially manufactured food, usually prepared from pelagic fish from wild stocks. Not all fish farmers should be lumped together but, unfortunately, in many countries there is not as much regulation as in others, and investors are looking to make much money in a short time. 'We need to introduce quality standards that take a variety of factors into account', says Simon Bush of Wageningen University in the Netherlands: 'For example, we need to label products with ecological and social product values, create independent audits and train qualified people to ensure that they are properly monitored.'

There are more problems, such as an increase in parasitism and disease as the animals become overcrowded, the introduction of alien species and inappropriate application of genetic resources that can alter the wild populations. On the other hand, we are now gradually beginning to apply protocols in which species are bred in a more or less regulated way, but until recently we could say that each farm met its needs as well as it might, and at its own convenience.

Aquaculture in general, and mariculture in particular, do therefore affect ecosystems, although it should be stressed that the effects are not uniform and cannot be generalized. There are problems, it is true, but there are also solutions. What happens is that many of them involve investments and changes in social perception, and have associated political costs. Ángel Borja of ATZI in Spain indicates that there is a good set of environmental indicators at hand to check whether or not a mariculture practice is appropriate: 'When studying the viability or otherwise of a given species, various factors must be taken into account, such as the currents and physical and chemical characteristics of the water in the area, the depth of the seabed, the composition of the benthic fauna and flora, the carrying capacity of the system and, of course, the species to be bred and under what conditions', explains Borja. As it stands, this seems so trivial an exercise that is often not carried out. In Europe, the protocols for action have laid down a tortuous path, yet in the end there are certain criteria that are gradually being met. However, in developing countries, the push for expansion and immediate profit, as well as the lack of scruples so often evident, have damaged many entire coastlines that now have difficulties in recovering.

But it's not the same everywhere. I was told by a retired colleague from the Alfred Wegener Institute, Wolf Arntz, that in the countless fjords of southern Chile aquaculture has experienced an unstoppable boom that is taking the area by force, destroying it. Although on paper there are a number of regulations, the land and its aquatic environment for salmon farming are sold to the highest bidder, and the degradation of the fjords is difficult to reverse. In this respect, I believe that aquaculture has done a great deal of damage, and the worst is that there are many areas, such as these fjords, where the balance is fragile and this impact can kill off a large number of species.

The social perception of sea farms is also important. 'Aquaculture has gained a low acceptance because of the poor forecasting of the environmental and landscape impact in areas where it has been done very extensively', says Maria Giovanna Palmieri of CEMARE at the University of Portsmouth in the United Kingdom: 'In Scotland, where there is a large continental and marine production, a balance is beginning to be made between the benefits obtained or potentially obtained from this practice and the costs for the ecosystem and the impoverishment of the very vision of natural spaces from a purely aesthetic point of view.' The only compensation that people seem to accept is the creation of jobs, but this is a double-edged sword, as too much degradation renders the area unusable for many other economic ventures. Another issue that has been dragging on in aquaculture is competition for resources in economies based on land-based agriculture or livestock. Resources that aquaculturalists see as partly their own are diverted into developing production that they obviously do not control. In developing countries, where resources are particularly scarce, this social and economic problem has become significant, especially in view of the explosion in aquaculture since the 1980s and 1990s.

But land-based produce, especially that derived from livestock, faces a threat from farmed fish: people perceive it as being healthier for them. Not surprisingly, fat levels in fish are lower, and the level of polyunsaturated fats (the famous omega-3 fatty acids) are higher than in pork, beef or chicken. Fish is considered a more complete food and is receiving such good press that competing with meat does not mean that it is left in the shadows. However, feeding these captive creatures, especially carnivorous fish, comes at a high price. Oils and meals from wild marine stocks, especially from small pelagic fish, are needed to feed fish in captivity (see below). The growth of aquaculture therefore depends on the stocks in our fishing grounds, which are already barely able to withstand the onslaught of commercial fishing.

However, there are solutions, although some of them are little studied, unviable or, from an environmental point of view, questionable. One that is being studied more seriously is to create substitute food derived from phytoplankton culture and terrestrial plants. More and more efficient feedstuffs are being achieved in terms of both their supply to animals and of their composition, obviating the need for fishmeal, thanks to enrichment with omega-3 fatty acids. We need to study both the acceptance of this food by fish and its efficiency in rearing them, and the popularization of these substitutes among people. It is clear that although the food requirements of fish, including the carnivores, are not as high as those of warm-blooded animals, diet is important and the products that replace those from wild stocks will need to have appropriate nutritional qualities if they are to compete with those that serve as their food.

Moreover, the price of the product needs to be competitive, although this does not seem to be a problem, given that the price per kilo of vegetable-based substitute is around 62 euro cents, compared to 89 euro cents for a kilo of fishmeal. 'It is clear that the best thing, from a food point of view, would be to raise animals that are predisposed to omnivory,' says Rosamond Naylor of Stanford University, 'that is, animals that are able to eat food of both plant and animal origin without problems.' Ransom Mayer goes on to say, 'A clear distinction must be made between aquaculture of molluscs, shrimps and omnivorous or herbivorous fish and carnivorous fish, the latter requiring economic, energy and technological effort that ultimately makes them a delicacy within the reach of a few'. Indeed, rearing large carnivores has a much higher cost in terms of food and energy than rearing other species.

Another solution that can go hand in hand with aquaculture feeding is to take advantage of the by-catch that we have discussed in previous chapters. The fish that are thrown overboard every trip could be destined for fish farms, properly treated. But there is a major lack of coordination in this regard, and introducing this venture will not be easy unless it is profitable after transporting and trading it: space, refrigeration and packaging are needed to land it and later use it as food for other fish. The possibility of using Antarctic krill as a food source for fish farms is also being considered. The problem here is mainly environmental (although there are still problems in exploiting this creature, which has a large amount of associated fluoride), because fishing for krill in greater quantities than are already caught to meet our demand for mariculture would mean interfering in the food chain of the southernmost seas of the planet.

Finally, I would like to explain something that has been applied systematically over the past fifteen years and which, as we will see right at the end, seems to be the most successful solution in the near future. FAO is increasingly promoting IMTA farms, which stands for Integrated Multi-Trophic Aquaculture. What is it about? Well, at its core, it is about emulating an ecosystem, about cultivating species that can make aquaculture sustainable from an environmental and, at the same time, production point of view. For example, there are combinations that work for fish farmers rearing carnivorous-omnivorous species of fish. As they would be unable to cope with all the organic matter and nutrients produced, the farmers supplement these carnivorous-omnivorous species with filter feeders such as oysters or mussels (which eat fish waste), together with polychaete worms or shrimp (which feed on suspended matter and zooplankton), which in turn provide the nutrients (along with ammonium and other excreta produced by the fish) for macroscopic algae, such as kelp, needed for their development. 'If we combine salmon and mussels,' says Jade Irisarri of the Vigo Marine Research Institute (CSIC), 'it increases the efficiency of both species in terms of reproduction, for example.' To complete the picture, some of these algae can be used as food by echinoderms such as sea urchins, especially in the breeding season. Of course, the ongoing experience (already a reality in countries such as Ireland, the United Kingdom, China, Chile or Canada) and the pilot ventures (in countries such as Spain, France or Portugal) must take into account many factors that make their implementation more complicated. 'It is important to make a good study in which all profits and losses are considered, bearing in mind that not all species have the same profitability or market share,' says Angélica Mendoza-Bertrán from the University of Leiden in the Netherlands: 'The different components must be very well combined and, in the end, we need to bet on something well structured in the long term.' The combination system is more expensive, long term, and may even in the short term be less profitable. Dr Martinez-Espiñeira of the Memorial University of Newfoundland in Canada stresses the fact that this type of aquaculture needs reinforcement through advertising that identifies it as being more suitable for the environment and viable in the long term.

Once again, it is shown that the way forward is complex—the way that takes into account many factors. In the case of IMTA, let's not forget that, in reality, we are simulating a highly simplified ecosystem so, apart from the economic viability and social acceptance of the derived products, we must carefully study whether or not any material is recycled and whether or not productivity is sufficient to carry out the project on a large scale. The Chinese are known to have made a great deal of progress in this respect, but they are coy with their results. Perhaps this is because they understand that they are facing the aquaculture of the future and, in a certain sense, it is the one who makes waves that will dominate the operation and succeed in putting it into practice on an industrial level. But for me it is paradoxical that, after decades of work, we are realizing that what we have to do, in the end, is to create multiple farms that simulate the functioning of an ecosystem. Here, matter is recycled, the organisms interact… that is to say, in the end, the success or otherwise of mankind in the domain of the sea in securing a viable food source that respects the environment will depend on our degree of simulation of nature itself.

Fishing for Fish?

Our need for food has come to create somewhat paradoxical situations. One of them is the exploitation of millions of tonnes of fish for use in feeding other fish, instead of being used directly as food. As we have seen, fishing for small pelagic fish (anchovies, sardines, mackerel, herring, etc.) represents a hugely important proportion of the world catch: around 30 million tonnes and, of these catches, aquaculture takes no less than 24 million tonnes, or 87%. 'Consumption for human use has been growing by 1.2% between 1970 and 2006,' analyses Albert Tacon of the University of Las Palmas of Gran Canaria, 'and, although non-human use has been decreasing due to greater efficiency in feeding the farms, the fact that aquaculture production has increased in the same period by 271% has put enormous pressure on pelagic stocks.' Aquaculture, therefore, is absolutely dependent on marine stocks, and without a doubt its growth depends on what the sea can yield.

The case of the Peruvian anchoveta serves as one of the clearest examples, in this sense. Only 0.7% of the annual catch reaches Peruvians themselves in the form of consumable fish, which means that, of the almost 7 million tonnes, only 43,000 tonnes reach the mouths of the inhabitants of this country that has sea on two shores and one of the greatest catch rates in the world. The chief problem is that there is already an established market for non-human grade anchovy, both for aquaculture (57%) and for animal oils and other derivatives (over 42%). Some countries have already realized the absurdity of this policy in which their own population is not enjoying the protein and energy on their doorstep, and that it travels across the seas, often to arrive back in the form of farmed fish. FAO's own guidelines promote the use of fish for direct human use, not for aquaculture, but it is a strong, growing and deep-seated business that desperately needs these calories to rear fish all over the world. Considering that some 23 children die of malnutrition every minute on our planet, isn't it incredible that so much of the fish goes into the aquaculture business rather than directly into people's mouths?

China

With more than 1.3 billion inhabitants, it is no wonder that China is the world's largest producer and consumer of fish. China alone consumes about a quarter of everything that is caught or grown in the world's aquatic environment. The figures for China are always set apart in the graphs presented by the FAO, especially since the early 1970s when the country decided to go in for this form of production, now seen as the key to its food survival: aquaculture. Some 72% of its own output comes from aquaculture, with a nine-fold increase in the amounts in this sector in just 22 years (Fig. 21.1).

Since 1990, China has been the world's leading aquaculture power, far ahead in both freshwater and mariculture. It provides protein, jobs and currency exchange for a country that today has sacrificed much of its lake water, rivers and coasts to create a huge factory from which more than 40 million tonnes of product are extracted. Considering that the production of aquatic biomass from farms worldwide is more than 70 million tonnes per year, we can only imagine its importance to the most populous country in the world. But the Chinese are concerned about the unbridled development of aquaculture in their territory. We have already discussed the problems that can arise from unreasonable and poorly managed aquaculture.

Fig. 21.1 Devastating effects of intensive aquaculture in a coastal area of China. Left image: the area in 1992; right image: the same zone in 2006. Note the hypersalinization effects close to the shore. *Source* NASA earthobservatory.nasa.gov/IOTD/=36953

In China, the problem is exacerbated by production and above all by the real lack of control and poor coordination between administrations. 'There are huge losses that we are already quantifying from an environmental point of view', explains Ling Cao of the College of Fisheries at Huazhong Agricultural University in China. In this great Asian country more than 1.5 million hectares are dedicated to this productive exercise, with almost 600,000 ha in the sea and more than 675,000 ha in shallow intertidal areas. 'There is so much crap that is produced in the system by the fish faeces or food that is not used that the animals end up eating their own waste', continues Cao.

'It is true that aquaculture has grown steadily, but organic aquaculture has grown the most, by 1,700% from 2003 to 2012,' explains Biao Xie of the Nanjing Normal University in China. 'We have grown from 5,000 tonnes to over 85,000 tonnes, using more than 400,000 ha for this purpose'. Efforts to improve the waste issue are real and very considerable, but they will not reach all places in the same way and, above all, not in time. Pollution accidents resulting from the activities of sea farms are continual, with an average of more than two thousand per year and at an official cost of over €110 million. In 2008, in the midst of the Olympic Games in Beijing, along the coast of Qingdao there was the largest population explosion or bloom of macroalgae ever recorded, in which *Porphyra yezoensis* in some cases covered an area of no less than 1,200 km of coastline in a total area of 40,000 km^2 (slightly less than the extent of Extremadura).

Right now, aquaculture in China faces a dilemma: people cannot live without it, because it is part of their plan for the progress of their nation, but it is destroying the habitat in many places, degrading it by simplifying its complexity, contaminating it and pushing it beyond its limits. 'We have been working with IMTA in China for decades,' says Jianguang Fang of the Chinese Ministry of Agriculture, 'taking advantage of natural bays like Sanggon Bay, which covers more than 100 square kilometres, where up to 30 different species are grown and feed on each other, producing almost 4 million tonnes each year.' In places like this, a balance between different organisms and a biogeochemical cycle in which pollutants are controlled has been achieved. Biotechnology is also one of the major challenges for China, where multiple solutions combine to increase yields and, at the same time, render the sustainability of the resource more viable. The path is the right one, but we'll see if they can maintain their high productivity while drastically reducing the impact on their own patch.

Shrimp and Mangrove Forests

The shrimp industry has experienced a production explosion since the 1970s, when several companies in tropical countries saw the immense potential for exploitation of the resource. No other aquaculture production system has had such growth, where from 1975 to 1985 it grew by 300% and from 1985 to 1995 by 250% around the world. Over a third of the shrimp consumed comes from shrimp farming areas, with more than 70% coming from Asian countries such as China, Vietnam, the Philippines or Indonesia, and, far behind, South America and Africa. But, as always, the 99% of production that comes from the third world goes almost entirely to the first world. The problem is that the environmental and social costs are taken on by the developing countries. 'What was originally an element to create foreign exchange for countries like Vietnam or the Philippines is now becoming a serious problem due to the environmental costs and the profound degradation that these areas are suffering', observes Marta Rivera-Ferre of the University of Vic in Spain. She, like other specialists, has seen at first hand what the exploitation of this crustacean is doing to the environment. 'The introduction of large operating companies,' Rivera-Ferrer adds, 'has been to the detriment of artisanal development in the exploitation of resources.' And there's no shortage of money in this industry. An estimated 14% of the world's total fisheries are based on wild or aquaculture fisheries for this type of invertebrate. There are dramatic cases, such as Myanmar, where production has risen from one tonne a year in 1990 to over 30,000 tonnes in 2004, or China, where it has risen from 185,000 tonnes to almost a million tonnes in the same period.

But what's the problem with cultivating this type of crustacean? Shrimp production has variations, but basically it is a matter of parcelling out areas, which are then grouped together to rear the crustaceans and then extract them. The problem is that most of these areas are in the middle of one of the most peculiar, interesting and diverse systems on the planet: tropical mangroves. Once cultivation begins, the mangrove swamp and its surroundings are degraded by excess food, organic matter from crustacean waste and other sources. These contaminants become entrenched through the restructuring of the system that becomes necessary to handle the product effectively; that is, the construction of shrimp farms. The salinization and acidification of the soil and putting certain pollutants into circulation cause not only that area, but also adjacent areas, to suffer damage that takes an indefinite period to heal. In this process, pathogens and parasites are introduced into the area, as well as alien species, in many cases rapidly degrading the ecosystem's complexity. The problem is that, once an area is no longer useful for shrimp farming, it is abandoned and another one used. A country like Ecuador has much experience in this regard, because it managed to degrade more than 90% of its mangrove areas in a very short time, leading to the total collapse of shrimp. When the shrimp business first started, it was not explained to the locals that in ten years they would have to abandon all production in the area as, by then, it would be useless.

The climatic and ecological conditions ideal for this type of exploitation involve minimal cost in certain tropical countries: it is enough to enclose the plots where the larvae are to be reared, without any major infrastructure. 'In Bangladesh, 6% of GDP comes from fishing or aquaculture products, and 85% of that production comes directly from farmed shrimp', explains Kalam Azad of the University of British Columbia in Canada. More than two billion wild larvae are moved to controlled plots where they are reared for market: 'A very large part of the wild stock is being lost, which in many places is unable to cope with such sustained exploitation', adds Kalam Azad. In fact, it is much like rearing tuna and other species: it is not real aquaculture so much as maintaining a species whose stock is already impoverished by fattening it up at will.

In recent years, production has increased by 100% per year in countries such as Vietnam and Bangladesh. The practice is untenable from any point of view. It is no better in the Philippines, where the 400 thousand hectares of mangroves recorded in 1918 have fallen to less than 100 thousand hectares, mainly because of intense shrimp pressure. But here, where they have been dealing with the problem for a long time, they are looking for a transdisciplinary agreement to solve the ecological, economic and social disaster that has led to this degradation. Here there is open talk of an entente between politicians, big business, social agents, aquatic farmers, technicians, non-governmental agencies and scientists to see how to solve the problem. 'The benefits of mangroves don't need to be overemphasized: they regulate greenhouse gases, provide ecological and social services to the community, are the cradle of biodiversity and exploitable species in adult or juvenile stages, recycle nutrients, have incalculable genetic resources.... Do I go on?' says Joshua Farley of the University of Vermont in the United States.

The fulminating development of shrimp aquaculture has eroded the resilience of the country's socio-economic system. 'The fishermen, the shell collectors, the small farmers, there are countless groups of people who are seeing their way of life seriously damaged by the incompatibility between this type of industrial business and the traditional way of treating the mangrove swamp', says Luciana Queiroz of the ICTA of the Autonomous University of Barcelona. Olivier Joffre of the University of Wageningen in the Netherlands also welcomes IMTA: 'We are looking for a more moderate and at the same time environmentally friendly production; it is not going to be easy because short-termism prevails especially in this type of practice; this new approach is much more distributive, which is going to make investors nervous.' Social rejection is already a fact, and it is taking its toll. We must bear in mind that the entire mangrove environment, anthropological, economic and social, has been transformed, and the common good has been lost in many cases (Fig. 21.2).

The solution, of course, is to make the type of exploitation much more rational and multiple, where several species, as we have said before, can be exploited at the same time. One example is tilapia, a cichlid. It has been shown that if these fish are reared in moderate amounts in the same enclosure

Fig. 21.2 Sometimes, a rough and ready approach serves to evaluate mangrove swamps. Left image: the positive relationship between fisheries yields and the extent of mangroves in Baja California, Mexico (*Source* Aburto-Oropenza et al. 2008). To evaluate a system sometimes you can take a rough but robust approach. In this case, the mangroves were evaluated, contrasting the fisheries yield with the mangrove extension (positively related) in Baja California, Mexico. On the right, an illustration of mangrove biodiversity (*Credits* ADOBE STOCK)

with shrimp, they cleanse the excess organic matter from the water, leading to much less degradation of the environment in which they are reared. In reality, as we have been insisting throughout this chapter, this is a mere shadow of a real ecosystem: several species, several different needs, several types of diet… More than forty developing countries have a serious environmental conflict with shrimp farming, places where coastal environmental degradation is extremely rapid and their recovery, once degraded, very slow or nil. This is yet another problem of the Blue Revolution, a misunderstood protocol of action in which, for a change, the environment and the very society that inhabits it have been relegated in countries' relentless pursuit of supposed modernization and securing resources from abroad.

Algae Fields

We haven't mentioned algae. Far from being the future (and only) food promised by the Soylent Green company in film of that name (1973), in which Charlton Heston faced up to the powers that be to try to reveal the secret of a world that was dying, and had only algae left to eat, macrophytes are a fundamental part of many human or other animal food processes. They are the basis for pharmaceutical, chemical and microbiological products. That is why they are a profitable business and are grown in various parts of the world. Algae production is well over 10 million tonnes per year, almost all of it farmed. Only kelp (the most widespread) has an annual production of as much as between 4 and 5 million tonnes. 'In coastal systems, a species like *Laminaria saccharina* can grow up to 14% daily', says Bela Buck of the German Alfred Wegener Institute. It is not difficult to grow algae, especially in areas with strong currents and waves where they can grow steadily, but the systems must be robust, and the technology used sometimes more sophisticated than that for algae grown within bays in sheltered areas. These systems, installed for example on wind farms far from the coast (as we will see later on), must be able to withstand currents of more than 2 m per second and waves that routinely exceed 6 m in height.

The algae industry has grown by almost 30% from 1993 to 2006, with a turnover of more than €4.7 billion per year, China and Japan being the main producers (and consumers) on the planet. Europe has only 6% of the world's algae production. The market for agar, used for cell culture, food stabilization or cosmetics, is worth more than €450 million per year alone. But algae have not been able to enter the Western food market strongly, although they are a condiment or base for many oriental dishes. 'If the planet is going to need another 40 million tons of food from the sea by 2030, algae will play a key role', says Aleksandre Maisashvili of Texas A&M University.

Until now, the cultivation of algae has been carried out on its own; that is, as a monoculture. However, as we have mentioned, a future opening is to combine different types of complementary aquaculture in order to take advantage of the facilities and create synergy between the various organisms. In this case, macroalgae would serve to capture the dissolved nutrients released by fish, shrimp, oysters or other animals reared in captivity. Up to 90% of these nutrients (nitrogen compounds or phosphates) can be used by certain algae species. 'Algae crops can become as dense or thicker than ter-restrial crops,' says Klaus Lüning of the Alfred Wegener Institute in Germany, 'so their profitability and ability to recycle nutrients that are a problem for farming systems are very high.' The other algae, the microscopic, unicellular ones, are not yet exploited from a direct food point of view. There are exceptions: in Japan, you can buy phytoplankton biscuits. So far, however, no one has taken Soylent Green's project seriously, as the sea's proteins have not run out. Several research groups have been working for more than a decade to create biscuits from phytoplankton in a profitable way, beyond this Japanese prototype. Who knows, at the rate we're going, we'll be left with just seaweed and zooplankton for food, accompanied by a nice, succulent jellyfish…

22

The Sense Behind Marine Protected Areas

When the first Marine Protected Areas (MPAs) began to be created back in the early twentieth century, few people would have believed that protecting a particular area of coastline from fishing would benefit primarily the fishermen themselves. At first sceptical of this type of regulation, fishermen (especially small scale) came increasingly to welcome the fact that specific areas are being limited so that neighbouring areas benefit and the number of catches (fish, crustaceans, molluscs) increase. In their ground-breaking work, Antoni García-Rubies and collaborators from the Centre for Advanced Studies of Blanes (CSIC) and the Department of Ecology (UB) had already demonstrated more than two decades earlier that even the most endangered species of fish could recover quickly if people took seriously the protection of a specific coastal area. But the fact is that, although the positive influence of protected areas had been widely demonstrated, for example for adult fish species vulnerable to overfishing (groupers, gilthead bream, imperial bream, etc.), it was less clear where their true role lay in the recovery of the coast in general. 'It is well established that MPAs generally have a number of fish larger than 25 cm in length that is double that in the adjacent areas,' explains Graham Edgar of the University of Tasmania, 'but sometimes it is not enough to house large animals inside.' First and foremost, marine reserves serve to increase habitat quality while providing a good place for various species to reproduce and release their eggs, sperm or seeds, as well as creating the conditions best suited for larvae and juveniles to survive. 'A net movement of adults of different species of fish, crustaceans or cephalopods out of the protected area is sought,' says Dennis Heinemann of the Ocean Conservancy in Washington, 'the so-called "spill-over" that depends heavily on the design of the reserve in question.'

© Springer Nature Switzerland AG 2019
S. Rossi, *Oceans in Decline*,
https://doi.org/10.1007/978-3-030-02514-4_22

It's clear to me that there are no more 'pristine' spaces. I can't help but think that in places like the Western Mediterranean, the coasts of the Maine or the outskirts of Shanghai or Tokyo the transformation has been so acute and there are so many people that creating reservations is an extremely delicate business. On the coast, apart from the 'indigenous' human population (who live mainly near the coast), the influx of 'non-native' people (tourists) increases every year in some areas by up to an order of magnitude. This is why the creation of MPAs is both a necessity and a social and legislative problem.

With the establishment of protected areas, not only are the fishery resources conserved but also the ecosystem that houses them, and this enhances the biodiversity of the region. This ensures the viability of many species, as well as the recovery of those that have been heavily exploited or that, indirectly, may have experienced heavy impact or mortality. In a recent compilation study, Heinemann shows that out of 44 areas, only four had a decrease in the abundance of different key organisms, while the other 40 had an increase in these species. 'Often the areas are not well designed or well monitored', says Heinemann.

Following several studies, we can see clear positive examples in this regard: an increase in biomass in the species studied of between 190 and 446%, an average increase in density of 166% and an average 20% increase in biodiversity, as well as 30% larger sizes—all in the relatively short time of five to ten years. In Kenya, for example, several protected areas have increased the catch outside the protected area by up to 110%, and in Florida by up to 75%. The concept of a marine area is simple: if we protect a given area from human influence, nature itself returns in a certain sense to its own course and 'regulates itself'.

But it's not so simple. Marc Rius of the University of Barcelona has been able to show that some species can be harmed if others benefit too much. 'Fish can negatively influence mussel populations within a protected area,' says Rius, 'at their most vulnerable stage: the larvae. Rius shows that within the area there may be a decrease in mussel larvae due to an excess of predators, in this case fish, which forage on the undersea cliffs where the bivalves are established. Too many hungry fish means that they go looking for food anywhere, and mussels are an appetizing snack. Thus, outside the protected area the mussels are under pressure from humans, and inside it is the excess of fish that makes them diminish or not increase to a desirable level. I suspect that something similar happens with other organisms like the gorgonians themselves or red coral in places like the Medes Islands, where an excess of fish can make the few youngsters that there are fail in their first stages of life, devoured by hungry shoals of seabream or *Obladas*.

A few years ago we conducted a red coral regeneration study and saw that the recruitment rate was abnormally low in the same place as Rius carried out his study, the Medes Islands. In the past, monk seals or a healthy population of dolphins had been in charge of keeping the fish down but, without any top predator to devour them, they would reproduce in a sustained manner to exceed the numbers allowing the viability of species vulnerable to their predatory action. Today, there are no monk seals in places like the Medes, as we have discussed in previous chapters, so it is clear that tools must be provided so that the fish population does not exceed the carrying capacity of the system.

Furthermore, in this context, it has yet to be demonstrated that protected areas receive the dispersed larvae and juveniles of various species, not just adults; that is, for the movement of new 'migrants' from areas of abundance to areas where there was far more active human pressure. Among other work, Dr Richard Cutney-Bueno's group at the University of Arizona's School of Natural Resources in the United States showed that larvae, in this case molluscs, spread from protected to unprotected areas much faster than they had appeared to: 'this implies that habitats can recover more quickly than we thought, at least for certain species', says Cutney-Bueno. 'We have taken into account the currents, the underwater topography, the original populations and a very intensive sampling for larvae, and we have seen that, although there is great variability between years and depending on the area, the effect is always positive.' The 'zero point' issue is something that obsesses the managers and scientists involved in monitoring MPAs. The zero point is a quantitative estimate of the status of the area to be protected before any restrictions on fishing or management of that area begin. 'We cannot make a model in which we can see progress if we do not have reliable data on what was in place before the ban on fishing or access to certain areas', comments Joachim Claudet, from the MAERHA Laboratory of the French IFREMER: 'The monitoring of certain key species is fundamental to obtain a realistic view of the evolution of the protected area (monitoring based on the BACI model: Before and After the Control Impact (the impact before and after the control exercised through a restriction or prohibition). But which species should be protected? And above all, which are the most appropriate to monitor the various impacts inside and outside the reserve?

We can give an example by focusing on the Mediterranean environment. Not all species have the same social weight, either aesthetically or productively. Red coral is an iconic seafloor species, yet perhaps its immense transformation has led it to play an ecological role secondary to that of the lesser-known red gorgonian (*Paramuricea clavata*). These gorgonians have an important role as structurers of the ecosystem, a role that the coral has now lost through shrinking to barely a few centimetres in height, as it cannot now act as an 'animal tree' or bio-engineer in current conditions. But no one knows about the red gorgon apart from divers and specialists. Therefore, in terms of the score that we would award to them, neither species would be awarded protection, and for different reasons.

The same goes for other species of fish, crustaceans, molluscs and, of course, plants. The specialists to be consulted are, without a doubt, marine biologists and ecologists. They must adequately describe the distribution, reproduction, growth and other parameters of the species of a protected area, and they have the casting vote when awarding specific roles within the ecosystem. 'Many people still don't understand what it means to apply these marine protected area concepts', says Tundi Agardi of Sound Seas in the United States. Indeed, we must not forget that around protected marine areas live humans who are accustomed to the dynamics of fishing, occupations, urban planning, and so on. There must be a balance between the protected area and the people who live around it, so the study must not be only biological or ecological, physical and geological, but above all, human.

The ideal approach to designing a protected area would be first to identify the role of the zone in the dynamics of the coastal system. Political, aesthetic and socio-economic factors can come up against purely scientific ones: 'In many cases, the choice of site and the design of the protected area have little scientific justification', complains Simonetta Fraschetti of the Zoology and Marine Biology Laboratory at the University of Salento in Italy. 'The socio-economic reality of the area can compromise the real viability of the area, and sometimes things that should go hand in hand, such as ecosystem conservation, the projection of an area for the good of fisheries and eco-tourism, clash head-on.' Among other things, little attention has been paid to the dispersal from these protected areas of the true 'engineers' of benthic ecosystems, namely the algae, corals, sponges and other organisms that structure the seabed. A research group from the Centre for Advanced Studies of Blanes (CSIC) decided to compare the so-called functional diversity within and around Marine Protected Areas in the Mediterranean. 'We are looking to see if the effect of the protection is restricted to the local area or has a more general effect on surrounding areas', says Adriana Villamar, researcher for the working group. The aim is to see not only the vulnerable species, but also the health of the entire ecosystem, including species that may go unnoticed but that also structure and give life to the habitat.

On the other hand, when it comes to protection, migratory species must be considered. For example, Kelly Pendoley, of the conservation company

Pendoley Environmental Ltd, has seen that Australia's largest protected area, the Commonwealth Marine Reserve (almost a million square kilometres), has serious 'holes' where its protection does not reach yet where, for example, turtles, sharks or whales may pass through: 'Sometimes up to 48% of the station travelled by these species is not protected, and it is there that these highly mobile species can be very vulnerable; If you don't protect with these kinds of factors in mind, it may not do much good to have such a big protection effort.'

But the political and social issue is what is on the rise. You have to involve people from the beginning. 'A protected area must be well planned from the outset; those who live in the area, especially fishermen and shell-fishermen, must have a voice and a vote, otherwise it will not work.' This reflection by Ratana Chuenpagdee of the Memorial University of Newfoundland in Canada is pivotal. Many researchers are of a similar opinion, because respect for people and the democratization of decision-making have proved to be fundamental to the proper use and success of these places. Elisabet de Santo, from Dalhousie University in Canada, goes further: 'It's a question of environmental justice. The political and social interface must be given greater priority than the ecological one.' It's not that the biology, ecology and conservation of the species are not important, but that we have to integrate the community into the decisions and make them part of that restoration plan. Among the effects to be taken into account are cultural and spiritual effects. We are natural, and it has been amply demonstrated that we all benefit from seeing and being in touch with landscape and nature in a good condition. This is taken into account by many indigenous peoples, who see these protected areas as an opportunity, at times, to preserve their own cultural heritage.

On the other hand, there is the question of the minimum area. How much space should be protected, asked Tundi Agardi and his collaborators in a review of the issue. They came to the conclusion that the 'magic number' of 20% of the coastal area would be sufficient to maintain the functioning of various areas of the planet. This number—it is not known where it came from—calls for the creation of fully protected marine areas on at least 20–30% of the coastline. Large-scale MPAs are a reality but, as Pierre Leenhardt of the CRIOBE Institute at the University of Perpignan in France says, they have become more political than conservationist. 'Since 2004, 80% of the protected area in our seas has been increased', says Leenhardt. These are political areas because they are under the umbrella of the 'precautionary principle', and in many cases they are intended to preserve without even knowing for certain whether they will meet the initial objectives, given the little real control that some of them have.

What about the other 70–80% that wouldn't be protected? Do you have to stop protecting yourself? Every area needs to be studied, without a strict and inflexible 20% preventing more coherent coastal management. Flexibility and, above all, the adaptation of laws on MPAs are essential for the system to work. In fact, in countries such as Chile, where small-scale fishing is fundamental to the subsistence of coastal human communities, this point is quite clear: 'The fishermen's perception of the problems and the education and transmission of information from the scientific community to the different stakeholders is essential for the management areas and protected areas to function properly,' conclude Stephan Gelcich and Juan Carlos Castilla of the Catholic University of Chile: 'Only in this way, by making the main protagonists of the exploitation co-participants in the decisions, are satisfactory results obtained.' In all forums, the question of participation comes up concerning the democratization of decision-making. To take people into account.

But we also insist that MPAs must serve to improve the relationship between the environment and human beings themselves, who need controlled areas to continue to exercise a viable economic policy with guarantees. The information required is particularly difficult to acquire when it comes to deep-sea areas, habitats outside the continental coastal strip.

We must be able to change the vision of a Marine Protected Area from one of aesthetics or merely political compensation to one of environmental preservation, ecosystem management and human–nature harmony. Focusing on one of these areas to be protected beyond the coastline (Cap de Creus, in the north of Catalonia, beyond its current protection as a Natural Park), we can see that the complexity of the decisions is high and the repercussions undoubtedly profound.

Let's start with the natural environment. Josep-María Gili's research team at the CSIC's Institute of Marine Sciences has been responsible for a series of in-depth studies on the area in which various factors have been analysed, such as biocenosis (habitat types and communities), biodiversity, the state of health of the populations, and mortality factors of species of bioconstructive organisms in the area in question. This is the first step if you want to preserve either a particular environment or a vast area. 'Due to the complexity and richness of the environment, this is a very important enclave for a number of typical Mediterranean communities', confirms Gili: 'The most interesting thing is that we have found very diverse and vulnerable communities from 50 to 200 m deep, affected mainly by trawling.'

Indeed, the protection will go far beyond the coastal communities closest to the coast, and it has been observed that large areas have been destroyed by trawlers that leave furrows more than half a metre deep and destroy the entire living three-dimensional structure (corals, sponges, gorgonians, etc.), the basis of the viability of the populations of a large number of organisms. 'In this wide area of Cap de Creus, from the shallowest area a few metres deep to the animal forests at more than 200 m deep, we have already found more than 1,600 different species, approaching the biodiversity numbers of the Mediterranean, which has more than 2,500 different species', adds Jordi Grinyó, also from the Gili team. And the number continues to rise, because Cap de Creus (like other sites along the Mediterranean or Atlantic coast) combines different morphologies, geological components and hydrodynamic factors that favour the creation of habitats.

But there is more, as the continental shelf (the area where most of the trawling and industrial longlining takes place) has a high density of communities and organisms due to the fertilizer coming from the continent itself down the streams, rivers and groundwater floods. 'We found something very important: biological corridors,' Gili concludes, 'that is to say, spaces in which vagrant organisms (fish, crustaceans or cephalopods) move from the deepest to the shallowest areas, such as terrestrial forests.' These biological corridors, composed of gorgonians, pennatulaceae (sea pens), crinoids and other sessile organisms, are a refuge for both juveniles and adults, where they can find shelter and food. But in this area they are much disturbed, because there is hardly a square kilometre that has not been visited by trawlers. The aim is to create a large protected area where fishing is severely restricted or impracticable due to deterrent measures such as surveillance and artificial reefs.

I firmly believe that the future lies in this direction. We have to manage the sea and lend far more importance to creating areas where human interference is either minimal or highly controlled, otherwise the capacity for regeneration will be increasingly diminished and we do not know which way the evolution of the ecosystem will go. On a human timescale, the people of the next generation need to see what we have been able to do so that they, too, can enjoy a rational exploitation of resources.

The real problem arises, once again, when you touch on the human side of the issue: what do you do about the fishing sector in the area? At the moment, only 10% of the Cap de Creus marine area has even partial levels of protection. If you close an area of, let's say, 70–80%, many, many people will be unable to fish. Industrial fishing involves a major commitment of both money and people. It is estimated that by closing the area to all types of fishing, more than 260 direct jobs would be lost and more than 1,500 indirect jobs eliminated. This is without taking into account fishermen from other areas who visit these waters at certain times of the year. Logically, if the area is to be closed to this type of activity, the next step is to create employment opportunities in the area for those who could no longer go out trawling or longlining, with a detailed study of the real possibilities linked to other sectors such as park infrastructure, surveillance, tourism or small-scale fishing. It is ridiculous to consider an MPA without providing management and monitoring tools for the area in question, and it is absurd to create an area that will be viewed with suspicion or even hatred because the needs of the people who have been working in the area for decades are not taken into account. Marine Protected Areas are essential for the future of our seas, but always remembering humans, and even more so in an area as stressed as the Mediterranean coast, one of the most transformed and influenced by humans on the planet.

No Monitoring, No Reservation

Over the past decade, people in several villages have realized that their standard of living is rising, thanks to the tourism attracted by the second largest coral reef in the world (after the Australian Barrier Reef). Paradoxically, they have also realized that this 30% of their country's gross domestic product that comes from tourism, let's call it 'ecological', may be endangered due to this very use of the reef, whether by overfishing, over-frequentation or pollution (and, as we have said in previous chapters, something as uncontrollable as climate change).

Belize depends on the conservation of its Marine Protected Areas and adjacent areas for its survival. 'People in towns and villages and small towns perceive change; they know that it is tourists who are making a positive difference in their lives', proclaims Dr Diedrich of the University of Rhode Island. We don't need to look so far afield. As has been commented earlier, in the largest marine natural park in Spain the problems of maintaining balance are similar. In Cap de Creus, the authorities have to deal with more than ten thousand pleasure boats in the ports and piers of the northernmost Costa Brava, plus those that come from France or other places from the summer. It involves anchoring, recreational fishing and pollutants. More than 70% of the skippers do not check where they drop anchor, and most of them do so in the areas of *Posidonia oceanica*, with consequent damage. But there's more: what is environmentally friendly tourism, in the sense of enjoying nature, can turn into a real headache for the ecosystem. Diving (more than 100,000 divers a year, 85% of them from diving centres) is a disturbance, especially when between 60 and 70% is concentrated in the area closest to Roses and

Empuriabrava (Cap de Creus), both because this is where there are more people and the shelter from the strong wind, the tramontane.

In order to see the effect of these and other disturbance factors, rigorous scientific monitoring must be carried out on certain species that may be considered important from several points of view. 'Any marine reserve has as its objective the conservation of the fauna and flora of a given place (biodiversity), but it also seeks the recovery of the fishing stocks of adjacent areas, environmental education, advances in scientific research and, of course, the improvement of the local economy, thanks to tourism or the enrichment of the natural heritage', says retired professor Joan Domenec Ros of the Department of Ecology of the University of Barcelona. It is therefore necessary to choose indicators that give us a precise idea of the ecosystems' state of health. A matrix is made in which all the values (abundance, ecological importance, aesthetics, ethics, etc.) are arranged against those species that can give us some indication of improvement, status quo or degradation of a system. Let's not fool ourselves. I am often the first to spot that some of the species or values are considered for reasons of a certain human subjectivity. I believe that at the moment it is more important to maintain marine animal or plant forests (spermatophytes, gorgonians, etc.) in good health than any animal that is no longer functional in certain areas yet is considered sacred due to fishing, tradition, culture, and so on.

In this context, we need to look for places inside and around the reserve to enable comparison, and the time series must be long (in fact, once the reserve is established, perpetual). On the Costa Brava, in the early 1990s several research groups were coordinated to monitor key species of the various protected areas extend along the coastline. Thanks to this type of monitoring, within the reserve it has been possible to see the positive evolution of *Posidonia oceanica*, red coral, lobsters and groupers, to name a few. In this last species, there has been an increase from about 50 individuals in 1991 around the Medes Islands Nature Reserve to over 250 in 2005. This has been more than just preserving the species, it has contributed to the export of biomass and, more subtly, has helped us to look again at some of the habits of this imposing creature.

The concept of monitoring has changed. Lorenso Bramanti, from the French CNRS in Banyuls sur Mer, believes that people should be involved in observing signs of health that are easy to distinguish in certain species: 'More than 80% of divers in Italy and on Italian territory will plan to see red coral.' Why not take advantage of this to give studies a helping hand? In this way, you not only raise people's awareness, promoting conservation, but you obtain semi-quantitative data that helps you. So-called citizen scientists now commonly monitor many animals and plants. Their observations have made it possible to carry out an extensive study on, for example, the migratory movement of birds and the effects of climate change. Once again, the key is involving people. It is a question of participation and budgets: scientific teams cannot cover everything, and women and men willing to collaborate can be the way forward in the conservation of many species, bearing in mind that going anywhere takes a great deal of money. Something as simple as our divers taking the water temperature at each dive is already being used by research groups to measure seas' warmth across the entire planet.

The monitoring also allows us to approach systems that are less influenced by man's hand (yet in no case 'pristine', as is proposed), so that we can evaluate these effects in areas that are influenced to a greater or lesser extent by our actions. But we must repeat something that the authorities do not seem to understand: without scientific monitoring (social, economic, biological, of ecosystems) there is no possibility of seeing what is happening, no rigour in the results, no rigorous conception of changes and disturbances or how factors may or may not influence the evolution of the reserve itself. And most importantly, laws, guidelines and management itself must evolve and adapt to what the health indicators find. They need to be much more flexible, otherwise we will be faced with a strictly political measure, something very proper on paper but that is far from fulfilling its real role as a cog in the recovery of the coast and the oceans.

How Much is an Ecosystem Worth?

The rapid loss of ecosystems is beginning to stimulate new ways for scientists and specialists to tackle the problem. Since some managers and politicians need reference to statistics to assess the real impact of degradation, some groups of ecologists, sociologists and economists are beginning to provide them. This is the case for the mangroves of Mexico. A recent joint study by Mexico (Autonomous University of Baja California), United States (Scripps Institution) and Spain (CEAB-CSIC), reported in the North American magazine *PNAS*, tries to put a realistic value on the loss of these valuable sites. Considering only the small-scale fishing in these places by the inhabitants of the area, the value of a mangrove swamp in Baja California is estimated at €24,000 per hectare per year from figures on the catches recorded during the years 2001 and 2005. 'Local catches are positively related to the health of the mangroves in this area. The larger the mangrove swamp,' explains the leading scientist, Octavio Aburto-Oropeza, 'the greater the benefit.' We have focused on a wide but specific area, so as not to mix up possible ecosystems that would confuse the study.

The calculation is very conservative, because it takes into account only something that can be valued with relative simplicity: fishing. However, mangroves have a much more complex function, ranging from stabilizing the system against the waves, recycling nutrients to capturing carbon dioxide and maintaining high biodiversity, or providing an almost magical space for its inhabitants. Other sectors, such as tourism or industry, should be taken into account more or less directly. On the one hand, tourism is greatly benefited by this type of coastal vegetation, preventing the erosion of the beaches and providing very diverse life to the surrounding areas for the enjoyment of visitors. On the other hand, industry needs this type of vegetation to mitigate greenhouse gas emissions. Aburto-Oropeza argues that giving a realistic monetary value to the ecosystem may be one of the mechanisms that make us think long term, not short term as we do now: 'In Mexico alone, 2% of the mangroves are lost every year, mainly due to the construction of tourist complexes and shrimp aquaculture. In the area of La Paz, for example, 23% of the mangroves disappeared between 1973 and 1981 and have not recovered again', adds the Mexican scientist, of both Scripps Oceanographic Institute and the Autonomous University of Baja California: 'In just 10 years, some €2,400,000 per hectare could be lost, taking into account only the fishing sector.'

In any case, it is increasingly common to assign a monetary value to ecosystems. Not without controversy, this materialistic way of approaching nature runs up against a broad sector that is against the understanding that nature is a good in itself, thus cannot be evaluated from a monetary point of view. The reality is that politicians are increasingly taking a better view of this approach, and since the 1970s more and more tools have been developed to determine whether ecosystem degradation pays off from an economic point of view.

In the United Kingdom, Nicola Beaumont and collaborators from the Plymouth Marine Laboratory studied at various locations what would happen if the marine ecosystems of the British Isles were degraded. The result was 'an uncertain future in terms of lost profits', a future in which income from fishing, tourism and human health would be lost. But the conclusion of all the studies is that the degradation of marine (and terrestrial) ecosystems' complexity and the drastic reduction in biodiversity lead to a deterioration in our quality of life in all senses: we all want our landscapes to be preserved and for the contemplation of nature to be a pleasure. At the 2009 Prague Conservation Congress, I heard much about this new approach to the problem. There was controversy, and the truth is that it is an issue that I consider to be highly complex. In the International Marine Conservation Congress in Newfoundland in 2016, the topic was even more relevant and there were many lectures on it. In these, the issue of not everything being valued in terms of money came up strongly.

I very much agree that there are aesthetic, cultural and spiritual values that cannot be translated into finance. And, in my opinion, they are the most important, because, in the end, the loss of connection with nature is the loss of well-preserved natural landscapes. I believe that it is a difficult chain of decisions in a world so materialized that it is forcing us to set a price for everything around us in order to value it, and that starts off a dangerous dynamic…

The Sea Ignored

Marine Protected Areas have sometimes served as a source of information to understand the potential changes in trends on the planet. However, there is a clear asymmetry between knowledge of terrestrial and marine systems. The 2007 IPCC recorded 28,586 significant changes in terrestrial systems due to climate change. In the same report, only 85 are collected at sea and in freshwater systems, almost all of which are also directly related to climate change. This imbalance shows a deep divergence between the spending on terrestrial science and on marine and aquatic science, as well as a lack of real knowledge of these ecosystems that cover more than 70% of our planet's surface. 'If we measure the research effort, both in engineering and in science, from 1996 to 2004, taking note of the publications in specialized magazines, we will see that only 11% are dedicated to the biology, ecology and conservation of the marine environment', says Elvira Poloczanska of the Australian CSIRO. 'There are only 30 long-term marine time series that collect physical or biological data (or both), while terrestrial systems have no fewer than 527', adds Anthony Richardson of the University of Queensland, also in Australia.

There is no doubt that studying the aquatic environment is more expensive due to its inaccessibility, even in shallow coastal systems. Ships, special devices and remote action robots are very expensive instruments, especially when great depths are involved. But the problem could be more complicated, 'because the anthropogenic signs of the system may have hidden the effects of climate change', adds Poloczanska—that is to say, the effects such as from overfishing, pollution and the profound transformation of the coast due to urbanization overlap with the effects of widespread temperature increase or acidification. In estimating the climatic changes that climate change itself may be causing, the biggest challenge now facing the scientists and technicians is that we are degrading marine systems two to three times faster than land-based systems: 'So we don't know the sea's ability to withstand such dramatic transformations', concludes Richardson.

There is too much ignorance about ocean systems and their regulatory role on the planet as a whole: the number of marine scientists involved in future IPCC expert commissions needs to be increased significantly. Protecting the sea should be more than 'intuitive'; politicians will not invest more money just because the sea is 'important'. But, as we have indicated throughout this book, our exploitation of the sea is blind, and there are no limits because in many cases it is common property. 'Destroying the services provided by the sea in an accelerated manner is a bad global strategy, especially because we still do not know what the real consequences of our actions might be', says Giovanni Bearzi, director of the Milan-based Tethys Research Institute. The powerlessness of seeing that the studies, recommendations, reports and scientific articles are never implemented by management makes men like Bearzi wonder what those who foreground our environmental problems are doing wrong: 'Sometimes the scientists' proposals are unrealistic. Our science may be very good, but I think it is often very theoretical and impractical.' In this sense, an open channel between citizens and scientists must be taken more seriously. 'Environmental education is an issue that has changed a lot in the last decade', says Juanita Zorrilla-Pujana from the Institute of Environmental Science and Technology of the Autonomous University of Barcelona: 'The more education, more awareness and involvement'. She adds, 'we must link the community to those who make decisions, explaining things clearly, but, above all, we must create indicators to understand how this information is being transmitted'.

We can no longer manage information just for experts. The spectrum needs to be much broader, involving more people at all levels. On the other hand, apart from science, socio-economic aspects and their importance in conservation processes are fundamental to the protection of our coasts and oceans. Perhaps, in the not too distant future, the role of marine scientists in interpreting the global changes that the planet is undergoing may be further strengthened. For the time being, it is still a matter of convincing politicians and managers of the importance of the eternally ignored key component of the planet: the sea.

23

Underwater Restoration

More than ten years ago, I participated in a (failed) project in which the vision of the company was to replant *Posidonia oceanica* (a Mediterranean submarine plant par excellence) on the Mediterranean seabed using robots directed by people on the land. The Anaphysis project (as it was called) was to help to repopulate the damaged seafloor where this plant had previously covered extensive meadows. The idea was not a bad one. What went wrong? Partly the incompetence of the company, no doubt, yet also the lack of adequate technology and, above all, the absence of interest by society itself…

It's been a long time, long enough for us to make some progress. Do we have the tools? I was recently in contact with Professor Tetsu Sato of the Research Institute for Humanity and Nature in Japan. This professional has been dedicated to following a community of fishermen and sea lovers in Okinawa for the past 25 years. There is a cooperative there that has managed to promote the protection and restoration of the tropical reefs off part of the island. The cooperative does two things. First of all, it promotes any initiative that can improve the state of health of the reefs. For example, it is dedicated to removing invasive starfish that otherwise eat the coral polyps, and monitoring the sediment discharges from rivers and streams so that the reef is not smothered. Scientists like Dr Sato are responsible for guiding these efforts, indicating the threats to the reef, how they can be avoided and how certain activities, such as fishing itself, tourism, urban expansion, and so on, are to be regulated. It is a one-on-one dialogue between scientists and fishermen, always listening to the other.

But perhaps most interesting is the active creation of the coral plantations in this project. This second component also demands technicians and

© Springer Nature Switzerland AG 2019
S. Rossi, *Oceans in Decline*,
https://doi.org/10.1007/978-3-030-02514-4_23

scientists, but it is the cooperative itself that has organized it. It finances the venture with a part of the profit from the sale of algae and an extremely active donor programme. To do this, much volunteer labour is used, coordinated by the cooperative itself and the scientists. Why is it restoring the reefs? Simply because the members of the cooperative have understood that the survival of the fishing stock depends on the health of the reefs and the condition of the Acropora coral and other species. The most interesting aspect for me is that they have done it themselves, creating the mechanisms and calling in the experts, all without administration, politicians or subsidies. They couldn't wait, and they knew it. They understood the relationship between the reef itself, biodiversity and the number of fish that they could catch. It is a step beyond conservation and the concept of a Marine Protected Area.

In some ways, Marine Protected Areas are a passive way of addressing the problem of environmental degradation. Can it be done actively? What are the mechanisms that will enable us in the future to restore some of our marine ecosystems, in part by banning certain practices, redirecting certain uses or restricting specific areas to disturb them as little as possible? There are two initiatives on the scene in this respect: creating artificial structures (reefs) and re-implementing eco-engineering species in degraded areas or where they have disappeared. Both Marine Protected Areas and artificial reefs are increasingly seen as complementary measures in restoring coastal marine ecosystems.

An artificial reef can be described as 'any material that is deliberately placed on the seabed and that did not previously exist under natural circumstances'. In general, these structures serve to protect, regenerate, concentrate or increase the populations of organisms living near the seabed, especially those that are bound to hard substrates. Artificial reefs provide a habitat and refuge for species where they have not previously been or have been degraded by direct or indirect human action. These structures need to be strategically placed so that fish and other organisms can actually use them for breeding, shelter and food, or remain in them for part of their life cycle. However, it is still debated whether they are merely attracting wandering wildlife (fish, cephalopods, etc.) rather than creating biomass (areas capable of producing new individuals of various species). What is clear is that putting a rigid structure in the middle of sandy or muddy environments promotes an increase in the biomass of species, either because they are attracted by its three-dimensionality and deflection of currents or because there is food around it. It should also be borne in mind that a reef, according to its complexity (today, this can be very sophisticated), increases diversity simply by creating a new environment aside from the soft substrate that surrounds it (Fig. 23.1).

Fig. 23.1 Coral reef restoration. Several tools are now available for areas in which restoration makes sense. *Source* Horozowski and Rinkevich (2017)

If well designed, artificial reefs can be managed as new fishing grounds for several species. Their implementation can be seen as an increase in the habitable area in particularly degraded or over-exploited areas, increasing the biomass by adding more living space. Some studies show that, when well thought out, an artificial reef can be self-sustaining. Since the early 1990s, artificial reefs have been neatly placed on Portugal's Algarve coast to increase the potential for small-scale spearfishing, trammel and longline fishing. Here, Fausto Leitão's team from the Portuguese National Institute of Biological Resources has shown considerable increases in the biomass of such a commercial genus as bream. 'The colonization was very rapid', says Leitão: 'The reefs serve as a nursery for new generations, a place for mating and release for adults and an ideal place to feed.' The reefs can therefore be seen as an alternative fishing site, reducing the pressure on coastal communities. 'We will have to be careful in the future not to over-exploit the artificial reef as well', concludes Leitão: 'What is clear is that we must not allow managers to make the same mistake, because the situation of small-scale fishing on the coast was limited.' In the Algarve, there are no less than 44 km^2 occupied by this type of reef, a big gamble that makes it one of the top places in Europe for this type of recovery strategy. Today, in the Mediterranean for example, reefs are very common, and there are over 300 structures along the coast.

A reef also provides protection for specific areas where trawling is prohibited. It is therefore a preventive measure, a warning to boats skippers that indicates a place where efforts are being made to try to regenerate the environment, to give it a new chance. They have often been installed as a physical barrier to try to discriminate between small-scale and trawl fishing, between coastal and shelf fishing, with varying success. Indeed, at first the vast majority of artificial reefs were off-putting, but it was soon realized that, well designed, they could be more. Artificial reefs may, in fact, have unsuspected economic benefits besides helping to regenerate or enhance an environment, and many are now being designed as recreational areas for both sport fishing and diving.

Are artificial reefs attractive to divers? In a place like the United States, where more than 2.8 million people participate in this activity, representing about 23 million dive days a year, I would say that this is an important question. 'Generally, people who have a natural reef next to them appreciate that more', says Chi-Ok Oh of Clemson University in South Carolina, USA. 'However, they are accepted by a large majority of divers, especially those who don't have a great deal of experience or a fixation with certain types of dives.' People may perceive the material as being 'out of place' because it is artificial, but it creates interest and there is a market for such reefs. However, it is not a substitute for natural communities. In the near future and at the rate at which diving is growing, it may be necessary to use our imagination and inventiveness to create alternative sub-aquatic excursions, and reefs can make an interesting contribution in this regard.

Therefore, before we dismantle those that have existed for decades and to which people have become accustomed, it is useful to make a detailed study of the pros and cons of dismantling existing artificial reefs. This is the case in the restructuring plan carried out in the Gulf of Mexico, one of the areas where systematic oil extraction has been taking place for decades and where there are many installations scattered along the coast, abandoned when oil exploitation ceased. Dolly Jorgensen of the Norwegian University of Science and Technology has been concerned about this phenomenon and has found that most people, accustomed to these structures since the 1980s, do not want them to go. 'The platforms have become an integral part of the landscape in this area, even though they are aesthetically frightening', says Jorgensen: 'The problem is that the companies themselves have to assess whether or not it is worthwhile to dismantle them or to prepare them for people to continue to frequent them.' Apart from the benefit for local fishermen, who see more opportunities for fishing around these structures, they have an importance as pockets of diversity and are areas where a significant increase in the biomass of many organisms has been noted. The Gulf of

Mexico has very little hard substrate in certain areas, so they have become 'islands' where the benthic organisms enjoy a hard substrate upon which to establish themselves and reproduce.

More than one company has taken seriously this type of approach, restoring the seafloor and, at the same time, creating an attractive business. I am struck by one approach in particular that aims to create an integrated project going far beyond the restoration of environments and provision of recreation for tourism. 'The aim is to involve people in an education, awareness and involvement project to restore large areas of potentially damaged seabed', says Marc García-Duran of Underwater Gardens International. In no case is it a question of submerging four cement structures and forgetting them in order to regenerate the biomass of fish (throwing four concrete blocks in has questionable results), but rather carrying out a step-by-step project of coastal regeneration in which a study of ecosystems' forms and functionality is essential for success. At the same time, complicity with society is sought, whether local or foreign, to complete the project and follow it up successfully.

The same concept is beginning to be considered with other macrostructures that we will discuss in the following chapter: offshore wind farms. 'Wind energy production plants can act as centres of diversity by adding an ideal platform for the settlement of new organisms', says Richard Inger of the School of Biosciences at the University of Exeter in the United Kingdom. The benefits and risks of these macrostructures are yet to be studied, as scientific results are limited. However, the loss of habitat for some communities, the diversion of marine currents, the noise, the harmful effect of electromagnetism and the erection of obstacles to avian or cetacean fauna (many of these are still under study) must be weighed against the benefits of creating new biotopes in the middle of the sea and providing a safe place where trawling ceases. If properly designed, offshore wind turbines can be helpful, as explained in more detail in the next chapter. Among the positive aspects is the creation of new artificial reefs around which to manage fisheries effectively, especially small-scale fisheries. Moreover, these can be used to restore previously damaged ecosystems, but it is clear that a good preliminary study must be made in which several simultaneous factors are taken into account, not only the ecological.

Reefs, however, can often give a misleading idea of a system's recovery. Around Dubai, there has been a shift from about 50 km of beach to more than 200 km of concrete structures in the form of dykes, jetties and retaining walls. The latest works have added no less than 1,600 km of 'coastline', in the form of sand, to the famous structures of Palmera and Mundo, designed to attract tourists and foreign capital. Quite apart from the problems discussed in the chapter on 'The Swamped Shoreline', those 200 km of hard substrate forming the reefs have been colonized by corals and other sessile organisms. 'There is an increase in coverage in artificial structures, because corals cover more than 50% of the hard substrate surface compared to just over 30% in areas where there is a natural reef,' says Dr Burt of Zayed University in the United Arab Emirates. But we must not be fooled: the structures have allowed the arrival and proliferation of some species over twenty-five years yet cannot emulate a natural reef, as not all species become established in the same way. Such reefs have the potential for increasing wealth and diversity, yet are not substitutes for the natural environment.

It is very important to take into account the longevity of the structures in order to make a proper evaluation. The oldest ones have contributed the most to the accumulation of sessile fauna, the creators of systems such as coral reefs. You need patience to see the results of building reefs, because it takes decades for the community to mature, something that politicians often find annoying when they can't see beyond the four years of mandate provided by the voters.

Another potential problem involved in large-scale artificial structures is dispersing unwanted species, such as aliens. As we have already mentioned in previous sections, a hard substrate is ideal for allowing sessile species to extend their populations along the coast. When it comes to an alga or an ascidia, bivalve or unwanted polychaete worm, the prospect of erecting large submerged structures where there is none raises issues that must be taken into account. Laura Airoldi of the University of Ancona in Italy was able to see this with an invasive algae (*Codium fragile* ssp *tomentosoides*) moving through the Adriatic Sea means of artificial structures.

In order to understand the phenomenon of contagion, seasonal factors and possible interference with other species must be taken into account, but humans distort, interrupt or promote the movement of many species in various places, directly or indirectly, thanks to this great extension of concrete. By putting hard substrate in the sand, we favour dispersion by giving a home to species that previously could not cross what was once a desert to them; that is, sand and mud, the soft-type substrates. Today, it is estimated that more than half of the coastline in built-up areas is artificial hard substrate, which has to be considered in species dispersion models if we are to be able to act effectively in the event of a proliferation of unwanted organisms.

I believe that one of the bravest bets involves another type of regeneration: the re-establishment of the creatures that are architects of the sea, where they have been damaged or have disappeared. It is more complex, requires more study and may fail yet, in my view, if taken seriously, it is our immediate future. 'Inspired by forestry, an approach to restoration emerges as coral gardens', says Baruch Rinkevich of the Israel Oceanographic and Limnological Research Institute: 'This kind of approach can bring ecological, economic and social benefits, but we must take it more seriously.'

As we have already seen, one of the systems most directly or indirectly affected by humans is that of the coral reef. One concept that has been gaining momentum for almost two decades is precisely the 'coral gardening' concept mentioned by Rinkevich. It should be borne in mind that, in developed countries, forests have long since ceased to be natural. In some places, up to 90% of the forests around us are manmade, and in general 4% of the world's tree mass comes from forestry practices. In Great Britain, for example, more than 68% of the forests can be considered human. It is a significantly lower figure in France, where it is estimated that only 13% can be considered to have been planted.

'Coral gardening' is an active measure in which small pieces of coral or larvae recruited from healthy colonies are transplanted. 'The restoration of our funds with this type of system will be commonplace in the 21st century,' says Baruch Rinkevich, 'the problem is that, just as we have been doing on land for decades, we are still starting to do so at sea.' That's not surprising. Everything is more complicated at sea, and the technology and methods to be used are always more complex than on land. The problem, as I mentioned earlier, is that we have never really taken it seriously, and we see the challenges of feasibility and scale. I agree with this point of view, I believe that it has never before been considered by the authorities and managers of our environment, although there is now a new trend.

Taking it seriously would involve plans at the local and global level, with entities like the World Bank itself committed to the cause. The costs calculated so far range from €800 to €8,000 per square metre, depending on the species, the place and its accessibility. This is a limiting factor, because in developing countries the cost can in no case exceed €500 to €700 per square metre, otherwise it is absurd. However, the advocates of this solution complain that it has always been done on a shoestring or as a pilot study, and has rarely been taken seriously.

It is true that to undertake 'coral gardening' you need to keep many concepts in your head at the same time. The clonal organisms that make up the reef (i.e. corals) are not all susceptible to transplantation, yet in most cases the percentage of small fragments that survive is high and the mother colonies do not lose the capacity to regenerate the lost tissue. It is necessary to take into account the health of the donors, as well as to choose the best time of year and the most suitable methodology for transplants. The few specialists in this type of biotechnology also insist on studying factors such as the distance between transplanted colonies, genetic flows, what kind of corals (or other species) are more resistant and the competitors that will be found in their new environment. Of course, the area where the transplant is going to be done must not be disturbed, otherwise the exercise would turn into an immense failure. It is therefore necessary to resort to the ecology of forests, as well as to principles that we know (which are many) of benthic ecology to create the appropriate conditions for this active transplant.

Coral reefs are declining at a rate of about 2% per year, more than terrestrial forest stands and occupying a much smaller area. Other coastal systems are also at risk of regression and irreparable degradation on a human timescale. 'Marine restoration is a young science', says Arigdar Abelson of Tel Aviv University in Israel: 'There are gaps in its applications, the science behind it and, above all, the kind of legal elements that need to be taken into account to avoid conflict.' However, if we can achieve restoration far beyond the ecological model, we will achieve something that is urgent: restoring the lost ecosystem services. Finding new ways to restore our seafloor is certainly an immediate challenge that we face. Active measures such as the implantation of artificial reefs or the farming approach of recovery must always be taken into account with rigour and in the hope that they will help us to alleviate the current situation of impoverishment that our oceans are experiencing.

I must say, if a group of Japanese fishermen realized more than two decades ago that by keeping their reefs healthy and transplanting coral their catches were maintained or increased, why can't this be done on a much larger scale in other parts of the world?

Growing Red Coral

Perhaps one of the crops that is giving most hope in the Mediterranean region is that of precious red coral. Recent studies carried out at the University of Pisa by Giovanni Santangelo and Lorenzo Bramanti of the French CNR have comprised implanting coral onto marble slabs near the mother colonies. Using this technique, young coral become established after the extensive process of ovogenesis, liberation and swimming that the larvae of this species undertake between July and August each year. It has been proven that the survival rate is high, and after several years of monitoring it seems that the colonies are growing robustly on the marble slabs.

Once the young corals are established on the slab they can be transported, along with the slab itself, to where red coral has disappeared or is in clear regression. Slabs have proven to be a good small-scale solution, but what about doing it on a large scale? Can medium or large populations be regenerated using this system? 'Maybe this kind of program would need volunteers', Bramanti explains: 'There are a lot of divers who would possibly be willing to collaborate on a large-scale program. This is not a crazy idea, it is possible to recover populations with well-designed transplants of plaques with new recruits.' In collaboration with my working group, Bramanti also tries to achieve the reproduction of larvae in the laboratory so that it is easier to eliminate any possible competitors from the slabs, promoting the growth of the established colonies. It would be like creating a coral garden in which the growth of only certain species is favoured. Another solution jointly tested by Bramanti and my working group is the reintroduction of adult colonies capable of creating young that would repopulate the area through transplants. However, this technique provokes controversy since it involves the extraction of red coral from one place to put it in another, partially depleting the original banks.

Perhaps the most interesting thing is to cultivate red coral. Although slow, coral growth can be promoted by several factors that are now better understood whereby we can make colonies grow useful coral polyps. This would involve taking the coral from small colonies (the really risky part) and replacing it in a coral 'pasture' under controlled conditions. It is a logical extension to think that in this way our traditional coral need not become extinct (nor does it have to be if it is done rationally), growing into 'luxury' coral of large size and collected from great depth.

Biological Corridors

One of the features of artificial reefs is their ability to protect certain areas that are vulnerable to trawling because of their wealth, strategic location or simply because there is believed to be a greater concentration of fish. As mentioned above, these reefs may deter trawlers, if they are large enough, because the vessels risk losing their nets in the tangle of large iron and concrete blocks that have been placed strategically to prevent their passage. And another of their features that is beginning to be recognized as a conservation tool is how they can act as a biological corridor. On land, wandering fauna are hindered by roads, railways, power lines, ski slopes, wind farms or large urbanized areas, as these obstruct their free passage from one side to the other of their range, either interfering or preventing them from passing to places that may provide food, shelter or a mating site. At sea, these barriers or human communication routes are provided by the action of trawling. I have no doubt that just a century ago there were extensive 'meadows' of pennatulaceans, gorgonians and sponges, areas where sea lilies were very abundant, which have been swept away by incessant dragging by our fishing vessels. Over and over again, trawlers destroy the fauna that structure the underwater shelters and food areas (as we have seen in the explanation of the area to be protected off Cap de Creus). Discontinuities are created that may interrupt the free passage of certain species of fish, crustaceans or cephalopods that move from deep into shallower areas (or vice versa) at certain times or throughout the year. In addition, the deep-sea populations of many organisms that are capable of living across a wide bathymetric range are more stable, and it is these that are the most likely to feed in shallower ones at times when they experience one or another setback.

On land, a biological corridor is a passage in which it is important to interfere as little as possible so that species can flow without impediment. The purpose of a corridor is always the same: to connect populations, to allow the movement of species. In the case of the sea, a corridor would be an area where any aggressive activity that attacks the seabed ceases so that both sessile and agile settlers can regroup to regenerate the marine forest described so many times in the pages of this book. But if there is little information about corridors on land, in the sea it is simply zero: it doesn't exist.

In the search for biological corridors in the sea it is crucial to conduct scrupulous mapping of the bottom to identify populations of agile and sessile organisms. The next step is to monitor the movements of certain species of fish or crustaceans with radio-tracking, that is to say a system in which a transmitter is placed on their bodies without damaging them. Once identified, corridors are protected by various types of deterrent systems. However, given the high environmental degradation resulting from trawler fishing, finding suitable areas where it is still possible to identify the passage of organisms from one place to another will be difficult. And it will not be easy to understand, at this stage, to what extent we have influenced the population dynamics and movements due to the profound transformation of the seafloor.

But there is another type of corridor that is not directly related to fishing and that I consider equally interesting: the evolutionary corridor. This is the case of the so-called Gondwana living corridors, extensive routes along which the dispersion of species can be traced throughout the history of the planet. While there are still many doubts about this type of corridor on land, the confusion and lack of information at sea (as always) is even greater. It is known, for example, that an important evolutionary corridor is the one that extends from Patagonia to Antarctica via the so-called Scottish Arc, but there are still many gaps to be filled before we can understand which organisms and at what times they have spread from one place to another. In order to understand this type of corridor we must take into account the history of the planet and the pathways of dispersion, following geological, environmental and ecosystem evolution over millions of years. What fascinates me about this second type of corridor is that it is unstoppable: nothing will ever obstruct it, not even the sixth extinction that we are already experiencing today because, while the routes are narrow, they always end up finding one channel or another to connect and disperse, giving life and diversity to our world, no matter how our obtuse vision of things tries to impede the way.

Replanting Posidonia and Other Spermatophytes

Within the discourse on habitat restoration, we cannot forget the contro-versial issue of replanting spermatophytes. These terrestrial plants that returned to the sea have experienced a strong regression. For more than twenty years, in view of this regression many groups have dedicated them-selves to replanting them in various areas to recover such an important habitat —with a greater or lesser degree of success. The restoration plans have been ambitious in some cases, but their success has in many cases been more than questionable. 'It is crucial that both the habitat to be recovered and the population that "donates" the transplant cuttings are adequate', explains Dr van Katwijk of Radboud University Nijmegen in detail. This scientist and his colleagues made an excellent compilation of past work done in the Wadden Sea, especially on the Dutch coast, where the genus *Zostera* has undergone a major setback along practically the entire coast. But van Katwijk warns that most of the 42 transplants had very poor long-term survival rates and failed to take root: 'Few have been really effective, in many cases not exceeding seven to eight years of life; then they disappear, with the conse-quent failure and the problem of having impoverished a healthy area.' First of all, it is important to transplant in an area where you can make sure that the effects that had depleted it no longer obtain, whether eutrophication, heavy metal pollution, excess sedimentation or turbidity or trawling. Otherwise, the result is a total failure, and unfortunately many times, due to uninformed goodwill or a perverse lack of previous research, some of these groups have replanted spermatophytes in places that are far from fully recovered from the disturbances that led to their earlier collapse.

The most serious thing is when transplants are performed on the pretext of maintaining a population that is known to have been exterminated. José Sánchez-Lizaso and his colleagues at the University of Alicante were able to see this problem at first hand. In the expansion of the Campomanes Marina in the Valencian community, under pressure from the law that is supposed to watch over the survival of the spermatophyte meadows, the authorities decided to carry out a large-scale transplant of an area that was clearly going to disappear with the development of the port. Sánchez-Lizaso had warned that a study should be carried out on a medium scale to verify the viability of this movement of *Posidonia oceanica* rhizome, and took pieces measuring a square metre and 40 cm deep to a safe place. However, they found that the mortality was very high and the live beam density decreased from about 360 to less than 50 per square metre (only 15% of the original density). 'Large-scale transplantation is not feasible', confirms Sánchez-Lizaso. The truth is that the data are quite conclusive: mortality rates of zero to 96%, showing variations that are too marked, thus serving only to impoverish one area and barely recover another. There are also positive results, such as those of *Posidonia australis* in Australia where, with a custom-designed planting machine (capable of moving clamps of 0.25 square metres and 50 cm deep) and a rigorous protocol, Eric Pauling's research team at Murdoch University in Australia achieved 77% survival after five years of follow-up.

Perhaps the best solution is to plant spermatophyte seeds in captivity and bring the seedlings to the affected sites. This is what Dr Balestri did with his collaborators at the CIBM, Interuniversity Center for Marine Biology and Applied Ecology in Livorno, Italy, with remarkable success. More than 70% of the seedlings took root and survived, thanks to a well-designed retention system for the young plants, designed to prevent the loss of individuals. 'The advantage is that you don't sacrifice plants, you don't break from one place to another', says Balestri. Adriana Alagna of the Institute for Coastal Marine Environment of Castellamare in Italy has gone further, finding a stony substrate to be the ideal place for small plants to take root. Seeds (or larvae, in the case of sponges, corals, gorgonians, etc.) are ideal because the 'mother' population is less affected. The biggest problem is finding the plant seeds, because only in certain places do they flower, pollinate and bear fruit. It is not necessary for me to insist that the best thing, as always, is to preserve the remaining populations of spermatophytes, but this last solution could be something to take into account as a possibility of recovery of disturbed grasslands that can be recovered after having re-establishing conditions suitable for their survival.

Once again, the issue of restoration needs to be part of the solution, taking it seriously. 'By making an active plan for the conservation and transplantation of marine plants such as spermatophytes,' says Carlos Duarte of King Abdulaziz University in Saudi Arabia, 'we can go from 177 to 1,337 tonnes of carbon dioxide sequestered per hectare per year.' Those famous carbon sinks that we need so much…. Maintaining what we have and regenerating what is missing will be the key, in the near future. We owe it to the sea, always mistreated and ignored.

24

Energy and Matter from the Sea

It is shocking, at least to me, to see how the words of an influential politician can sometimes make available huge amounts of resources, generating a dormant or latent interest in issues that were waiting for an opportunity to become viable. In one of his initial speeches, US President Barak Obama turned the direction of Bush's energy policy, which until then had not been in support of renewable energies, towards greater independence for his own country and greater consistency with climate change. Hundreds of small, medium and large companies engaged in the renewable energy business in its most diverse facets witnessed exponential expansion. In fact, it had been more than a decade since the search for a solution to the many energy problems had started underway, with substantial monetary investment. But many of the energies that had been ignored until then, or that had been kept in technicians' and engineers' drawers to await an opportunity, began to appear on the tables of politicians, businessmen and managers of public entities. They were searching for solutions, and various projects that had so far been regarded as science fiction or coming from respectable dreamers wanting a utopia now became seen as attractive and surprisingly efficient.

Now, with an incoming President Trump, this trend is more than likely to be truncated. A whole energy plan will be reshaped and, in a short time, we will see the defenders of energy derived from fossil fuels emerge from their dark corner once again. We must not forget that 90% of the energy on the entire planet comes from non-renewable sources, with strong asymmetries between countries, especially those that have made a firm commitment to renewables, where this percentage may be greatly reduced.

© Springer Nature Switzerland AG 2019
S. Rossi, *Oceans in Decline*,
https://doi.org/10.1007/978-3-030-02514-4_24

Whatever happens, and it doesn't look like anything good is going to happen, the reality stamps its crudeness on our faces without any consideration. We are a species that has gone from consuming about 100 W per person per day when we were just Australopithecines to more than 12,000 W in modern times. We need an amount of external energy to live in an absurd way spurred on in the past by the apparent 'inexhaustibility' of fossil resources. But we also know that this is coming to an end, and that we are already in the so-called peak-oil. We look for oil and other fossil fuels in the most unlikely places: in an Arctic Ocean that lacks ice, in deep areas, sometimes thousands of metres away, in the subsoil in the form of fracking… a flight forward that only makes things worse. The powerful lobby for non-renewable fuels (perhaps the most powerful on the planet) will not allow its arm to be twisted: it denies or ignores climate change; ignores the effects of pollution; stonewalls the lobby for renewables; immobilizes proposals for a carbon dioxide tax…. The only way out is to renew the energy model completely, to consume far less energy and optimize the renewable sources that already far exceed the capacity of many fossil fuel-based installations. It is a difficult route, because it will mean the democratization of energy and reduction in its cost (always a key factor in human history), but it is inevitable.

For many reasons, the new energies (nuclear, wind, solar, second- and third-generation biodiesel, large-scale ocean currents or geothermal energy, to give just a few examples) have no way back. I am not going to go any further into the debate on 'how much oil or natural gas is left' (although I do remember, once again, that these are finite energies and the alarm was raised decades ago about their extent). On the other hand, I have already insisted on the fact that using the amounts of fuel we do (and will) is undoubtedly responsible for some of the changes that we are undergoing and will undergo over the next few years. The change in energy policy and alternative sources for this market is approaching and the sea has much to do with this future. Here are a number of examples that are already being developed that give hope for change and independence in the next two decades because they are feasible, not exclusive to any part of the planet and, although some of them are not at present entirely profitable in view of the 'everlasting' oil, coal, natural gas and uranium of today, they will be so in the short term and may help us to resolve the problem of dependence on third countries.

The most important thing is to act with prudence and with a series of serious elements of criteria in hand to rule out hasty solutions, as was the case with first-generation biodiesel. We know that renewable fuel is a possible alternative (as a patch, until such time as energy from nuclear fusion arrives) to fossil fuels. We cannot forget that trucks, tractors, cars and many other machines need these fuels, and replacing them with electric motors is not about to happen in a couple of short weeks (especially in certain places). At present, huge amounts of money, farmland and human effort have been invested in creating biodiesel and bioethanol as renewable alternatives to our current fuels, but the failure is considerable. 'The various types of first-generation biofuel have proven to be environmentally counterproductive, even more so than gasoline itself,' says Jörn Schalermann of the Smithsonian Tropical Research Institute in Panama. Land plantations are not viable, as is currently demonstrated, partly because a significant proportion of agricultural produce for human consumption is now being burnt by cars and partly because producing something that even approaches replacing petrol and diesel from trucks, tractors and cars would require a disproportionate amount of land, endangering already battered ecosystems, especially the tropical ones where the production is being spurred on. 'One of the factors to be taken into account when assessing this type of energy is undoubtedly the real ecological impact', adds Schalermann: '21 out of 26 crops studied for this purpose reduce the greenhouse effect by 30%, but 12 of these 26 have unacceptable environmental and food costs.' These crops include soybeans, palm oil and sugar cane, constituting three of the best for this type of conversion to fuel.

Let's look at some figures to understand the problem. Palm oil is one of the most profitable terrestrial vegetables from the point of view of energy that is convertible into biodiesel. One hectare of this vegetable produces about 5,950 litres of oil. A country like the United States needs about 530 million cubic metres of diesel per year to meet its various needs for this fuel, especially for road traffic. Therefore, it would need, from approximate calculations, some 111 million hectares to produce enough biodiesel to meet current needs, or 61% of the country's arable land. Unacceptable. Another example: to power 6% of the US population, this country has to invest all the corn that it produces in biofuel. Simply intolerable. With soya it would be worse, because one hectare of soya produces about 400 litres of fuel a year, an order of magnitude less than palm oil. Calculate how many hectares that would take....

Second-generation biodiesel seeks to optimize processes and convert vegetable and animal oils to generate energy. However, it is far from cost effective and faces problems similar to those of first-generation diesel. Another diesel, third-generation, is made from plant micro-organisms (microscopic algae) and is making a decisive leap. It may serve to patch up the precarious situation in which we will find ourselves before too long. It is a question of using microalgae, especially those of marine origin, which have a high yield of oils in their interior, are relatively easy to cultivate and occupy much less land than terrestrial vegetables, apart from being inedible by humans directly, therefore not creating a conflict of interest, in this sense. 'Biodiesel from algae seems to be the only fuel that can really replace the current diesel fuel without such a great harm to the environment', says Yusuf Chisti of Massey University in New Zealand. In reality, the equation for any biodiesel is often simple. Vegetation captures carbon dioxide from the atmosphere and, once converted into fuel, generates carbon dioxide that re-enters the cycle. However, this equation is much more complicated than it seems.

We have to take into account that in every process there is not only the plant. There is also everything that surrounds its production—maintenance, transport, and so on—that generates carbon dioxide either directly or indirectly. In the case of microscopic algae, if the power plants are well designed, the balance between carbon dioxide used and carbon dioxide absorbed could actually be zero. Algae can contain between 30 and 80% lipids. Unlike palm oil, an algae crop can yield up to 240,000 litres per hectare per year of fuel. Their proportion of protein matter and carbohydrates is much lower than that of terrestrial plants, so a series of well-designed bioreactors (transparent channels or pipes in which algae are grown and then harvested) can provide a great deal of oil if the environmental conditions are right. 'We have to look for the ideal conditions of temperature, pH, nutrients, light and carbon dioxide concentration for its cultivation', says Claudio Fuentes-Grünewald of the Institute of Environmental Science and Technology of the Autonomous University of Barcelona: 'Marine biodiesel is easy to grow, produces a lot of oil and takes up little space.' And it does not consume drinking water, because it is produced by seaweed.

In tropical areas, ideal for light and temperature, a well-designed plant can produce about 1,535 kg of biomass per cubic metre per day. With 30% of its weight in oil, one hectare can produce about 123 cubic metres during 90% of the year. In other words, we have an average of 98.3 cubic metres per hectare per year, far more than with any terrestrial plant species and occupying a space of only 3% of the cultivable area, 58% less than the most optimized terrestrial crops. Even those algae that have lower yields produce an order of magnitude more oil than the best terrestrial crop.

We must choose the most suitable strains and algae, and optimize the production process to the maximum. In addition, these microalgae can be used to extract other types of protein molecules and can be used far more efficiently than terrestrial plants. 'One of the most interesting applications is to take what is left of the algae, the solid non-lipidic waste, and transform it into biogas', says David Santos-Ballardo from the Autonomous University of

Sinaloa. Indeed, David, other colleagues and I undertook a project in which we showed that the biodiesel production cycle could be optimized, because the biogas produced from the digestion of the waste could feed the biodiesel production plant. In other words, everything is used and a self-managed plant is created that does not depend on external energy for its operation.

Ideally, outdoor crops should be harvested with minimal energy expenditure in the form of electricity, transport or fertilizer. The algae are very fast growing, and the cells of the normal strain used for this purpose can achieve cell division every two days. 'We have to do a good study of the materials and energy needed for the plant to operate,' concludes Eva Sevigné-Itoiz of the ICTA-UAB in Barcelona, 'but by making calculations we can see that open-air crops are even more productive than those grown in industrial warehouses behind closed doors.'

The complex process of optimization is already underway, with the oil companies on the frontline trying to achieve large production plants in a few years. Isn't that so? In reality, oil companies see this type of biodiesel more as a threat. Why? Well, for the simple reason that this type of cultivation can actually be done by anyone. Producing algae is easy and cheap. Imagine a small coastal town of about 200,000 inhabitants. If it were possible for half of its energy needs to be absorbed by this type of biodiesel and the other half by other renewable energies, it would be a catastrophe for the big monopolies. They'd lose a great deal of control. And there would be no need to renew its entire fleet of vehicles with electric machines, as some of them could continue to use diesel in this transition to totally clean energy.

Another alternative source of energy that has entered the content and comes from the sea is wind power. Europe, with Britain at the forefront, went from having a cumulative capacity of 700–800 MW of energy from offshore wind turbines in 2007 to more than 2,000 MW in 2009, just two years later. And this is rising, because to this day these projects continue to grow without ceasing. Offshore wind farms, once a utopia, are now a very tangible reality. Wind parks with columns supporting fifty to five hundred turbines have been installed along the North Sea and the Baltic to produce electricity from the strong winds offshore in areas especially exposed to this type of energy. Wind farms have different problems from land-based ones, although the aesthetic and bird life aspects seem to be the most critical. However, if the site has been carefully selected and the facilities have been moved away from areas that intercept migratory routes or are make abundant or delicate bird life (with protected species) vulnerable, the number of collisions can be low, from 0.01 to 23 per year per turbine, which is a risk requiring study yet can be solved. Marine mammals may also be displaced from the area by noise or excessive interference in their resting, fishing or travelling areas, but the effects (the few times that they have been observed) are minimal. The underwater wiring and above all the electromagnetic fields generated by the transmission of energy from the sea to the land are another of the problems that were cited as potentially dangerous for ecosystems, but the reality is that the few serious studies that exist (and I insist on the few, because there is still not much going on, in this sense) have shown a negligible effect on fish fauna.

These structures have more interference on navigation, which is restricted and, in many cases, totally blocked. Although these problems are on the table, the offshore wind solution seems one of the most acceptable and, in the medium term, can generate a large amount of energy. 'There is still a lot of doubt among people, entrepreneurs or local governments themselves', says Brian Snyder of the LSU Center for Energy Studies in the United States: 'The specialists and ecologists who have worked on the subject over the past decade have not seen any major environmental problems.' However, everyone agrees that, at least for the time being, a strong injection of state capital is needed to make these projects a reality. One hundred turbines of five megawatts each (i.e. 500 MW of energy generated) placed some 15 nautical miles from the coast cost some €270 million to build, some €80 million to cement them in and another €126 million to install underwater cabling and electricity receivers. Estro gives us a cost of about €43 per megawatt hour, a price at the limit of profitability, to which must be added the costs of maintenance, environmental monitoring, and so on.

There are places where this technology is being developed exponentially, because it is very suitable from a logistical and energy perspective. As we mentioned earlier, Britain is a country that has always tried to maintain a certain amount of energy independence, with the help of its North Sea oil exploitation, but now it is finding that its reserves are diminishing and the exploitation of both this resource and coal is becoming scarce. By 2020 (just around the corner), Britain estimates that, if this is not remedied, it could become up to 75% energy dependent on the outside world. That is why there is an ambitious renewable energy plan to provide up to 20% of the energy needed by the British urban, industrial and agricultural machine over the next two decades. Among the energies already being put into operation are that of offshore wind farms, especially in the eastern part of Scotland. Prior to 1995, there were no wind farms in Scotland, then in 2008 some 59 operating wind farms were installed, 65 more are now approved and 103 more are under study. Wind energy, especially at sea, is a strong contender in this area, where winds have an average speed of 7.5 m per second (only matched by the

Norwegian coast and some areas of Denmark and Ireland), with considerable strength and continuity, allowing wind farms to be seen as a feasible reality.

As in other places, and as we mentioned before, wind energy awakens suspicion in the landscape: nobody wants wind turbines to be visible, because they are bulky, unsightly and noisy, although technology, especially in this last area, has improved drastically. Here, as elsewhere, we have gone from turbine blades of about 50 m in diameter (with productions of 1 MW) to more than 125 m in diameter (capable of generating up to 5 MW). The new generations of offshore wind turbines, especially those that are more than ten nautical miles away, are likely to have a diameter of more than 200 m and a correspondingly significantly larger production. Returning to the previous reflection, the landscape problem is one of the most difficult to solve, especially in the face of tourism that is does not favour a sea horizon covered in turbines. However, overall, wind turbines are a benefit because, as we have already said, the environmental benefits are considerable compared to other forms of energy production. 'When you look at the studies on people's acceptance of this kind of impact, up to 75% support this kind of energy, even though they don't know about its impacts', says Bouke Wiersma of the University of Exeter in the United Kingdom. However, if you share this energy with people, that is to say you give them direct control, the thing changes in a radical way. In this case, the possibility of creating energy and being able to supply it makes up to 85% of people accept its visual impact.

If you don't want to see turbines, why not submerge them in water? This is another possibility that is already being considered by different companies in the United States, Great Britain and Germany: turbines driven by sea currents. Just a few years ago, a small Scottish company started up a project of turbines driven by the energy of the currents, and it is now valued at almost €4.5 million: a tiny investment that grew to almost ten times in a short period of time. In this case, tidal currents (which were already spoken of more than three decades ago as a source of energy) are the ones that provide electricity through the movement of turbines. There are no large-scale operations, but there are generators capable of producing more than 1.2 MW with only a handful of small generators. Projects to create generators from ocean currents are more real than ever. 'The ocean currents are very interesting from an energy point of view', says Charles Finkl of Coastal Planning and Engineering in Boca Raton, United States of America: 'About 30 million cubic metres of water per second are moving in the Florida Strait; turbines could generate a lot of energy.'

When this scheme began to be spoken of, sketches and prototypes had already been appearing for decades to replace oil and other non-renewable energy elements. At a very approximate and possibly underestimated figure, they could generate about 450,000 MW (or about 450 GW), equivalent to about €434 billion worldwide. Obviously, this estimate is considering a series of generators that would intercept more than half of the ocean currents, something that Finkl's team does not even consider. As we have already mentioned regarding the Scottish venture, there are dozens of serious projects that consider mainstream energy as a safe bet. Near Florida, the currents are particularly strong and turbines at between 100 and 500 m deep could be placed to generate electricity near the Miami conurbation. The generators have to be close to urban centres in order to lose as little electricity as possible in their transport from the bottom of the sea to the coast. 'The currents have to exceed an average of one metre per second for the turbines to be efficient', explains a researcher on a recent study by the National Sun Yat-Sen University of Taiwan. Undersea turbines are slower than wind turbines, but the kinetic energy, due to the much higher viscosity of the water compared to air, is much greater and generates more electricity: a current of 5 knots can provide energy equivalent to an air speed of 350 km per hour.

The placement of current-driven turbines has its problems, like everything else. The largest of these is undoubtedly so-called fouling; that is, the plant and animal organisms that settle on anything that is submerged (port dykes, ships' keels, buoys, etc.). This problem is undoubtedly one of the greatest impediments, but by placing the turbines at a depth of more than 100 m the fouling would be reduced as the plankton would be more concentrated in areas where the light would reach and microscopic algae grow. However, the maintenance operation would be costly. The interference with marine life is minimal—fish, cetaceans and birds would not be affected by the turbines if they were well designed and placed in areas where there are no migratory phenomena, quite aside from the fact that at those depths the presence of many organisms, such as birds, may be discounted as it is too deep for them to dive. The most critical part is the installation of the turbines and the wiring, a little like the wind turbines, where we commented that one of the chief issues stems from the very installation of the macrostructures and from the cables that pass go to the coast to carry the electricity. Some specialists even talk about putting the turbines in specific areas for a dual purpose: generating electricity in the Gulf of Mexico and mitigating the effect of hurricanes there. A series of turbines placed in strategic areas of the Antilles could supply energy to industry and reduce the currents that accelerate their hurricane formation, according to a study by Richard LaRosa (died in 2018) of SeaLevel Control in New York.

Finally, I'm not going to forget about waves. When I was a child, in the 1970s, I had already seen drawings and pilot projects in which waves supplied energy. They are now a reality, although still much underestimated. Why? I compare renewable energy sources and I am stunned: while wind produces about 0.4–0.6 KW per square metre and solar energy between 0.1 and 0.2 KW in the same area, waves can produce up to 3 KW per square metre. Wave energy is available 90% of the year, unlike wind and sun, which range from 20 to 30% of the year at full capacity; and 37% of the human population lives less than 90 km from the coast, which means that it could supply a great deal of energy to many people. Their environmental impact is minimal and technical problems (conversion of mechanical energy into electrical energy, variation in wave height and adaptation of the generator system, changing wave direction and impact on mechanisms, etc.) are being solved. Why don't we use it more? In my humble opinion, this simply because it is very cheap, very affordable and, once again, within everyone's reach.

Now that we have a real explosion of alternative energies, we must be aware of the real possibilities and their profitability, avoiding systematic application without serious prior study of the possible impacts on the environment and their efficiency (unless in some cases the remedy is worse than the complaint). In a changing world, energy is going to be one of the key themes of our will to change. I will not tire of saying that the path exists, and we just have to follow it.

Deep Mining

We cannot ignore another type of resource in the sea that will also grow enormously in a short time: oil and minerals. Until now, exploitation has been shallow, due to the associated technical problems, but over the past two decades South African diamond mines, for example, have penetrated more than 8 km offshore from the coast to obtain, from hundreds of metres deep, valuable loot. The exploitation of the sea in terms of minerals is not new, but we are increasingly prospecting and exploring deeper and deeper. Between the 1970s and 1980s, a term that I remember hearing first in Cousteau's underwater programmes became fashionable: manganese nodules. We could see these images from hundreds of metres deep, grey balls that submarines had collected from the abyssal plains to be analysed on the surface. The analyses showed significant concentrations of minerals: 11% copper, 4% zinc and, in certain areas, up to 150 g of silver and 15 g of gold per tonne of material. It was the future, but so, so far away... mining was going to develop at legendary depths, with science fiction gadgets. The various companies involved in this adventure lost some €500 million in prospecting and exploitation attempts at that time and, when in 1982 the submarine *Alvin* discovered new mineral sources in the underwater chimneys of the Atlantic Ocean, engineers and marine geologists did not want to hear about the possibility of opening a mining concern at a depth of thousands of metres.

Until now. Because, as they say in Spain, '*a la fuerza ahorcan*'—we've been forced into it. Obviously, the technology that we use today is infinitely superior to that of three or four decades ago, but the crux of the matter is that we are beginning to have problems in meeting our needs, and the seabed is immense… and there is much to exploit. For more than a decade, the Brazilians have had oil platforms that operate to depths of more than 1,500 m, and in the Gulf of Mexico (one of the richest areas on the planet in terms of underwater oil resources) they are already extracting oil from more than 2,500 m down, to give two examples of deep-sea underwater mining (as in the chapter on chemical pollution in the example of BP).

It is therefore hardly surprising that two Canadian-based mining companies (Nautilus Minerals and Neptune Minerals) are already systematically mining minerals just over 60 nautical miles off the coast of Papua New Guinea. There, robots of about 180 tonnes extract different types of nodules. They suck them with powerful turbines to the surface using nozzles coupled to a complex remote system from a depth of more than 1,500 m. Across the Pacific (New Zealand, Vanuatu, Papua New Guinea), there are many areas susceptible to systematic exploitation, and the few companies capable of doing so already see real benefits in this. 'There is increasing pressure on science and technology to find new resources and extract them profitably from the seabed', says geologist Bramley Murton of the National Oceanography Institute in Southampton in the United Kingdom.

Currently, of the more than two hundred areas of underwater exploration, ten have ores to exploit profitably, each concentrating deposits of potentially more than 100 million tonnes of ore. There are more than 200,000 seamounts around the world so, in fact, more than 155 countries could benefit directly from this beneficial business, which will soon be systematic. What is more, in many cases no one needs to be asked for permission, because many of them are located in international waters and, as we have already seen with precious corals or trawling, the question is who comes first and takes the spoils. 'Some 400 sites have already been successfully explored, 165 of which have exploitable material that could be exploited for profit', concludes Murton. Many of these mining operations would be complicated if the mountains or veins are more than 2,000 m deep, because the technology required there is really extremely sophisticated and it probably would not be a profitable undertaking. But the European Union, for example, produced a report in which it considered that no less than 5% of minerals such as cobalt, copper or zinc would come from the bottom of the sea in 2020, rising to 10% in 2030. That would mean about €10 billion a year, a not inconsiderable slice of the pie.

The main problem is that legislation on underwater mining stems from active mining of shallow areas or from land-based mining, with little or no provision for deep-water mining. Oil platforms, polymetallic nodules, carbon dioxide sinks—there is nothing written on how to conduct environmental impact studies, their consequences for deep-water communities or the degree of susceptibility of areas to being prospected, exploited or used. 'Seamounts face many dangers', says Malcolm Clark of New Zealand's National Institute of Water and Atmospheric Research. Clark and his colleagues have recently reviewed potential assaults on the still-unknown deep seabed at great depths and warn of the little work that has been done on how communities operate at those depths and how it would affect them if they began to move huge amounts of rock around them: 'Perhaps the effect is even worse than that of trawling', says Clark.

Seamounts are highly complex environments, where currents, relief, connectivity between communities and many other factors yet to be studied play a role. Being unknown environments, it is necessary to undertake in-depth (and never better said) studies of their capacity for resistance to the impact of such aggressive manipulation. The idea is to put semi-permanent stations on the surface, through which the material to be extracted is channelled. But there is no consensus on the matter. One of the pioneers in this type of exploration and an enthusiastic promoter of deep mining, Dr Scott of the Scotiabank Marine Geology Research Laboratory, believes that there are several incentives to carry it out without major problems for the deep-water communities: 'To begin with, the environment is not acidified the same as in land-based mining, because the exploitation is surrounded by an alkaline environment such as the sea.' But in addition, in most cases the extraction is not carried out by drilling but by collecting from the bottom, far from the communities that could be affected by these subaqueous operations.

Other specialists in the field, such as Jochen Halfar of the Mississauga University of Toronto, are cautious about this: 'The prospects for regulating underwater mining in general and deep-water mining in particular are not good.' Yet others, such as Dr Boschen of New Zealand, see dangers even in environmentally friendly places such as their country: 'Licences have been granted in New Zealand for exploitation in places like the deep hydrothermal vents in New Zealand before they even know what is there and what the implications for the communities would be.' For this country, current legislation, weak or non-existent, would be fully exploited by companies seeking these resources, taking advantage of the legal vacuum and lack of knowledge. For these companies, it is a bonanza to find strategic minerals in a no man's land, where there are no troublesome, unstable governments to be financed, in an underhand way, to sustain the exploitation of a range of resources so important to the functioning of our current social and economic model. I wrote a novel on the subject, *The Iceberg Cemetery*, where the place where the exploitation was to take place was the Antarctic seafloor: I am sure that, if it were profitable, there would be immense pressure on politicians and society to gain access to the treasure.

For me, the problem is simple. Don't we realize that we have to dig more and more to maintain production for the absurd way of life that we lead? Wouldn't it be logical to think that before we start an assault on the bottom of the oceans we need to carry out a more detailed study of how to do so without disturbing deep and surface ecosystems? We are fleeing forward without thinking about the consequences of our actions.

Integrated Future

As we move forward with the construction of wind farms, we need to optimize the land that we are basically gaining from the sea. Back in the 1970s, I was so keen on collecting a particular science magazine that I went to out find it every week with religious punctuality. It was called *View to Know*, with magnificent illustrations and a ground-breaking layout, for its time. I saw this integration between sea farms and wind turbines as something for the remote future. Today, more than thirty years later, it is already becoming a reality, and part of our future may be to integrate ideas like this one, in which energy and food come out of a complex built in the middle of the sea.

How do we make the most of our gigantic installations dedicated to transforming wind energy into electricity? Dr Buck, of the Alfred Wegener Institute in Germany, and his collaborators make up one of the groups that devise the optimum use of space for the operation of offshore wind farms, those beyond 10 or 12 nautical miles, as are already beginning to operate in various parts of Northern Germany, Denmark or Great Britain. 'We are proposing to create infrastructure with certain types of mariculture products', explains Buck. 'Wind farms have very limited access, so it would be very feasible to set up sea farms using a large area between the perennial structures of the mills.' The first thing to consider in this integrative equation is what kind of species might be susceptible to cultivation. 'The crops would be exposed to very harsh ocean conditions,' continues Buck, 'so the facilities have to be very robust, equipped with technology designed to withstand wind and waves of several metres high.'

This is especially true in areas such as the North Sea, where, without the shelter of the coast and shallower areas, waves can easily reach 6 or 7 m in height. However, the turbines themselves could serve as perfect anchors and solid safeguards for these crops. The other factor is choosing the right species. Most species would not withstand such exposed conditions, and on the other hand they must be easy to maintain. The ideal ones are undoubtedly algae such as various types of laminaria or rhodophytes and active or passive filter feeders. Oysters and mussels, for example, would require minimal maintenance costs, as they would not have to be fed (they eat suspended particles). 'We have shown that the combination of wind power and mussels is not only viable but also very profitable', argues Robert Griffin of Stanford University: 'What we need to do is to achieve synergy between entrepreneurs with different interests and technical expertise.'

Indeed, the most fundamental aspect of these projects is their acceptance by those who have to invest in them. Despite the fact that in the area under German influence (both the North Sea and the Baltic), where there are already 47 offshore wind farms with some 80 to 500 turbines each, shellfish farmers are not entirely convinced of the profitability of this business. There are studies on carrying out this type of supplement, but they are still scarce (most of them developed by the Alfred Wegener Institute). Studies are needed to understand the degree of stress of the organisms exposed to this dynamic, as well as to understand, with multicriteria analysis (ecological, social, economic, etc.), if the market price for the species, their demand and the power to market them makes the business profitable overall.

However, everyone welcomes this possibility, because it is takes advantage of an space already 'doomed' for practices such as commercial or private vessels, fishing or recreation to produce more than just energy. 'Today, one of the major problems for aquaculture, which is becoming a real substitute for commercial fishing, is finding coastal areas where it can set up sea farms', concludes Buck. 'Wind farms are a unique opportunity to produce protein or other sea products on a very large scale, which are to be used to clear coastal areas compromised by natural parks, small-scale fishing, etc.'

25

Anything to Add?

Throughout the book I have tried to give some clues to a possible future model of management of our oceans. However, I cannot neglect to comment on some aspects that I have not dealt with, in addition to highlighting others that I believe must be stressed in order to make them clear. It is obvious to me that the problem of the oceans comes largely from our great ignorance of the subject, especially from the fact that we do not 'see' what lies beneath the surface of the sea. Degrading something without seeing what is being degraded is easier than it looks, because it's not 'bad': you don't worry so much. A few years ago when I saw a shopping trolley submerged at a depth of more than 20 m I realized that, in a sense, we had reached a state of passivity with regard to the marine environment, which was too large. How was it possible that someone could have intentionally thrown it in there, near the Cap de Begur, this element that was so foreign to the seabed? There was no evidence that it could have been transported by a flood, or anything like that. Therefore, anything that we say about ocean management must be based on one premise: the seas are not a sinkhole. They are not a place where everything disappears and life goes on. They have a limit.

It is true that scientists and technicians themselves have not contributed much to a clear vision of the seas, but neither have we. We lack a great deal of information. What happens is that, when we obtain it, it is difficult for us to spread it through society and to those who make decisions from a social and political point of view. For example, in the area of fishing, there is a clear lack of good lubrication between the various parties to increase (by a great deal) the necessary agility in decision-making. The mechanism is highly bureaucratic, and it works awkwardly through annual reports, grey literature and

© Springer Nature Switzerland AG 2019
S. Rossi, *Oceans in Decline*,
https://doi.org/10.1007/978-3-030-02514-4_25

scientific articles that have sometimes become obsolete by the time they come out. When this information concerns fishing quotas, apart from being deficient in many respects it is already out of date. In many cases, some of the landings or catches registered in ports do not take into account of certain sizes that are discarded from the by-catch or do not consider the by-kill, which are those organisms (including those that are the object of fishing) that die without ever reaching the vessel. 'In 40 years of herring exploitation in the North Sea, the average catch has been around 140,000 tonnes, but the by-catch has been around 87,000 tonnes… What a waste!' complains Robert Callum of the Environmental Department of New York University.

Our problem has been (and is) that we see the sea as an inexhaustible system, a place that regenerates everything. Until a couple of decades ago, the degradation of the same system through aggressive techniques of different types was not contemplated, nor were the changes in the availability of food and the transformation of food chains experiencing unbearable pressure. Only the data handled recently have made it clear that this is no longer the case, although even the best models of fishing or of the functioning of ocean ecosystems are only a pale reflection of the reality. Among the gambles for the future that have to be adopted in the shortest possible time are the reduction of fishing, avoiding fishing for what we do not know about (and therefore taking undue risks), making management programmes much more flexible, avoiding the most aggressive fishing gear or human interference, and not permitting huge by-catch or the degradation of systems when we have more than enough tools to prevent these (Fig. 25.1).

Those are fine words, but applying them is a different matter. As we have seen, the growth of Marine Protected Areas is inevitable and highly desirable, but we must remember that this has been proposed for more than a century and no one has paid much attention to it so far. As early as 1912, Marc Hérufel, an illustrious French scientist, spoke of the importance of managing the sea, creating zones of exclusion where human interference was minimal. If his advice had been followed, we would not be as we are. We are rediscovering his ideas a century later. We have a long way to go before we accept this perception and really implement it.

The people involved in fishing believe that, in the end, their models will work and we will be able to redress the situation. But time is running out. The sea can regenerate itself, a fact intrinsic to nature itself, but this recovery (which would benefit us all) must be done by weighing matters differently. Some scientists have already made it clear that 20–40% of all oceans should be protected without further ado by several types of legal mechanisms, especially in productive areas. In the end, the pyramid must be reversed,

Fig. 25.1 How we are transforming and impoverishing our oceans. Overfishing, seafloor destruction and the disappearance of both living three-dimensional structures (corals, plants, etc.) and big predators degrade the system and make it less resilient. *Source* Daniel Pauly

leaving little room for unprotected and unmanaged sites and much more for those that we want to preserve.

Returning to the concept that opened the chapter, the biggest problem is, without a doubt, our only partial knowledge of the workings of the sea's various ecosystems that we need to undertake rational discourse. 'Chaos theory has shown that complex systems (such as marine ecosystems) are by definition unpredictable', says Ferdinando Boero of the University of Salento in Italy. Perhaps we should not be so pessimistic, but the truth is that many of the models applied lack a solid database, which is essential to be able to execute the mathematics and statistics that they incorporate.

Today, the compartmentalization of science is taking us down a dangerous path, because there is a great deal of data but sometimes it is difficult to interconnect them. It is difficult to base a whole model only on what we see, because there are many of those compartments of the ecosystem that certainly play an important role but are still unknown or full of unknowns. Until not so long ago, we ignored the role of bacteria and other micro-organisms. We neglected the role of viruses, pathologies and parasites in the functioning of ecosystems, and we avoided 'hidden' phases of life cycles such as the existence of cysts or latent states of organisms. These are just some of the examples that lead us to the conclusion that there is a long way to go.

We have fewer and fewer excuses. Although we are still better able to carry out an analysis of what happens above the water than what happens under the it, because of obvious physical and chemical barriers, the technology exists. Take some examples. We have the technique and protocols to make acceptable maps of the seabed through multibeam probe surveys (devices that, through sound, make relief maps of the marine topography), Remotely Operated Vehicle (ROV) robots and manned submarines. These are expensive instruments, no doubt, but are increasingly affordable. However, there is a lack of coordination at the local, regional and national levels to make good seafloor map, especially in areas more than 50 m below the surface. We can also follow surface currents with oceanographic buoys, or at depth with current meters arranged on different levels from the bottom. We know how much matter falls to the bottom and take out evidence from the sediment to reconstruct the history of the sea and the planet. We can even apply satellite data to observe changes in sea and land temperatures, quantify soil moisture, provide data on chlorophyll and salinity of the oceans at the planetary level… this information can be (and is) processed in computer, chemical and image analysis laboratories. We are able to create fish-shaped robots that analyse certain elements or molecules such as oxygen to give us an idea of pollution, disturbances and seasonal changes in water. These robots are self-contained, can roam the sea for a specific period of time, occasionally surface to transmit data through a wireless information system and return to port when they feel that their batteries are running low…

Then why do we know so little? Why the hell aren't we applying it? I am going to give an example that I saw at a congress in 2009 and then let you draw your own conclusions. Speaker Beatriz Morales-Nin from the IEO of Mallorca presented a series of data on marine science in Spain. For a period, she was the area coordinator of the marine projects that were developed in our country both at national and European level. Well, despite the fact that 10% of Spain's gross national products comes directly from activities related to the sea (tourism, fishing, transport, etc.), the investment in finding out about the sea is, if I may say so, ridiculously small. Between 2000 and 2003, some €15 million was invested in different research and development projects distributed across 294 projects, while this figure fell to around €12.4 million in the period from 2003 to 2007. Projects related to diversity in the that period amounted to just under €2 million. With the economic crisis, the amount of money was decimated, and only now it seems that, very slowly, it is recovering...

Lack of support personnel, lack of resources, a poor attitude towards small to medium enterprises and large companies, lack of data structuring, lack of visibility... I know, we all complain about the shortcomings in our own field, but if we consider that between 2000 and 2007 less than €28 million had been invested (about €4 million a year for the whole of Spain, including universities, CSIC, IEO, etc.), the amount borders on the ridiculous. So, when we become full of talk about putting forward a plan for the 'salvation' of the oceans, we should not forget to speak about a monetary injection at the same time, one commensurate with the vastness of the ocean.

Anyway, let's not get our hopes up. There's many people out there who aren't up to the job. Environmental issues may be in vogue, but if they are not dealt with properly, if they are not given the rigour that they deserve, they come up against obstructions that are designed to put a spoke in the wheels. In the early 1970s and 1980s, the United States was at the forefront in addressing the planet's environmental problems, including those of those whose livelihoods depend on the sea. Today, the world's leading power is reluctant to sign up to any measure that will globally slow down its growth model. We could talk and talk about it, but the reality is that people don't want to hear about changes if it leads to 'trauma' to their acquired habits and well-being. From the mid-1970s onwards, dissenting voices began to emerge on various environmental issues, questioning concepts, results and proposals.

Interestingly, more than 90% of the books and articles critical of these environmentalist theses between 1975 and 2005 were linked to what is known as 'conservative think tanks', that is, thinkers or working groups linked to ideas deeply rooted in the more 'liberal economic' system. Doubt is planted in an implicit way, because science and scientists are not perfect. The 'think tanks' took advantage of these scientific doubts to plant the seed of negativity and encouraged people to ignore the ecological warnings. The strategy has worked, because it has achieved its goal. Now, with so many accumulated problems, many people are beginning to be immobilized with doubts about this type of issue, which is not going to help us at all when it comes to solving them.

On the one hand, this attitude puts scientists in an uncomfortable position, a situation in which they must become involved— 'get wet'—where they must express their opinions and make choices. It is then that the Cartesian mind of the academic must swim in unconventional waters, accustomed as it is to trial and error, to working with probabilities, to always doubting everything (a scientific character has snags). On the other hand, this strategy that we were talking about attracts those who want to hear what is best for them. It has been shown that people hear what they want to hear, and if a scientist explains something that, because of ideology, religious belief or any other premise, we do not agree with, we will tend to be sceptical.

A clear example of what I am trying to say is the climate change controversy, especially the awkward stance that made such a fuss days before the 2009 Copenhagen conference (which, by the way, was another failure). Contradictory emails, information taken out of context, data not properly validated... and sceptics having a picnic. Creating a little more confusion. Ideal! The reality is that, like all sciences, those who follow climate change have dark areas, gaps in knowledge and a lack of conclusive data in certain areas.

Is there climate change? Of course. Of course. I believe that, more than just looking at the past and making predictions with sometimes abstract and unintelligible models, it is enough to observe what we specialists are seeing: the present. A few sceptics base their criticism on the barely discernible effects of climate change compared to other human interference. To be honest, I think that this is stupid. The first thing is to acknowledge that there is a problem, then to separate the parts. The issue, in my view, is that many of the scientists who have come to the forefront on an issue such as climate change, where politicians in particular and society in general are so concerned, have tried to swim in the turbulent waters of the media and power brokers, of the factual powers.

No, gentlemen! What we have to do is to put on the table what we have, with all its faults and shortcomings, explain it well and discuss it in depth. But without wasting time. Because we don't have time to spare. If we still want to redress the situation, if we want to go for a different model in which we are an integral part of the system—and not just a parasite that feeds on it to the point of exhaustion and transforms it in a way that makes it unrecognizable even to our eyes—we must act immediately, without further hesitation.

The Value of the Naturalist

In part, I chose my profession for the pleasure of being in contact with the sea and the landscapes that the seabed offers. The years go by and one's daily research, report and article writing and miscellaneous bureaucracy mean that one has fewer and fewer opportunities to put on an air tank, grab the regulator and put on one's fins to descend in the water to track and observe the animals and plants under study. For me, it is a pleasure to be there, under-water, observing coral or counting fish, but the decrease in my contact with nature is, unfortunately, a process of natural succession. In this process I give way to the new generations to do what I did between the ages of twenty and thirty years old: to make the most of the sea from a point of view that is not only scientific but for the pure pleasure of observation. I have not given up this pleasure, and I hope to be able to spend many years enjoying seeing, live, what I am told through scientific articles: observing.

The naturalist, the observer, is much discredited in a society in which haste, the concreteness of results, the desire for answers and competitiveness prevail. Anything superfluous, anything that does not go 'straight to the point' and, if possible, 'right now' is considered useless or a hindrance. However, as many specialists agree, the vision of management through natural history is essential to address our current problems. 'In order to make a proper (and necessary) simplification of ecosystems and habitats, we have to make use of observation and a much more holistic view of nature', explains Paul Dayton of the Scripps Institution of Oceanography of the University of California: 'A model is needed in which the role of the naturalist is revalued, with the application of long-term observations and with the development of certain disciplines such as taxonomy or the biology and ecology of species that are currently very discredited.'

There are many scientists who do not even see what they are studying, who are dedicated to laboratory experiments or to developing complex models in which the direct observation of nature is conspicuous by its absence. During the last decades, branches such as genetics or theoretical ecology have acquired much strength (and resources) and, although they are essential in the social and scientific context in which we operate, they have tended to bury disciplines such as the analysis and classification of species. It is curious to see how the word biodiversity fills our ears when only about two million species have been adequately described, given that there may be an estimated five to fifty million species on the planet. An example of coordination between these latest generation genetic techniques and the concern to know the biodiversity of the planet is the discovery of hundreds of variants in bacterial groups of which until now little or nothing was known. The marine microbiological environment itself has seen a profound revision both in the number of new species and in the function of the different groups.

What is clear is that we have to learn to appreciate this kind of science again. There is a growing shortage of specialists capable of carrying out a holistic, comprehensive analysis of the system. The new generations are often marked by such a restricted line of research that it is difficult for them to understand beyond their small area of study. How can we distinguish natural disturbances from anthropogenic ones if we are not clear about how nature works? We talk about recovery and stability of habitats, but do we really know when they are stable, and when and how they can be recovered? In order to measure stress, we must be very clear about certain basic concepts and look at the system as a whole, not base our conclusions on box experiments or laboratory manipulations. 'We have to be able to ask the right questions', Dayton concludes: 'That way we'll be able to deal with problems with confidence.' Those questions always come from a deepening of the naturalists' observation of the environment. The new generations must be able to appreciate this type of approach, this way of approaching science, otherwise it will be difficult for us to manage our natural environment properly.

Educate, Transmit, Raise Awareness

I think that almost everyone in my age group (between forty and fifty years old) remembers with special affection those Sunday afternoons when we were in front of the television, towards the end of the 1970s, the chapters of the *Underwater World of Jacques Cousteau*. From the lack of decent programming in Spain (and with no margin for zapping, except with the famous UHF), this and similar programmes took the lead. We were hypnotized by jumping salmon, coral forests or the harsh conditions of Antarctica that a fragile *Calypso* had to weather, all intoned in the first person by the French Commander, with added voiceover, explaining the adventures of his team around the world. Cousteau, although it sounds like a cliché, introduced us to the sea, as Felix Rodríguez de la Fuente did with Iberian fauna, Carl Sagan with the principles of astronomy, David Attenborough with the various ecosystems of the planet and Gerald Durrell with the patient vision of the accomplished naturalist. It is not a question of claiming any place in this canon, I am just trying to reflect a little on how to help the better management of the oceans from an educational point of view. Many of the programmes that I mentioned were broadcast on prime-time television, while most of them are now relegated to secondary (and possibly loss-making) television channels and antisocial viewing hours. Nothing, in my opinion, is going to change the view of current television producers on programmes that don't sell, such as documentaries. However, explaining things in a correct yet entertaining way and with a solid knowledge base is not such a complex task and can attract audiences.

You have to know and show the problems so that they can be understood, and therefore solved. The first step is, without a doubt, the involvement of the scientist in the task of disseminating information. You don't have to write novels or become a science journalist like me, and you don't have to write essays about transformed seas, but society has the right to demand that you make an effort to make yourself understood in the best possible way. It is fundamental to speak directly and give clear explanations to the media, and the person speaking must be a valid interlocutor, someone capable of understanding what they are being told, someone with standards for not screwing up the orders of magnitude or giving the wrong units, someone who knows how to filter the information properly and then make it known in a level-headed, understandable way to a carpenter, a bus driver or a lawyer. On the one hand, scientists fail if they do not want to waste time talking to this person, or who believe that they already play a certain role and should not have to give so many explanations about their work: for this reason, they write in specialized journals and produce reports, don't they? This attitude, partly because of arrogance and partly because of a work overload, is not conducive to the transcendence of their specialism. In many cases, this does not facilitate their acceptance by society, which may not understand why it should continue to finance them on that basis. It is difficult to demonstrate that understanding the reproduction of gorgonians is essential in obtaining a full picture of their population dynamics and, therefore, of their ability to withstand natural or human disturbances affecting the population. And that these populations are essential, for example to maintain a state of health capable of hosting the fish or crustaceans that we eat. Therefore, even what we think is the most difficult thing to explain—what we are sure people give a damn about—must be made available to everyone in one way or another.

But I am not forgetting the other side, the journalist who has specialized in science and technology. On many occasions, there is a lack of rigorous preparation that allows him or her to go deeper into the subject, to sift through the contents, and to present an article that is both attractive and rigorous. Their lack of preparation can lead to misinterpretations, to the drawing of figures and conclusions out of context, to the proposal of alarmist titles. They are not alone: editors and subeditors may be responsible for the headline or edit that distorts the very truth of the written column for lack of space, either to attract the reader or because of their ignorance about the subject. Few newspapers, radio stations or television programmes are dedicated to science, and those that do so try to do so with great dignity, even under pressure in a market where they sell more with a visual or scandalous impact.

All people are targeted for disclosure, but who are the preferred ones? In my opinion, there are three areas in which more effort is needed, from different perspectives and with different methods: those who will live with the consequences of the problems that we seem to be failing to address (children and young people); those who pay taxes; and those who make political, social and economic decisions. I mean, everyone. But in different ways. I am not going to dwell on the first two groups: there are many initiatives, working groups and efforts that seem to be gradually changing the course of things. Ministry, regional and local administrations or the European Union are devoting more and more effort and budget to these initiatives. And, moreover, there is a civic and solidarity consciousness that is increasingly seeking another way of life, outside this capitalist and consumerist world that devours us at every step we take.

Yes, I would like to comment on something about politicians, businessmen, economists and sociologists. In my opinion, a simple course in ecology and biology (not just ecology) would make them understand why we can no longer function as we are. The assumption that the only possible economic and social model is that of continuous growth or expansion is based on a very human and unnatural conception of our environment. Nothing grows indefinitely. Explaining how an ecosystem works would make one understand why there can be no depletion of resources or management that degrades the environment so that the species (we) who inhabit it has to look for a supplementary one in a finite world. I am convinced that if an economist understood this, he or she would realize why the current model is simply destined for the most absolute failure, especially when transformations are happening so fast.

This is a far from simple task that will ultimately lead to a change of model that will seem to us radical and in many ways unpopular yet which we will have no choice but to adopt in the near future. But to be able to do this, the first to assimilate it must be the ones who pull the strings—the ones who make the decisions—and this can only be done through an adequate transmission of knowledge.

Epilogue

> '*We are going to the collapse...*'
> (Robert O'Neill, research professor at the
> Oak Ridge National Laboratory,
> United States, at the 1998
> INTECOL conference)

Why are we such fools? What leads us to destroy our own base of life? The answer is much simpler than it seems: as we have never stopped being animals, we follow the same mechanism as those around us. A lynx, in a simple predator–prey model, will eat as many rabbits as it is possible for it to capture. It does not think at any time that the prey will become extinct. When they do, the lynx goes hungry until the hare stock recovers. But sometimes it doesn't, and the lynx dies or migrates, just disappears from the place. We, despite our capacity for reason, our thousands of years of cultural experience, our hundreds of philosophers and thinkers, are applying the same mechanism. This is because we are animals, animals in a natural context. The difference lies in what was said at the beginning: we are able to change the planet, to govern its energy and matter flows, to alter the carbon or nitrogen cycle, to be decisive in Gaia's course. But we are not yet fully aware of this.

As much as eminent scientists like Simon Thrush of NIWA New Zealand insist that 'there is an urgent need to expand our understanding of our oceans in order to be able to manage them properly at the ecosystem level and not just at the species level to change the current destructive trend', our real vision is limited. The vast majority of us live in large cities and are increasingly far removed from the changes taking place around us. Things are seen from a distance, changes always affect others and, if we're talk about the sea, the concept is even further away.

Things are not going well, and we have to face the change that has been taking place in our natural systems for decades now as something to be faced

© Springer Nature Switzerland AG 2019
S. Rossi, *Oceans in Decline*,
https://doi.org/10.1007/978-3-030-02514-4

with difficult and courageous decisions. We must wait for no further results: if we want to maintain something of what we have and give regeneration a chance, we have no choice but to take a decisive step forward and radically change our concept of the future, going through a considerable transformation of our needs, our economy and our society. Among other things, conservationists themselves would have to lead by example, changing their habits, because otherwise the opportunities for regeneration will become more and more distant. 'We can only guess what kind of organisms will benefit from these profound changes due to both over-exploitation of species and climate change', says Jeremy Jackson, but it is also true that if there is a set of ecosystems that is able to regenerate, it is ocean ecosystems.

In order to change, we will have to stop and reverse many processes and trends, which will require a change in the model of fishing, agricultural activity and greenhouse gas emissions, but we must not forget that we are the only species that can affect the fate of the entire planet, either negatively or positively.

Synergy is the word that, in my opinion, best defines what is happening to our planet: a series of problems, of bad actions, of excesses, that work in a synergic way, never in an isolated way. More coordination between politicians, institutions and science, no longer afraid of a decisive change for our future, will mean that we can finally survive in a dignified way. Daniel Pauly of the Sea Around Us Project of Canada, like the overwhelming majority of fisheries and ocean conservation specialists, adds to this view and comments that the current crisis in fisheries and the degradation of marine ecosystems should be an opportunity to renew our vision of what is and is not exploitable, and how to do it properly. Because it is not that we will not survive it, but that the future looming on the horizon is grey, lacking in variety, poor in nuances, monotonous and not without dangers. We are not going to become extinct, but I have the impression that we will live worse and worse, more and more locked up in a bubble of resistance that prevents us from enjoying who we are: part of the environment.

We may become a sad and miserable species on a planet made exclusively for us and, in that case, it would be no surprise that future generations if cursed us for having left such a mess, to which they will have no choice but to adapt. When I analyse the steps to be taken, I realize that it is not as complicated as it seems. A colleague of mine at the Institute of Environmental Sciences, Jeroen van den Bergh, once explained to me that 'making the right environmental policy to correct mistakes is not as expensive as people think; there are many ways to change trends'. What really scares me is when the three E's cross in synergy, as that can lead to a real collapse: creating an Energy, Ecological and Economic (EEA) crisis at the same time. At that moment, when it starts to squeeze us to the point of drowning, will we be able to use our capacity for reasoning to agree on something as essential as maintaining a dignified life?

Bibliography

General

Barnes, R. S. K., & Hughes, R. N. (1999). *An introduction to marine ecology*. Oxford: Blackwell Science Ltd.

Broad, W. J. (1997). *The universe below*. New York: Touchstone.

Clover, C. (2006). *The end of the line*. California, US: University of California Press.

Dudley, N., et al. (1996). *Bad harvest?*. London: Earthscan Publications Ltd.

Kinderley, D. (2007). *Océanos*. Madrid, Spain: Pearson Education, SA.

Koslow, T. (2007). *The silent deep*. Chicago: University of Chicago Press.

Kurlansky, M. (2008). *The last fish tale*. England: Penguin Books Ltd.

Margalef, R. (1997). *Our biosphere*. Oldendorf/Luhe, Germany: Ecology Institute.

Pimm, S. L. (1991). *Balance of nature?*. Chicago: University of Chicago Press.

Rossi, S. (2013). The destruction of the 'animal forests' in the oceans: Towards an over-simplification of the benthic ecosystems. *Ocean & Coastal Management, 84*, 77–85.

Chapter 1

Beerling, D. (2007). *The emerald planet*. Oxford, England: Oxford University Press.

Chyba, C. F. (2005). Rethinking Earth's early atmosphere. *Science, 308*, 962–963.

Gili, J. M., et al. (2006). A unique assemblage of epibenthic sessile suspension feeders with archaic features in the high-Antarctic. *Deep-Sea Research Part II, 53*, 1029–1052.

Hurtgen, T. M. (2003). Ancient oceans and oxygen. *Nature, 423*, 592–593.

Kasting, J. F., & Siefert, J. L. (2002). Life and the evolution of Earth's Atmosphere. *Science, 296*, 1066–1068.

Kiene, R. P. (2008). Genes in the glass house. *Nature, 456*, 180–181.

Redfern, R. (2001). *Origins*. Oklahoma, US: University of Oklahoma Press & Cassel-Co.

Saito, M. A. (2009). Less nickel for more oxygen. *Nature, 458*, 714–715.

Veizer, J. (2005). Celestial climate driver: A perspective from four billion years of the carbon cycle. *Geoscience Canada, 32*, 13–28.

Chapter 2

Hughes, T. P. (1994). Catastrophes, phase shifts, and large-scale degradation of a Caribbean coral reef. *Science, 265,* 1547–1551.

Jackson, J. B. C. (1997). Reefs since Columbus. *Coral Reefs, 16,* S23–S32.

Jackson, J. B. C. (2001). What was natural in the coastal oceans? *PNAS, 98,* 5411–5418.

Jackson, J. B. C., Budd, A. F., & Pandofi, J. M. (1996). The shifting balance of natural communities? In D. Jablouski, D. H. Erwin, & J. H. Lipps (Eds.), *Evolutionary paleobiology* (pp. 89–122). Chicago: University of Chicago Press.

Jackson, J. B. C., Kirby, M. X., Berger, W. H., et al. (2001). Historical overfishing and the recent collapse of coastal ecosystems. *Science, 293,* 629–638.

Jiménez, C., & Orejas, C. (2017). The builders of the oceans—Corals from the past to the present: The stone from the sea. In S. Rossi, L. Bramanti, A. Gori, & C. Orejas (Eds.), *Marine animal forests.* Switzerland: Springer International Publishing. https://doi.org/10.1007/978-3-319-17001-5_31-1.

Knowlton, N. (1992). Thresholds and multiple stable states in coral reef community dynamics. *American Zoologist, 32,* 674–682.

Lotze, H. K. (2005). Radical changes in the Wadden Sea fauna and flora over the last 2000 years. *Helgoland Marine Research, 59,* 71–83.

Officer, C. B., Biggs, R. B., Taft, J. L., Cronin, L. E., Tyler, M. A., & Boynton, W. R. (1984). Chesapeake bay anoxia: Origin, development, and significance. *Science, 223,* 22–27.

Osborne, A. H., Vance, D., Roholing, E. J., Barton, N., Rogerson, M., & Fello, N. (2008). A humid corridor across the Sahara for the migration of early modern humans out of Africa 120,000 years ago. *PNAS, 105,* 16444–16447.

Pitcher, T. J. (2001). Fisheries managed to rebuild ecosystems? Reconstructing the past to salvage the future. *Ecological Applications, 11,* 601–617.

Roberts, C. (2007). *The unnatural history of the sea.* Washington, US: Island Press.

Sandin, S. A., Smith, J. E., De Martin, E. E., et al. (2008). Baselines and degradation of coral reefs in the Northern Line Islands. *PlosOne, 3,* 1548.

Thurstan, R. H., Pandolfi, J. M., & zu Ermgassen, P. S. E. (2017). Animal forests through time: Historical data to understand present changes in marine ecosystems. In S. Rossi, L. Bramanti, A. Gori, & C. Orejas (Eds.), *Marine animal forests* (pp. 947–964). Switzerland: Springer International Publishing. https://doi.org/10.1007/978-3-319-17001-5_31-1.

Chapter 3

Carlton, J. T., Geller, J. B., & Reaka-Kuda, M. L. (1999). Historical extinctions in the sea. *Annual Review of Ecology and Systematics, 30,* 515–538.

Estes, J. A., & Palmisano, J. F. (1974). Sea otters: Their role in structuring nearshore communities. *Science, 185,* 1058–1060.

Estes, J. A., Tinker, M. T., Williams, T. M., & Doak, D. F. (1998). Killer whale predation on sea otters linking oceanic and nearshore ecosystems. *Science, 282,* 473–476.

Güçlüsoy, H. (2008). Damage by monkey seals to gear of the artisanal fishery in the Foça Monk seal pilot conservation area, Turkey. *Fisheries Research, 90,* 70–77.

Parrish, F. (2009). Do monk seals exert top-down pressure in subphotic ecosystems? *Marine Mammal Science, 25,* 91–106.

Serjeantson, D. (2001). The great auk and the gannet: A prehistoric perspective on the extinction of the great auk. *International Journal of Osteoarchaeology, 11,* 43–55.

Chapter 4

Assenburg, C., Harwood, J., Matthiopoulos, J., & Smout, D. (2006). The functional response of generalist predators and its implications for the monitoring of marine ecosystems. In I. L. Boyd & S. Wanless (Eds.), *Management of marine ecosystems: Monitoring changes in upper trophic levels* (pp. 262–274). London, UK: Zoological Society of London.

Bundy, A., Heymans, J. J., Morissette, L., & Savenkoff, C. (2009). Seals, cod and forage fish: A comparative exploration of variations in the theme of stock collapse and ecosystem change in four Northwest Atlantic ecosystems. *Progress in Oceanography, 81*, 188–206.

Kurlansky, M. (1997). *Cod*. England: Penguin Books Ltd.

Myers, R. A., Hutchings, J. A., & Barrowman, N. J. (1996). Hypothesis for the decline of cod in the North Atlantic. *Marine Ecology Progress Series, 138*, 293–308.

Myers, R. A., Hutchings, J. A., & Barrowman, N. J. (1997). Why do fish stocks collapse? The example of cod in Atlantic Canada. *Ecological Applications, 7*, 91–106.

Pauly, D. (1995). Anecdotes and the shifting baseline syndrome of fisheries. *Trends in Ecology & Evolution, 10*, 430.

Punt, A. E., & Butterworth, D. S. (1995). The effects of future consumption by the Cape fur seal on catches and catch rates of the Cape hakes. 4. Modelling the biological interaction between Cape fur seals *Arctocephalus pusillus pusillus* and the Cape hakes *Merluccius capensis* and *M. paradoxus*. *South African Journal of Marine Science, 16*, 255–285.

Rose, G. A. (2004). Reconciling overfishing and climate change with stock dynamics of Atlantic cod (*Gadus morhua*) over 500 years. *Canadian Journal of Fisheries and Aquatic Sciences, 61*, 1553–1557.

Rose, G. A., & O'Driscoll, R. L. (2002). Capelin are good for cod: Can the northern stock rebuild without them? *ICES Journal of Marine Science, 59*, 1018–1026.

Sinclair, A. F., & Murawski, S. A. (1997). Why have groundfish stocks declined? Chapter 3. In J. Boreman, B. S. Nakashima, J. A. Wilson, & R. L. Kendall (Eds.), *Northwest Atlantic groundfish: Perspectives on a fishery collapse* (pp. 71–93). Bethesda, Maryland: American Fisheries Society.

Trzinski, M. K., Mohn, R., & Bowen, W. D. (2006). Continued decline of an Atlantic cod population: How important is gray seal predation? *Ecological Applications, 16*, 2276–2292.

Yoneda, M., & Wright, P. J. (2004). Temporal and spatial variation in reproductive investment of Atlantic cod *Gadus morhua* in the northern North Sea and Scottish west coast. *Marine Ecology Progress Series, 276*, 237–248.

Chapter 5

Clapham, P. J., Young, S. B., & Brownell, R. L., Jr. (1999). Baleen whales: Conservation issues and the status of the most endangered populations. *Mammal Review, 29*, 35–60.

Fowler, C. W. (1987). A review of density dependence in populations of large mammals. *Current Mammalogy, 1*, 401–441.

Frantzis, A. (1998). Does acoustic testing strand whales? *Nature, 392*, 29.

Harvell, C. D., et al. (2002). Climate warming and disease risks for terrestrial and marine biota. *Science, 296*, 2159–2162.

Josephson, E., Smith, T. D., & Reeves, R. R. (2008). Historical distribution of right whales in the North Pacific. *Fish and Fisheries, 9*, 155–168.

Kasuya, T. (2008). Cetacean biology and conservation: A Japanese scientist's perspective spanning 46 years. *Marine Mammal Science, 24*, 749–773.

Konishi, K., Tamura, T., Zenitani, R., Bando, T., Kato, H., & Walloe, L. (2008). Decline in energy storage in the Antarctic minke whale (*Balaenoptera bonaerensis*) in the Southern Ocean. *Polar Biology, 31,* 1509–1520.

Lee, M. Y. (2010). Economic tradeoffs in the Gulf of Maine ecosystem: Herring and whale watching. *Marine Policy, 34,* 156–162.

Lukoschek, V., Funahashi, N., Lavery, S., Dalebout, M. L., Cipriano, F., & Baker, C. S. (2009). High proportion of protected minke whales sold on Japanese markets due to illegal, unreported or unregulated exploitation. *Animal Conservation, 12,* 385–395.

Morell, V. (2009). Mystery of the missing humpbacks solved by Soviet data. *Science, 324,* 1132.

Ragen, T. J., Huntington, H. P., & Hovelsrud, G. K. (2008). Conservation of Arctic marine mammals faced with climate change. *Ecological Applications, 18,* S166–S174.

Robinson, R. A., et al. (2008). Travelling through a warming world: Climate change and migratory species. *Endangered Species Research.* Published online 17 June 2008.

Simmonds, M. P., & Eliott, W. J. (2009). Climate change and the cetaceans: Concerns and recent developments. *Journal of the Marine Biological Association of the United Kingdom, 89,* 203–210.

Springer, A. M., Estes, J. A., van Vliet, G. B., Williams, T. M., et al. (2003). Sequential megafaunal collapse in the North Pacific Ocean: An ongoing legacy of industrial whaling? *PNAS, 100,* 12223–12228.

Tormosov, D. D., et al. (1998). Soviet catches of southern right whales, *Eubalaena australis,* 1951–1971: Biological data and conservation implications. *Biological Conservation, 86,* 185–197.

Whipple, A. B. C. (1979). *The Whalers* (176 pp). Alexandria, VA: Time-Life Books.

Chapter 6

Arntz, W. E., et al. (2006). El Niño and similar perturbation effects on the benthos of the Humboldt, California, and Benguela current upwelling ecosystems. *Advances in Geosciences, 6,* 243–265.

Asencio, C. (2008). Pesca. In Ministerio de Medio Ambiente, Medio Rural y Marino-Secretaría General del Mar-Dirección General de Sostenibilidad de la Costa y del Mar (Eds.), *Actividades Humanas en los mares de España.* Madrid, Spain.

Bertrand, S., Dewitte, B., Tam, J., Díaz, E., & Bertrand, A. (2008). Impacts of Kelvin wave forcing in the Perú Humboldt current system: Scenarios of spatial reorganizations from physics to fishers. *Progress in Oceanography, 79,* 278–289.

Borja, A., Fontán, A., Sáez, J., & Valencia, V. (2008). Climate, oceanography, and recruitment: The case of the Bay of Biscay anchovy (*Engraulis encrasicolus*). *Fisheries Oceanography, 17,* 477–493.

Botsford, L. W., Castilla, J. C., & Peterson, C. H. (1997). The management of fisheries and marine ecosystems. *Science, 277,* 509–515.

Cahuin, S. M., Cubillos, L. A., Ñiquen, M., & Escribano, R. (2009). Climatic regimes and the recruitment rate of anchoveta, *Engraulis ringens,* off Peru. *Estuarine, Coastal and Shelf Science, 84,* 591–597.

Coll, M., Palomera, I., Tudela, S., & Sarda, F. (2006). Trophic flows, ecosystem structure and fishing impact in the South Catalan Sea, North Western Mediterranean. *Journal of Marine Systems, 59,* 63–96.

FAO. (2016). *The state of world fisheries and aquaculture 2014.* Rome, Italy: FAO Fisheries Department.

Heino, M., & Dieckmann, U. (2009). Fisheries-induced evolution. In *Encyclopedia of life sciences (ELS)* (pp. 1–7). Wiley: Chichester.

Hibberd, T., & Pecl, G. (2007). Effects of commercial fishing on the population structure of spawning southern calamary (*Sepioteuthis australis*). *Reviews in Fish Biology and Fisheries, 17*, 207–221.

Hutchings, J. A. (2000). Collapse and recovery of marine fishes. *Nature, 406*, 882–885.

Le Floc'h, P., Poulard, J. C., Thébaud, O., Blanchard, F., Bihel, J., & Steinmetz, F. (2008). Analyzing the market position of fish species subject to impact of long-term changes: A case study of French fisheries in the Bay of Biscay. *Aquatic Living Resources, 21*, 307–316.

Lleonart, J., & Maynou, F. (2003). Fish stock assessments in the Mediterranean: State of the art. *Scientia Marina, 67*, 37–49.

Lloret, J., Palomera, I., Salat, J., & Sole, I. (2004). Impact of freshwater input and wind on landings of anchovy (*Engraulis encrasicolus*) and sardine (*Sardina pilchardus*) in shelf waters surrounding the Ebre River delta (North Western Mediterranean). *Fisheries Oceanography, 13*(2), 102–110.

Myers, R. A., & Mertz, G. (1997). The limits of exploitation: A precautionary approach. *Ecological Applications, 8*, S165–S169.

Palomera, I., Olivar, M. P., Salat, J., Coll, M., García, A., & Morales-Nin, B. (2007). Small pelagic fish in the NW Mediterranean Sea: An ecological review. *Progress in Oceanography, 74*, 377–396.

Pecl, G. T., & Jackson, G. D. (2008). The potential impacts of climate change on inshore squid: Biology, ecology and fisheries. *Reviews in Fish Biology and Fisheries, 18*, 373–385.

Pierce, G. J., et al. (2008). A review of cephalopod-environment interactions in European Seas. *Hydrobiologia, 612*, 49–70.

Rice, J., & Smith, D. M. (2017). Ecosystem-based management: Opportunities and challenges for application in the ocean forest. In S. Rossi, L. Bramanti, A. Gori, & C. Orejas (Eds.), *Marine animal forests* (pp. 965–988). Switzerland: Springer International Publishing. https://doi.org/10.1007/978-3-319-17001-5_26-1.

Tacon, G. C., & Metian, M. (2009). Fishing for feed or fishing for food: Increasing global competition for small pelagic forage fish. *Ambio, 38*, 294–302.

Chapter 7

Arrizabalaga, H., Restrepo, V. R., Maunder, M. N., & Majkowski, J. (2009). Using stock assessment information to assess fishing capacity of tuna fisheries. *ICES Journal of Marine Science, 66*, 1959–1966.

Baum, J. K., et al. (2003). Collapse and conservation of shark populations in the Northwest Atlantic. *Science, 299*, 389–392.

Dulvy, N. K., et al. (2008). You can swim but you can't hide: The global status and conservation of oceanic pelagic sharks and rays. *Aquatic Conservation: Marine and Freshwater Ecosystems, 18*, 459–482.

Ellis, R. (2008). *Tuna: A love story.* New York, US: Knopf, Random House inc.

Fromentin, J. M. (2009). Lessons from the past: Investigating historical data from bluefin tuna fisheries. *Fish and Fisheries, 10*, 197–216.

Game, E. T., et al. (2009). Pelagic protected areas: The missing dimension in ocean conservation. *TREE, 24*, 360–369.

Ganzedo, U., et al. (2009). What drove tuna catches between 1525 and 1756 in southern Europe? *ICES Journal of Marine Science, 66*, 1595–1604.

ICES (International Council for the Exploration of the Sea). (2002). *Report of the Working Group on Environmental Interactions of Mariculture.* ICES CM 2002/F:04, Copenhagen, Denmark, 105 pp.

Masuma, S., Miyashita, S., Yamamoto, H., & Kumai, H. (2008). Status of bluefin tuna farming, broodstock management, breeding and fingerling production in Japan. *Reviews in Fisheries Science, 16,* 385–390.

Miller, A. M. M., & Bush, S. R. (2015). Authority without credibility? Competition and conflict between ecolabels in tuna fisheries. *Journal of Cleaner Production, 107,* 137–145.

Myers, R. A., & Worm, B. (2003). Rapid worldwide depletion of predatory fish communities. *Nature, 423,* 280–283.

Myers, R. A., & Worm, B. (2005). Extinction, survival or recovery of large predatory fishes. *Philosophical Transactions of the Royal Society B, 360,* 13–20.

Myers, R. A., & Worm, B. (2009). Extinction, survival or recovery of large predatory fishes. *Philosophical Transactions of the Royal Society of London B: Biological Sciences, 360,* 13–20.

Pala, C. (2009). Protecting the last great tuna stocks. *Science, 324,* 1133.

Parkers, R. W. R., & Tyedmers, P. H. (2015). Fuel consumption of global fishing fleets: Current understanding and knowledge gaps. *Fish and Fisheries, 16,* 684–696.

Pereda, P., & Báez, J. C. (2008). Sobreexplotación de recursos pesqueros. In Ministerio de Medio Ambiente, Medio Rural y Marino-Secretaría General del Mar-Dirección General de Sostenibilidad de la Costa y del Mar (Eds.), *Actividades humanas en los mares de España.* Madrid, Spain.

Rott, E., Pipp, E., Pfister, P., Van Dam, H., Ortler, K., Binder, N., & Pall, K. (1999). *Indikationslisten für Aufwuchsalgen in österreichischen Fliessgewässern. part 2: Trophieindikation (sowie geochemische Präferenzen, taxonomische und toxikologische Anmerkungen).* Wasserwirtschaftskataster, Bundesministerium f. Land- u. Forstwirtschaft, Vienna.

Shiffman, D. S., & Hammerschlag, N. (2015). Preferred conservation policies of shark researchers. *Conservation Biology, 30,* 805–815.

Siskey, M. R., Wilberg, M. J., Allman, R. J., Barnett, B. K., & Secor, D. H. (2016). Forty years of fishing: Changes in age structure and stock mixing in northwestern Atlantic bluefin tuna (*Thunnus thynnus*) associated with size-selective and long-term exploitation. *ICES Journal of Marine Science.* https://doi.org/10.1093/icesjms/fsw115.

Stevens, J. D., Bonfil, R., Dulvy, N. K., & Walker, P. A. (2000). The effects of fishing on sharks, rays, and chimaeras (chondricthyans), and the implications for marine ecosystems. *ICES Journal of Marine Science, 57,* 476–494.

Walker, P. A., & Heessen, H. J. L. (1996). Long-term changes in ray populations in the North Sea. *ICES Journal of Marine Science, 53,* 1085–1093.

Walker, T. I. (1998). Can shark resources be harvested sustainably? A question revisited with a review of shark fisheries. *Marine and Freshwater Research, 49*(7), 553–572.

Ward, P., Myers, R. A., & Blanchard, W. (2004). Fish lost at sea: The effect of soak time on pelagic longline catches. *Fishery Bulletin, 102,* 179–195.

Watson, J. T., Essington, T. E., Lennert-Cody, C. E., & Hall, M. A. (2008). Trade-offs in the design of fishery closures: Management of silky shark bycatch in the eastern Pacific Ocean tuna fisheries. *Conservation Biology, 23,* 626–635.

Wilson, S. G., et al. (2005). Movements of bluefin tuna (*Thunnus thynnus*) in the northwestern Atlantic Ocean recorded by pop-up satellite archival tags. *Marine Biology, 146,* 409–423.

Wu, R. S. S. (1995). The environmental impact of marine fish culture: Towards a sustainable future. *Marine Pollution Bulletin, 31,* 159–166.

WWF Report. (2008). *Race for the last bluefin tuna.* WWF for a living planet. Report published by WWF Mediterranean, March, 126 pp.

Chapter 8

Aguilar, R., Perry, A., & López, J. (2017). Conservation of animal forests: Slowly on the way, though only just beginning. In S. Rossi, L. Bramanti, A. Gori, & C. Orejas (Eds.), *Marine animal forests* (pp. 1165–1208). Switzerland: Springer International Publishing.

Althaus, F., et al. (2009). Impacts of bottom trawling on deep-coral ecosystems of seamounts are long-lasting. *Marine Ecology Progress Series, 397,* 279–294.

Baillon, S., Hamen, J. F., Wareham, V. E., & Mercier, A. (2012). Deep cold-water corals as nurseries for fish larvae. *Frontiers in Ecology and the Environment.* https://dx.doi.org/10.1890/120022.

Bouderesque, C. F., et al. (2009). Regression of Mediterranean seagrasses caused by natural processes and anthropogenic disturbances and stress: A critical review. *Botanica Marina, 52,* 395–418.

Cau, A., Follesa, M. C., Moccia, D., Bellodi, A., Mulas, A., Bo, M., et al. (2016). *Leiopathes glaberrima* millennial forest from SW Sardinia as nursery ground for the small spotted catshark *Scyliorhinus canicula*. *Aquatic Conservation: Marine and Freshwater Ecosystems.* https://doi.org/10.1002/aqc.2717.

Coll, M., Lotze, H. K., & Romanuk, T. (2008). Structural degradation in Mediterranean Sea food webs: Testing ecological hypotheses using stochastic and mass balance modelling. *Ecoystems, 11,* 939–960.

Coll, M., Palomera, I., Tudela, S., & Dowd, M. (2008). Food-web dynamics in the South Catalan Sea ecosystem (NW Mediterranean) for 1978–2003. *Ecological Modelling, 217,* 95–116.

Company, J. B., et al. (2008). Climate control on deep-sea fisheries. *PloS One, 1*(e1431), 1–8.

Costa, M. E., Erzini, K., & Borges, T. C. (2008). Bycatch of crustacean and fish bottom trawl fishes from southern Portugal (Algarve). *Scientia Marina, 72,* 801–814.

Garstang, W. (1904). The impoverishment of the sea. *Journal of the Marine Biological Association of the United Kingdom, 6*(1), 1–57.

Hall-Spencer, J., Allain, V., & Fosså, J. H. (2002). Trawling damage to Northeast Atlantic ancient coral reefs. *Proceedings of the Royal Society of London B: Biological Sciences, 269,* 507–511.

Henry, L. A., & Roberts, M. (2017). Global biodiversity in cold-water coral reef ecosystems. In S. Rossi, L. Bramanti, A. Gori, & C. Orejas (Eds.), *Marine animal forests* (pp. 235–256). Switzerland: Springer International Publishing. https://doi.org/10.1007/978-3-319-17001-5_6-1.

Hinz, H. (2017). Impact of bottom fishing on animal forests: Science, conservation, and fisheries management. In S. Rossi, L. Bramanti, A. Gori, & C. Orejas (Eds.), *Marine animal forests* (pp. 1041–1060). Switzerland: Springer International Publishing. https://doi.org/10.1007/978-3-319-17001-5_37-1.

Jennings, S., Dinmore, T. A., Duplisea, D. E., Warr, K. J., & Lancaster, J. E. (2001). Trawling disturbance can modify benthic production processes. *Journal of Animal Ecology, 70,* 459–475.

Kaiser, M. J. (1998). Significance of bottom-fishing disturbance. *Conservation Biology, 12,* 1230–1235.

Koslow, J. A., et al. (2001). Seamount benthic macrofauna off southern Tasmania: Community structure and impacts of trawling. *Marine Ecology Progress Series, 213,* 111–125.

Lumsden, S. E., Hourigan, T. F., Brickner, A. W., & Dorr, G. (Eds.). (2007). *The state of deep coral ecosystems of the United States.* NOAA Technical Memorandum CRCP-3. Silver Spring MD.

Marliave, J. B., Conway, K. W., Gibbs, D. M., Lamb, A., & Gibbs, C. (2009). Biodiversity and rockfish recruitment in sponge gardens and bioherms of southern British Columbia, Canada. *Marine Biology, 156,* 2247–2254.

Morisette, L., Pedersen, T., & Nilsen, M. (2009). Comparing pristine and depleted ecosystems: The Sorfjord, Norway versus the Gulf of St. Lawrence, Canada. Effects of intense fisheries on marine ecosystems. *Progress in Oceanography, 81,* 174–187.

Mortensen, P. B., et al. (2005). Effects of fisheries on deepwater gorgonian corals in the Northeast Channel, Nova Scotia. *American Fisheries Society Symposium, 41,* 369–382 (P. W. Barnes & J. P. Thomas (Eds.), *Benthic habitats and the effects of fishing*).

Orejas, C., et al. (2009). Cold-water corals in the Cap de Creus canyon, northwestern Mediterranean: Spatial distribution, density and anthropogenic impact. *Marine Ecology Progress Series, 397,* 37–51.

Pauly, D., et al. (1998). Fishing down marine foodwebs. *Science, 279,* 860–863.

Pilskaln, C. H., Churchill, J. H., & Mayer, L. M. (1998). Resuspension of sediment by bottom trawling in the Gulf of Maine and potential geochemical consequences. *Conservation Biology, 12,* 1223–1229.

Pitcher, T. J. (2001). Fisheries managed to rebuild ecosystems? Reconstructing the past to salvage the future. *Ecological Applications, 11,* 601–617.

Prena, J., et al. (1999). Experimental otter trawling on a sandy bottom ecosystem of the Grand Banks of Newfoundland: Analysis of trawl bycatch and effects on epifauna. *Marine Ecology Progress Series, 181,* 107–124.

Puig, P., Canals, M., Martín, J., Amblas, D., Lastras, G., Palanques, A., & Calafat, A. M. (2012). Ploughing the deep sea floor. *Nature, 489,* 286–289.

Puig, P., Martín, J., Masqué, P., & Palanques, A. (2015). Increasing sediment accumulation rates in La Fonera (Palamós) submarine canyon axis and their relationship with bottom trawling activities. *Geophysical Research Letters, 42,* 8106–8113.

Rossi, S. (2013). The destruction of the 'animal forests' in the oceans: Towards an over-simplification of the benthic ecosystems. *Ocean & Coastal Management, 84,* 77–85.

Sánchez, P., Demestre, M., & Martín, P. (2004). Characterization of the discards generated by bottom trawling in the northwestern Mediterranean. *Fisheries Research, 67,* 71–80.

Sardà, F., & Maynou, F. (1998). Assessing perceptions: Do Catalan fishermen catch more shrimp on Fridays? *Fisheries Research, 36,* 149–157.

Thrush, S. F., et al. (2001). Fishing disturbance and marine biodiversity: The role of habitat structure in simple soft-sediment systems. *Marine Ecology Progress Series, 221,* 255–264.

Thrush, S. F., & Dayton, P. K. (2002). Disturbance to marine habitats by trawling and dredging: Implications for marine biodiversity. *Annual Review of Ecology and Systematics, 33,* 449–473.

Walting, L., & Norse, E. A. (1998). Disturbance of the seabed by mobile fishing gear: A comparison to forest clearcutting. *Conservation Biology, 12,* 1180–1197.

Wouters, N., & Cabral, H. N. (2009). Are flatfish nursery grounds richer in benthic prey? *Estuarine, Coastal and Shelf Science, 83,* 613–620.

Chapter 9

Arlinghaus, R., Matsumura, S., & Dieckmann, U. (2009). Quantifying selection differentials caused by recreational fishing: Development of modeling framework and application to reproductive investment in pike (*Esox lucius*). *Evolutionary Applications, 2,* 335–355.

Brechtol, W. R., & Kruse, G. H. (2009). Reconstruction of historical abundance and recruitment of red king crab during 1960–2004 around Kodiak, Alaska. *Fisheries Research, 100,* 86–98.

Cinner, J. E., et al. (2009). Linking social and ecological systems to sustain coral reef fisheries. *Current Biology, 19,* 206–212.

Coleman, F., et al. (2004). The impact of United States recreational fisheries on marine fish populations. *Science, 305,* 1958–1959.

Coll, J., Linde, M., García-Rubies, A., Riera, F., & Grau, A. M. (2004). Spear fishing in the Balearic Islands (west central Mediterranean): Species affected and catch evolution during the period 1975–2001. *Fisheries Research, 70,* 97–111.

Colloca, F., Crespi, V., Cerasi, S., & Coppola, S. R. (2004). Structure and evolution of the artisanal fishery in a southern Italian coastal area. *Fisheries Research, 69,* 359–369.

Dalzell, P., Adams, T. J. H., & Polunin, N. V. C. (1996). Coastal fisheries in the Pacific Islands. *Oceanography and Marine Biology: An Annual Review, 34,* 395–531.

Fox, H. E. (2004). Coral recruitment in blasted and unblasted sites in Indonesia: Assessing rehabilitation potential. *Marine Ecology Progress Series, 269,* 131–139.

Guidetti, P., Fraschetti, S., Terlizzi, A., & Boero, F. (2003). Distribution patterns of sea urchins and barrens in shallow Mediterranean rocky reefs impacted by the illegal fishery of the rock-boring mollusc *Lithophaga lithophaga. Marine Biology, 143,* 1135–1142.

Guillemot, N., Léopold, M., Cuif, M., & Chabanet, P. (2009). Characterization and management of informal fisheries confronted with socio-economic changes in New Caledonia (South Pacific). *Fisheries Research, 98,* 51–61.

Hardt, M. J. (2009). Lessons from the past: The collapse of Jamaican coral reefs. *Fish and Fisheries, 10,* 143–158.

Kronen, M. (2004). Fishing for fortunes? A socio-economic assessment of Tonga's artisanal fisheries. *Fisheries Research, 70,* 121–134.

Merino, G., Morales-Nin, B., Maynou, F., & Grau, A. M. (2008). Assessment and bioeconomic analysis of the Majorca (NW Mediterranean) trammel net fishery. *Aquatic Living Resources, 21,* 99–107.

Morales-Nin, B., et al. (2005). The recreational fishery off Majorca Island (western Mediterranean): Some implications for coastal resource management. *ICES Journal of Marine Science, 62,* 727–739.

Pais, A., et al. (2007). The impact of commercial and recreational harvesting for *Paracentrotus lividus* on shallow rocky reef sea urchin communities in North-Western Sardinia, Italy. *Estuarine, Coastal and Shelf Science, 73,* 589–597.

Queiroz, L., et al. (2017). Neglected ecosystem services: Highlighting the socio-cultural perception of mangroves in decision-making processes. *Ecosystem Services, 26,* 137–145.

Reglero, P., & Morales-Nin, B. (2008). Relationship between first sale price, body size and total catch of trammel net target species in Majorca (NW Mediterranean). *Fisheries Research, 92,* 102–106.

Wells, S. (2009). Dynamite fishing in northern Tanzania-Pervasive, problematic and yet preventable. *Marine Pollution Bulletin, 58,* 20–23.

Zeller, D., et al. (2008). What about recreational catch? Potential impact on stock assessment for Hawaii's bottom fish fisheries. *Fisheries Research, 91,* 88–97.

Chapter 10

Arai, M. N. (2001). Pelagic coelenterates and eutrophication: A review. *Hydrobiologia, 451,* 69–87.

Condon, R. H., Lucas, C. H., Pitt, K. A., & Uye, S.-I. (2014). Jellyfish blooms and ecological interactions. *Marine Ecology Progress Series, 510,* 109–110.

Graham, W. M., Pagés, F., & Hammer, W. M. (2001). A physical context for gelatinous zooplankton aggregations: A review. *Hydrobiologia, 451,* 199–212.

Hay, S. (2006). Marine ecology: Gelatinous bells may bring change in marine ecosystems. *Current Biology, 16,* R679–R682.

Johnston, B. D., et al. (2005). Ecological toxicology and effects of mass occurrences of gelatinous zooplankton on teleosts in enclosed coastal ecosystems. Society of Limnology and Oceanography, June 2005, Abstract.

Leone, A., Lecci, R. M., Durante, M., Meli, F., & Piraino, S. (2015). The bright side of gelatinous blooms: Nutraceutical value and antioxidant properties of three Mediterranean jellyfish (*Scyphozoa*). *Marine Drugs, 13*, 4654–4681.

Milisenda, G., Martínez-Quintana, A., Fuentes, V. L., Bosch-Belmar, M., Aglieri, G., Boero, F, & Piraino, S. (2016). Reproductive and bloom patterns of *Pelagia noctiluca* in the Strait of Messina, Italy. *Estuarine, Coastal and Shelf Science*. https://dx.doi.org/10.1016/j.ecss.2016.01.002.

Mills, C. E. (1995). Medusae, siphonophores, and ctenophores as planktivorous predators in changing global ecosystems. *ICES Journal of Marine Science, 52*, 575–581.

Mills, C. E. (2001). Jellyfish blooms: Are populations increasing globally in response to changing ocean conditions? *Hydrobiologia, 451*, 55–68.

Pagés, F., et al. (2001). Gelatinous zooplankton assemblages associated with water masses in the Humboldt Current System, and potential predatory impact by *Bassia bassensis* (*Siphonophora: Calycophorae*). *Marine Ecology Progress Series, 210*, 13–24.

Pérez-Ruzafa, A., et al. (2002). Evidence of a planktonic food web response to changes in nutrient input dynamics in the Mar Menor coastal lagoon, Spain. *Hydrobiologia, 475*(476), 359–369.

Purcell, J. E., & Arai, M. N. (2001). Interactions of pelagic cnidarians and ctenophores with fishes: A review. *Hydrobiologia, 451*, 27–44.

Purcell, J. E., Milisenda, G., Rizzo, A., Carrion, S., et al. (2015). Digestion and predation rates of zooplankton by the pleustonic hydrozoan *Velella velella* and widespread blooms in 2013 and 2014. *Journal of Plankton Research, 37*, 1056–1067.

Purcell, J. E., Uye, S. I., & Lo, W. T. (2007). Anthropogenic causes of jellyfish blooms and their direct consequences for humans: A review. *Marine Ecology Progress Series, 350*, 153–174.

Sun, M., Dong, J., Purcell, J. E., Li, Y., Duan, Y., Wang, A., & Wang, B. (2015). Testing the influence of previous-year temperature and food supply on development of *Nemopilema nomurai* blooms. *Hydrobiologia, 754*, 85–96.

Uye, S. I. (2008). Blooms of the giant jellyfish *Nemopilema nomurai*: A threat to the fisheries sustainability of the East Asian Marginal Seas. *Plankton and Benthos Research, 3*(Supp), 125–131.

Xian, W., Kang, B., & Liu, R. (2005). Jellyfish blooms in the Yangtze Estuary. *Science, 307*, 41.

Yoon, W. D., Yang, J. Y., Shim, M. B., & Kang, H. K. (2008). Physical processes influencing the occurrence of the giant jellyfish *Nemopilema nomurai* (*Scyphozoa: Rhizostomeae*) around Jeju Island, Korea. *Journal of Plankton Research, 30*, 251–260.

Chapter 11

Bramanti, L., et al. (2005). Recruitment, early survival and growth of the Mediterranean red coral *Corallium rubrum* (L 1758), a four-year study. *Journal of Experimental Marine Biology and Ecology, 314*, 69–78.

Bramanti, L., et al. (2013). Detrimental effects of ocean acidification on the economically important Mediterranean red coral (*Corallium rubrum*). *Global Change Biology, 19*, 1897–1908.

Bramanti, L., et al. (2014). Demographic parameters of two populations of red coral (*Corallium rubrum L.* 1758) in the north western Mediterranean. *Marine Biology, 161*, 1015–1026.

Cattaneo-Vietti, R., et al. (2016). An overexploited Italian treasure: Past and present distribution and exploitation of the precious red coral *Corallium rubrum* (L., 1758) (Cnidaria: Anthozoa). *Italian Journal of Zoology, 83*, 443–455.

García-Rodríguez, M., & Massò, C. (1986). Modelo de exploitación por buceo del coral rojo (*Corallium rubrum* L.) del Mediterráneo. *Boletin Instituto Español de Oceanografia, 3,* 75–82.

Garrabou, J., et al. (2001). Mass mortality event in red coral (*Corallium rubrum*, Cnidaria, Anthozoa, Octocorallia) populations in the Provence region (France NW Mediterranean). *Marine Ecology Progress Series, 207,* 263–272.

Garrabou, J. M., & Harmelin, J. G. (2002). A 20-year study on life-history traits of a harvested long-lived temperate coral in the NW Mediterranean: Insights into conservation and management needs. *Journal of Animal Ecology, 71,* 966–978.

Grigg, R. W. (1965). Ecological studies of black coral in Hawaii. *Pacific Science, 19,* 244–260.

Grigg, R. W. (1994). History of the precious coral fishery in Hawaii. *Precious Corals & Octocoral Research, 3,* 1–18.

Grigg, R. W. (2001). Black coral: History of a sustainable fishery in Hawaii. *Pacific Science, 55,* 291–299.

Hall-Spencer, J., Allain, V., & Fossa, J. H. (2002). Trawling damage to northeast Atlantic ancient coral reefs. *Philosophical Transactions of the Royal Society B: Biological Sciences, 269,* 507–511.

Roark, E. B., Guilderson, T. P., Dumbar, R. B., & Ingram, B. L. (2006). Radiocarbon-Based ages and growth rates of Hawaiian deep-sea corals. *Marine Ecology Progress Series, 327,* 1–14.

Rossi, S., et al. (2008). Survey of deep-dwelling red coral (*Corallium rubrum*) populations at Cap de Creus (NW Mediterranean). *Marine Biology, 154,* 533–545.

Santangelo, G., & Abbiati, M. (2001). Red coral: Conservation and management of an over-exploited Mediterranean species. *Aquatic Conservation of Marine Freshwater Ecosystems, 11,* 253–259.

Santangelo, G., et al. (2003). Reproduction and population sexual structure of the overexploited Mediterranean red coral *Corallium rubrum. Marine Ecology Progress Series, 248,* 99–108.

Tescione, G. (1965). *Il corallo nella storia e nell'arte.* Napoli, Italy: Montanino Editore.

Tescione, G. (1973). *The Italians and their coral fishing.* Naples: Fausto Fiorentino.

True, M. A. (1970). Étude quantitative de quatre peuplements sciaphiles sur substrat rocheux dans la région marseillaise. *Bulletin Institute océanographique Monaco, 69,* 1–48.

Tsounis, G., et al. (2006). Population structure of an exploited benthic cnidarian: The red coral case study. *Marine Biology, 149,* 1059–1070.

Tsounis, G., et al. (2007). Red coral fishery at the Costa Brava (NW Mediterranean): Case study for an over harvested precious coral. *Ecosystems, 10,* 975–986.

Tsounis, G., et al. (2010). Precious coral exploitation and conservation. *Oceanography & Marine Biology: An Annual Review, 48,* 161–212.

Tsounis, G., et al. (2013). Management hurdles in the sustainable harvesting plans of *Corallium rubrum. Marine Policy, 39,* 361–364.

Chapter 12

Borja, A., Ranasinghe, A., & Weisberg, S. B. (2009). Assessing ecological integrity in marine waters, using multiple indices and ecosystem components: Challenges for the future. *Marine Pollution Bulletin, 59,* 1–4.

Cadée, G. (2002). Seabirds and floating plastic debris. *Marine Pollution Bulletin, 44,* 1294–1295.

Cózar, A., et al. (2014). Plastic debris in the open ocean. *PNAS, 111,* 10239–10244.

Feder, H. M., & Blanchard, A. (1998). The deep benthos of Prince William Sound, Alaska. 16 months after the *Exxon Valdez* oil spill. *Marine Pollution Bulletin, 36,* 118–130.

Gray, J. S. (1979). Pollution-induced changes in populations. *Philosophical Transactions of the Royal Society of London Bulletin, 286,* 545–561.

Hoag, H. (2008). Arctic sentinels. *PloS Biology, 6*(e259), 2085–2088.

Jambeck, J. R., et al. (2015). Plastic waste inputs from land into the ocean. *Science, 347,* 768–771.

Jewett, S. C., et al. (1999). *Exxon Valdez* oil spill: Impacts and recovery in the soft-bottom benthic community in an adjacent to eelgrass beds. *Marine Ecology Progress Series, 185,* 59–83.

Kelso, D. D., & Kendziorek, M. (1991). Alaska's response to the *Exxon Valdez* oil spill. *Environmental Science & Technology, 25,* 16–23.

Laist, D. W. (1987). Overview of the biological effects of lost and discarded plastic debris in the marine environment. *Marine Pollution Bulletin, 18,* 319–326.

Law, K. L., et al. (2014). Distribution of surface plastic debris in the eastern Pacific Ocean from an 11-Year Data Set. *Environmental Science and Technology, 48,* 4732–4738.

Maki, H., Sasaki, T., & Harayama, S. (2001). Photo-oxidation of biodegraded crude oil and toxicity of the photo-oxidized products. *Chemosphere, 44,* 1145–1151.

Markham, A. (1994). *A brief history of pollution.* England: Earthscan Publications Ltd.

McDermid, K. J., & McMullen, T. L. (2004). Quantitative analysis of small plastic debris on beaches in the Hawaiian archipelago. *Marine Pollution Bulletin, 48,* 790–794.

Michel, J., et al. (2005). Potentially polluting wrecks in marine waters. In *2005 International Oil Spill Conference,* 84 pp.

Moore, C. J., et al. (2001). A comparison of plastic and plankton in the North Pacific Central Gyre. *Marine Pollution Bulletin, 42*(12), 1297–1300.

Okoh, A. I. (2006). Biodegradation alternative in the cleanup of petroleum hydrocarbon pollutants. *Biotechnology and Molecular Biology Review, 1*(2), 38–50.

Peterson, C. H. (2001). The 'Exxon Valdez' oil spill in Alaska: Acute, indirect and chronic effects on the ecosystem. *Advances in Marine Biology, 39,* 1–84.

Peterson, C. H., et al. (2003). Long-term ecosystem response to *Exxon Valdez* oil spill. *Science, 302,* 2082–2086.

Prouty, N. G., Fisher, C. R., Demopoulos, A. W. J., & Druffel, E. R. M. (2016). Growth rates and ages of deep-sea corals impacted by the Deepwater Horizon oil spill. *Deep-Sea Research II, 129,* 196–212.

Rios, L. M., Moore, C., & Jones, P. R. (2007). Persistent organic pollutants carried by synthetic polymers in the ocean environment. *Marine Pollution Bulletin, 54,* 1230–1237.

Sánchez, F., et al. (2006). Monitoring the *Prestige* oil spill impacts on some key species of the Northern Iberian shelf. *Marine Pollution Bulletin, 53,* 332–349.

Schwarzenbach, R. P., et al. (2003). *Environmental organic chemistry* (2nd ed.). Wiley: New Jersey.

Silliman, B. R., et al. (2012). Degradation and resilience in Louisiana salt marshes after the BP–Deepwater Horizon oil spill. *PNAS, 109,* 11234–11239.

Varela, M., et al. (2006). The effect of the *Prestige* oil spill on the plankton of the N-NW Spanish coast. *Marine Pollution Bulletin, 53,* 272–286.

White, H. K., et al. (2012). Impact of the Deepwater Horizon oil spill on a deep-water coral community in the Gulf of Mexico. *PNAS, 109,* 20303–20308.

Wright, S. L., Thompson, R. C., & Galloway, T. S. (2013). The physical impacts of microplastics on marine organisms: A review. *Environmental Pollution, 178,* 483–492.

Yamasita, R., & Tanimura, A. (2006). Floating plastic in the Kuroshio Current area, western North Pacific Ocean. *Marine Pollution Bulletin, 54,* 464–488.

Chapter 13

Altieri, A., & Gedan, K. B. (2015). Climate change and dead zones. *Global Change Biology, 21,* 1395–1406.

Alve, E., Lepland, A., Magnusson, J., & Backer-Owe, K. (2009). Monitoring strategies for re-establishment of ecological reference conditions: Possibilities and limitations. *Marine Pollution Bulletin, 59,* 297–310.

Arias-González, J. E., et al. (2017). The animal forest and its socio-ecological connections to land and coastal ecosystems. In S. Rossi, L. Bramanti, A. Gori, & C. Orejas (Eds.), *Marine animal forests* (pp. 1209–1240). Switzerland: Springer International Publishing. https://doi.org/10.1007/978-3-319-17001-5_33-1.

Blanco, R. (2008). Turismo de sol y playa. In Ministerio de Medio Ambiente, Medio Rural y Marino-Secretaría General del Mar-Dirección General de Sostenibilidad de la Costa y del Mar (Eds.), *Actividades humanas en los mares de España*. Madrid, Spain.

Borja, A., et al. (2009). Using multiple ecosystem components, in assessing ecological status in Spanish (Basque Country) Atlantic marine waters. *Marine Pollution Bulletin, 59*, 54–64.

Brizon-Portugal, A., et al. (2016). Anthropogenic pressure decreases species richness in intertidal reefs. *Marine Environmental Research, 120*, 44–54.

Conley, D. J., et al. (2009). Controlling eutrophication: Nitrogen and phosphorus. *Science, 323*, 1014–1015.

Dauvin, J. C. (2008). The main characteristics, problems, and prospects for Western European coastal seas. *Marine Pollution Bulletin, 57*, 22–40.

Diaz, R. J., & Rosenberg, R. (2008). Spreading dead zones and consequences for marine ecosystems. *Science, 321*, 926–929.

Esteban, V. (2008). Navegación de recreo y otras actividades deportivas. In Ministerio de Medio Ambiente, Medio Rural y Marino-Secretaría General del Mar-Dirección General de Sostenibilidad de la Costa y del Mar (Eds.), *Actividades humanas en los mares de España*. Madrid, Spain.

Foden, J., Rogers, S. I., & Jones, A. P. (2008). A critical review of approaches to aquatic environmental assessment. *Marine Pollution Bulletin, 56*, 1825–1833.

Goldberg, E. D. (1997). Plasticizing the seafloor: An overview. *Environmental Technology, 18*, 195–201.

Grantham, B. A., et al. (2004). Upwelling-driven nearshore hypoxia signals ecosystem and oceanographic changes in the northeast Pacific. *Nature, 427*, 749–754.

Knowlton, N., & Jackson, J. B. C. (2008). Shifting baselines, local impacts, and global change on coral reefs. *PloS Biology, 6*(2), e54.

Lawler, A. (2005). Reviving Iraq's wetlands. *Science, 307*, 1186–1189.

Melzner, F., et al. (2013). Future ocean acidification will be amplified by hypoxia in coastal habitats. *Marine Biology, 160*, 1875–1888.

Pimm, S. L., et al. (2001). Can we defy nature's end? *Science, 293*, 2207–2208.

Sardá, R., Avila, C., & Mora, J. (2005). A methodological approach to be used in integrated coastal zone management process: The case of the Catalan Coast (Catalonia, Spain). *Estuarine, Coastal and Shelf Science, 62*, 427–439.

Sardá, R., et al. (2012). Marine benthic cartography of the Cap de Creus (NE Catalan Coast, Mediterranean Sea). *Scientia Marina, 76*(1), 159–171.

Sheppard, C., et al. (2010). The Gulf: A young sea in decline. *Marine Pollution Bulletin, 60*, 13–38.

Spalding, M. D., et al. (2014). The role of ecosystems in coastal protection: Adapting to climate change and coastal hazards. *Ocean & Coastal Management, 90*, 50–57.

Stein, E. D., & Cadien, D. B. (2009). Ecosystem response to regulatory and management actions: The southern California experience in long-term monitoring. *Marine Pollution Bulletin, 59*, 91–100.

Willis, K. J., & Birks, H. J. B. (2006). What is natural? The need for a long-term perspective in biodiversity conservation. *Science, 314*, 1261–1265.

Chapter 14

Al Shehhi, M. R., Gherboudj, I., & Ghedira, H. (2014). An overview of historical harmful algae blooms outbreaks in the Arabian Seas. *Marine Pollution Bulletin, 86*, 314–324.

Anderson, A. D., & Keafer, B. A. (1987). An endogenous annual clock in the toxic marine dinoflagellate *Gonyaulax tamarensis*. *Nature, 325*, 616–617.

Bricelij, V. M., et al. (2005). Sodium channel mutation leading to saxitoxin resistance in clams increases risk of PSP. *Nature, 434*, 763–767.

Cembella, A. D., et al. (2002). The toxigenic marine dinoflagellate *Alexandrium tamarense* as probable cause of mortality of caged salmon in Nova Scotia. *Harmful Algae, 1*, 313–325.

Cembella, A. D., & John, U. (2006). Molecular physiology of toxin production and growth regulation in harmful algae. In E. Granéli & J. T. Turner (Eds.), *Ecology of Harmful Algae* (339 pp). Berlin: Springer.

Fire, S. E., & Flewelling, L. J. (2008). Florida red tide and brevetoxins: Associations and exposure in live resident bottlenose dolphins (*Tursiops truncatus*) in the eastern Gulf of Mexico, USA. *Marine Mammal Science, 24*, 831–844.

Fleming, L. E., et al. (2011). Review of Florida red tide and human health effects. *Harmful Algae, 10*, 224–233.

Garcés, E., Masó, M., Vila, M., & Camp, J. (2000). HABs events in the Mediterranean Sea: Are they increasing? A case study the last decade in the NW Mediterranean and the genus *Alexandrium*. *Harmful Algal News, 20*, 1–11.

Giuliani, S., Lamberti, C. V., Sonni, C., & Pellegrini, D. (2005). Mucilage impact on gorgonians in the Tyrrhenian sea. *Science of the Total Environment, 353*, 340–349.

Kirkpatrick, B., et al. (2014). Florida red tide knowledge and risk perception: Is there a need for tailored messaging. *Harmful Algae, 32*, 27–32.

Masó, M., & Garcés, E. (2006). Harmful microalgae blooms (HAB); problematic and conditions that induce them. *Marine Pollution Bulletin, 53*, 620–630.

Park, T. G., et al. (2013). Economic impact, management and mitigation of red tides in Korea. *Harmful Algae, 30*(S), S131–S143.

Rossi, S., & Fiorillo, I. (2010). Biochemical features of a *Protoceratium reticulatum* red tide in Chipana Bay (Northern Chile) in summer conditions. *Scientia Marina, 74*(4), 633–642.

Schiapparelli, S., et al. (2007). A benthic mucilage event in North-Western Mediterranean Sea and its possible relationships with the summer 2003 European heatwave: Short term effects on littoral rocky assemblages. *Marine Ecology, 28*, 341–353.

Schrope, M. (2008). Red tide rising. *Nature, 452*, 24–26.

Vila, M., Garcés, E., Masó, M., & Camp, J. (2001). Is the distribution of the toxic dinoflagellate *Alexandrium catenella* expanding along the NW Mediterranean coast? *Marine Ecology Progress Series, 222*, 73–83.

Chapter 15

Ballesteros, E. (2008). Especies invasoras. In Ministerio de Medio Ambiente, Medio Rural y Marino -Secretaría General del Mar-Dirección General de Sostenibilidad de la Costa y del Mar (Eds.), *Actividades humanas en los mares de España*. Madrid, Spain.

Carlton, J. T. (1987). Patterns of transoceanic marine biological invasions in the Pacific ocean. *Bulletin of Marine Science, 41*, 452–465.

Cohen, A. N., & Carlton, J. T. (1998). Accelerating invasion rate in a highly invaded estuary. *Science, 279*, 555–557.

Courtenay, W. R., et al. (2009). Risks of introduction of marine fishes. *Fisheries, 34*, 181–186.

Felline, S., et al. (2012). Subtle effects of biological invasions: Cellular and physiological responses of fish eating the exotic pest *Caulerpa racemosa*. *Plos One*, e38763.

Galil, B. S. (2000). A sea under siege-alien species in the Mediterranean. *Biological Invasions, 2*, 177–186.

Galil, B. S. (2007). Seeing Red: Alien species along the Mediterranean coast of Israel. *Aquatic Invasions, 2*, 281–312.

Grosholz, E. (2002). Ecological and evolutionary consequences of coastal invasions. *Trends in Ecology & Evolution, 17*, 22–27.

Hayes, K. R. (2002). Identifying hazards in complex ecological Systems. Part 1: Fault-tree analysis for biological invasions. *Biological Invasions, 4*, 235–249.

Jaubert, J. M., et al. (2003). Re-evaluation of the extent of *Caulerpa taxifolia* development in the northestern Mediterranean using airborne spectrographic sensing. *Marine Ecology Progress Series, 263*, 75–82.

Jørgensen, L. L. (2005). Impact scenario for an introduced decapod on Arctic epibenthic communities. *Biological Invasion, 7*, 949–957.

Jørgensen, L. L., et al. (2005). The introduction of the marine red king crab *Paralithodes camtschaticus* into the southern Barents Sea. *ICES Cooperative Research Report No. 277*.

Markert, A., Wehrmann, A., & Krönke, I. (2010). Recently established Crassostrea-reefs versus native Mytilus beds: Differences in ecosystem engineering affects the macrofaunal communities (Wadden Sea Lower Saxony, southern German Bight). *Biological Invasions, 12*, 15–32.

Mollo, E., Cimino, G., & Ghiselin, M. T. (2015). Alien biomolecules: A new challenge for natural product chemists. *Biological Invasions, 17*, 941–950.

Montefalocone, M., et al. (2015). A tale of two invaders: Divergent spreading kinetics of the alien green algae *Caulerpa taxifolia* and *Caulerpa cylindracea*. *Biological Invasions, 17*, 2717–2728.

Ojaveer, H., et al. (2014). Ten recommendations for advancing the assessment and management of non-indigenous species in marine ecosystems. *Marine Policy, 44*, 160–165.

Piazzi, L., & Balata, D. (2009). Invasion of alien macroalgae in different Mediterranean habitats. *Biological Invasions, 11*, 193–204.

Riesink, J. L., & Collado-Vides, L. (2006). Modeling the increase and control of *Caulerpa taxifolia*, an invasive marine macroalga. *Biological Invasions, 8*, 309–325.

Rius, M., Turon, X., & Marshall, D. J. (2009). Non-lethal effects of an invasive species in the marine environment: The importance of early life-history stages. *Oecologia, 159*, 873–882.

Secord, D. (2003). Biological control of marine invasive species: Cautionary tales and land-based lessons. *Biological Invasions, 5*, 117–131.

Sundet, J. H., & Hoel, A. H. (2016). The Norwegian management of an introduced species: The Arctic red crab fishery. *Marine Policy, 72*, 278–284.

Thresher, R. E., & Kuris, A. M. (2004). Options for managing invasive marine species. *Biological Invasions, 6*, 295–300.

Chapter 16

Brassell, S. C., et al. (1986). Molecular stratigraphy: A new tool for climatic assessment. *Nature, 320*(6058), 129–133.

Ciais, P. C., et al. (2013). Carbon and other biogeochemical cycles. In T. F. Stocker, D. Qin, G.-K. Plattner, M. Tignor, S. K. Allen, J. Boschung, A. Nauels, Y. Xia, V. Bex, & P. M. Midgley (Eds.), *Climate change 2013: The physical science basis* (pp. 465–570). Contribution of Working Group I to 5th Assessment Report of the Intergovernmental Panel on Climate Change. Cambridge, United Kingdom: Cambridge University Press. https://doi.org/10.1017/cbo9781107415324.015.

Clark, P. U., Alley, R. B., & Pollard, D. (1999). Climatology—Northern hemisphere ice-sheet influences on global climate change. *Science, 286*(5442), 1104–1111.

IPCC. (1997). *Impactos regionales del cambio climático: Evaluación de la vulnerabilidad* (p. 27). Grupo Intergubernamental de Expertos sobre Cambio Climático. ISBN: 92-9169-104-7.

IPCC. (2002). *Cambio climático y biodiversidad* (p. 93). Grupo Intergubernamental de Expertos sobre Cambio Climático. ISBN: 92-9169-104-7.

IPCC. (2007). *Climate change 2007: The physical science basis* (p. 996). In S. Solomon, D. Qin, M. Manning, Z. Chen, M. Marquis, K. B. Averyt, M. Tignor, & H. L. Miller (Eds.), Contribution of Working Group I to the 4th Assessment Report of the Intergovernmental Panel on Climate Change.

IPCC. (2013). *Climate change 2013: The physical science basis.* In T. F. Stocker, D. Qin, G.-K. Plattner, M. Tignor, S. K. Allen, J. Boschung, A. Nauels, Y. Xia, V. Bex, & P. M. Midgley (Eds.), Contribution of Working Group I to the Fifth Assessment Report of the Intergovernmental Panel on Climate Change. Cambridge: Cambridge University Press (1535 pp).

Martínez-Garcia, A., et al. (2009). Links between iron supply, marine productivity, sea surface temperature, and CO_2 over the last 1.1 Ma. *Paleoceanography, 24*, PA1207.

Martínez-García, A., et al. (2010). Subpolar link to the emergence of the modern Equatorial Pacific cold tongue. *Science, 328*, 1550–1553.

Masson-Delmotte, V. M., et al. (2013). Information from paleoclimate archives. In T. F. Stocker, D. Qin, G.-K. Plattner, M. Tignor, S. K. Allen, J. Boschung, A. Nauels, Y. Xia, V. Bex, & P. M. Midgley (Eds.), *Climate change 2013: The physical science basis* (pp. 383–464). Contribution of Working Group I to 5th Assessment Report of the Intergovernmental Panel on Climate Change. Cambridge: Cambridge University Press. https://doi.org/10.1017/cbo9781107415324.013.

Monastersky, R. (2009). A burden beyond bearing. *Nature, 458*, 1091–1094.

Parrenin, F., et al. (2007). The EDC3 chronology for the EPICA Dome C ice core. *Climate of the Past, 3*, 485–497.

Parry, M., Lowe, J., & Hanson, C. (2009). Overshoot, adapt and recover. *Nature, 458*, 1102–1103.

Pongratz, J., Reick, C. H., Raddatz, T., & Claussen, M. (2009). Effects of anthropogenic land cover change on the carbon cycle of the last millennium. *Global Biogeochemical Cycles, 23* (GB4001), 1–13.

Raymo, M. E. (1997). The timing of major climate terminations. *Paleoceanography, 12*(4), 577–585.

Raymo, M. E. (1998). Glacial puzzles. *Science, 281*(5382), 1467–1468.

Rosell-Melé, A. (2003). Biomarkers as proxies of climate change. In A. W. Mackay, R. W. Battarbee, J. B. Birks, & F. Oldfield (Eds.), *Global change in the Holocene* (pp. 358–372). London: Arnold.

Running, S. W. (2009). Ecosystem disturbance, carbon, and climate. *Science, 321*, 652–653.

Schneider, S. (2009). The worst-case scenario. *Nature, 458*, 1104–1105.

Thompson, D. W. J., Kennedy, J. J., Wallace, J. M., & Jones, P. D. (2008). A large discontinuity in the mid-twentieth Century in observed global-mean surface temperature. *Nature, 453*, 646–650.

Zahn, R. (2009). Beyond the CO_2 connection. *Nature, 460*, 335–336.

Chapter 17

Amstrup, S. C., et al. (2006). Recent observations of intraspecific predation and cannibalism among polar bears in the southern Beaufort Sea. *Polar Biology, 29,* 997–1002.

Atkinson, A., et al. (2004). Long-term decline in krill stock and increase in salps within the Southern Ocean. *Nature, 432,* 100–103.

Barnes, D. K. A. (2016). Iceberg killing fields limit huge potential for benthic blue carbon in Antarctic shallows. *Global Change Biology.* https://doi.org/10.1111/gcb.13523.

Bond, G., et al. (1992). Evidence for massive discharges of icebergs into the North-Atlantic Ocean during the Last Glacial period. *Nature, 360*(6401), 245–249.

Cherry, S. G., et al. (2009). Fasting physiology of polar bears in relation to environmental change and breeding behavior in the Beaufort Sea. *Polar Biology, 32,* 383–391.

Croxall, J. P., & Nicol, S. (2004). Management of Southern Ocean fisheries: Global forces and future sustainability. *Antarctic Science, 16,* 569–584.

Domack, E., et al. (2005). Stability of the Larsen B ice shelf on the Antarctic Peninsula during the Holocene epoch. *Nature, 436,* 681–685.

Durner, G. M., et al. (2009). Predicting 21st-century polar bear habitat distribution from global climate models. *Ecological Monographs, 79,* 25–58.

Forcada, J., & Trathan, P. N. (2009). Penguin responses to climate change in the Southern Ocean. *Global Change Biology, 15,* 1618–1630.

Gutt, J., et al. (2013). Shifts in Antarctic megabenthic structure after ice-shelf disintegration in the Larsen area east of the Antarctic Peninsula. *Polar Biology, 36,* 895–906.

Kawaguchi, S., Nicol, S., & Press, A. J. (2009). Direct effects of climate change on the Antarctic krill fishery. *Fisheries Management and Ecology, 16,* 424–427.

McGrath, D. (2014). The structure and effect of suture zones in the Larsen C Ice Shelf, Antarctica. *Journal of Geophysical Research: Earth Surface, 119,* 588–602.

Merino, N., et al. (2016). Antarctic icebergs melt over the Southern Ocean: Climatology and impact on sea ice. *Ocean Modelling, 104,* 99–110.

Moline, M. A., et al. (2008). High latitude changes in ice dynamics and their impact on Polar Marine ecosystems. *Annals of the New York Academy of Sciences, 1134,* 267–319.

Prop, J., et al. (2015). Climate change and the increasing impact of polar bears on bird populations. *Frontiers in Ecology and Evolution, 3,* 33.

Rebesco, M., et al. (2014). Boundary condition of grounding lines prior to collapse, Larsen-B Ice Shelf, Antarctica. *Science, 345,* 1354–1358.

Regehr, E. V., et al. (2010). Survival and breeding of polar bears in the southern Beaufort Sea in relation to sea ice. *Journal of Animal Ecology, 79,* 117–127.

Rockwell, R. F., & Gormezano, L. J. (2009). The early bear gets the goose: Climate change, polar bears and lesser snow geese in western Hudson Bay. *Polar Biology, 32,* 539–547.

Rühlemann, C., et al. (1999). Warming of the tropical Atlantic Ocean and slowdown of thermohaline circulation during the last deglaciation. *Nature, 402,* 511–514.

SCAR. (2009). *Antarctic climate change and the environment.* In J. Turner, R. Bindschadler, P. Convey, G. di Prisco, E. Fahrbach, J. Gutt, D. Hodgson, P. Mayewsky, & C. Summerhayes (Eds.). Cambridge, UK: Scott Polar Research Institute.

Smetacek, V., & Nicol, S. (2005). Polar ocean ecosystems in a changing world. *Nature, 437,* 362–368.

Smith, K. L., Jr., et al. (2007). Free drifting icebergs: Hot spots of chemical and biological enrichment in the Weddell Sea. *Science, 317,* 478–482.

Stirlong, I., & Derocher, A. E. (2012). Effects of climate warming on polar bears: A review of the evidence. *Global Change Biology, 18,* 2694–2706.

Thomas, D. (2004). *Frozen oceans*. London, England: Natural History Museum.

Thomas, D., et al. (2008). *The biology of polar habitats*. Oxford, England: Oxford University Press.

Vermeij, G. J., & Roopnarine, P. D. (2008). The coming Arctic invasion. *Science, 321,* 780–781.

Walsh, J. E. (2008). Climate of the Arctic marine environment. *Ecological Applications, 18,* S3–S22.

Walsh, J. E. (2009). A comparison of Arctic and Antarctic climate change, present and future. *Antarctic Science, 21,* 179–188.

Zahn, R. (1999). Polar-tropical and interhemispheric linkages. In F. Abrantes & A. Mix (Eds.), *Reconstructing ocean history—A window into the future* (pp. 1–6). Kluwer Academics/Plenum Publishers.

Zahn, R., et al. (1997). Thermohaline instability in the North Atlantic during meltwater events: Stable isotope and ice-rafted detritus records from Core SO75-26KL. *Portuguese Margin, Paleoceanography, 12,* 696–710.

Chapter 18

Andersson, A., et al. (2015). Projected future climate change and Baltic Sea ecosystem management. *AMBIO, 44,* S345–S356.

Azzurro, E., et al. (2007). Reproductive features of the non-native *Siganus luridus* (Teleostei, Siganidae) during the early colonization at Linosa Island (Sicily Strait, Mediterranean Sea). *Journal of Applied Ichthyology, 23,* 640–645.

Bamber, J. L., & Aspinall, W. P. (2013). An expert judgement assessment of future sea level rise from the ice sheets. *Nature Climate Change, 3,* 424–427.

Belkin, I. M. (2009). Rapid warming of large marine ecosystems. *Progress in Oceanography, 81,* 207–213.

Bell, J. D., et al. (2013). Mixed responses of tropical Pacific fisheries and aquaculture to climate change. *Nature Climate Change, 3,* 591–599.

Berggren, A., et al. (2009). The distribution and abundance of animal populations in a climate of uncertainty. *Oikos, 118,* 1121–1126.

Bernal, P. A. (1991). Consequences of global change for oceans: A review. *Climatic Change, 18,* 339–359.

Bianchi, C. N. (2007). Biodiversity issues for the forthcoming tropical Mediterranean Sea. *Hydrobiologia, 580,* 7–21.

Blum, M. D., & Roberts, H. B. (2009). Drowning of the Mississippi delta due to insufficient sediment supply and global sea-level rise. *Nature Geoscience, 2,* 488–491.

Boero, F. (2015). The future of the Mediterranean Sea ecosystem: Towards a different tomorrow. *Rend. Fis. Acc. Lincei, 26,* 3–12.

Brochier, T., et al. (2013). Climate change scenarios experiments predict a future reduction in small pelagic fish recruitment in the Humboldt current system. *Global Change Biology, 19,* 1841–1853.

Doney, S. C. (2006). Plankton in a warmer world. *Nature, 444,* 695–696.

Faezeh, M. N., et al. (2013). Future sea-level rise from Greenland's main outlet glaciers in a warming climate. *Nature, 497,* 235–238.

Hátún, H., et al. (2009). Large bio-geographical shifts in the north-eastern Atlantic Ocean: From the subpolar gyre, via plankton, to blue whiting and pilot whales. *Progress in Oceanography, 80,* 149–162.

Holland, G., & Bruyère, C. L. (2014). Recent intense hurricane response to global climate change. *Climate Dynamics, 42,* 617–627.

Hyndes, G. A., et al. (2016). Accelerating tropicalization and the transformation of temperate seagrass meadows. *BioScience, 66,* 938–948.

Kletou, D., Hall-Spencer, J. M., & Kleitou, P. (2016). A lionfish (*Pterois miles*) invasion has begun in the Mediterranean Sea. *Marine Biodiversity Records, 9,* 46.

Lefort, S., et al. (2015). Spatial and body-size dependent response of marine pelagic communities to projected global climate change. *Global Change Biology, 21,* 154–164.

Lejeusne, C., et al. (2010). Climate change effects on a miniature ocean: The highly diverse, highly impacted Mediterranean Sea. *TREE, 1204,* 1–11.

MacKenzie, B., & Köster, F. (2004). Fish production and climate: Sprat in the Baltic Sea. *Ecology, 85,* 784–794.

McIlgorn, A., et al. (2010). How will climate change alter fishery governance? Insights from seven international case studies. *Marine Policy, 34,* 170–177.

Mendelsohn, R., et al. (2012). The impact of climate change on global tropical cyclone damage. *Nature Climate Change, 2,* 205–209.

Montes-Hugo, M., et al. (2009). Recent changes in phytoplankton communities associated with rapid regional climate change along the Western *Antarctic peninsula. Science, 323,* 1470–1473.

Morton, O. (2009). Great white hope. *Nature, 458,* 1097–1100.

Paerl, H. W., & Paul, V. J. (2012). Climate change: Links to global expansion of harmful cyanobacteria. *Water Research, 46,* 1349–1363.

Poloczanzka, E. S. (2013). Global imprint of climate change on marine life. *Nature Climate Change, 3,* 919–925.

Portner, H. O., & Farrell, A. P. (2008). Ecology: Physiology and climate change. *Science, 322,* 690–692.

Rice, E., Dam, H. G., & Stewart, G. (2015). Impact of climate change on estuarine zooplankton: Surface water warming in Long Island Sound is associated with changes in copepod size and community structure. *Estuaries and Coasts, 38,* 13–23.

Romano, J. C., et al. (2000). Anomalie thermique dans les eaux du golfe de Marseille durant l'été 1999. Une explication partielle de la mortalité d'invertébreés fixes? Comptesrendus de l'Academie de Sciences de Paris. *Sciences de la Vie, 323,* 415–427.

Sabatés, A., et al. (2006). Sea warming and fish distribution: The case of the small pelagic fish, *Sardinella aurita,* in the Western Mediterranean. *Global Change Biology, 12,* 2209–2219.

Sallenger, A. H., Doran, K. S., & Howd, P. A. (2012). Hotspot of accelerated sea-level rise on the Atlantic coast of North America. *Nature Climate Change, 2,* 884–888.

Sarmiento, J. L., et al. (2004). Response of ocean ecosystems to climate warming. *Global Biogeochemical Cycles, 18*(3).

Smetacek, V., & Cloern, J. E. (2008). On phytoplankton trends. *Science, 319,* 1346–1348.

Tebaldi, C., Strauss, B. H., & Zervas, C. E. (2012). Modelling sea level rise impacts on storm surges along US coasts. *Environmental Research Letters, 7,* 014032.

Toseland, A., et al. (2013). The impact of temperature on marine phytoplankton resource allocation and metabolism. *Nature Climate Change, 3,* 979–984.

Winder, M., & Sommer, U. (2012). Phytoplankton response to a changing climate. *Hydrobiologia, 698,* 5–16.

Chapter 19

Albright, R., et al. (2016). Reversal of ocean acidification enhances net coral reef calcification. *Nature, 531,* 362–365.

Balch, W. M., & Utgoff, P. E. (2009). Potential interactions among ocean acidification, coccolithophores, and the optical properties of seawater. *Oceanography, 22,* 146–159.

Branch, T. A., et al. (2012). Impacts of ocean acidification on marine seafood. *Trends in Ecology and Evolution, 28,* 178–186.

Checkley, D. M., et al. (2009). Elevated CO_2 enhances otolith growth in young fish. *Science, 324,* 1683.

Comeau, S., & Cornwall, C. E. (2017). Contrasting effects of ocean acidification on coral reef 'animal forests' versus seaweed 'kelp forests'. In S. Rossi, L. Bramanti, A. Gori, & C. Orejas (Eds.), *Marine animal forests* (pp. 1083–1108). Switzerland: Springer International Publishing. https://doi.org/10.1007/978-3-319-17001-5_29-1.

Doney, S. C. (2006). The dangers of ocean acidification. *Scientific American, 294,* 58–65.

Doney, S. C., et al. (2009). Ocean acidification: A critical emerging problem for ocean sciences. *Oceanography, 22,* 16–25.

Dupont, S., et al. (2013). Long-term and trans-life-cycle effects of exposure to ocean acidification in the green sea urchin *Strongylocentrotus droebachiensis. Marine Biology, 160,* 1835–1843.

Feely, R. A., et al. (2004). Impact of anthropogenic CO_2 on the $CaCO_3$ system in the oceans. *Science, 305,* 362–366.

Fine, M., & Tchernov, D. (2007). Scleractinian coral species survive and recover from decalcification. *Science, 315,* 1811.

Gazeau, F., et al. (2013). Impacts of ocean acidification on marine shelled molluscs. *Marine Biology, 160,* 2207–2245.

Gibbs, S., et al. (2016). Ocean warming, not acidification, controlled coccolithophore response during past greenhouse climate change. *Geology, 44,* 59–62.

Godbold, J. A., & Calosi, P. (2013). Ocean acidification and climate change: Advances in ecology and evolution. *Philosophical Transactions of the Royal Society B, 368,* 20120448.

Hall-Spencer, J. M., et al. (2008). Volcanic carbon dioxide vents show ecosystem effects of ocean acidification. *Nature, 454,* 96–99.

Hoegh-Guldberg, O., et al. (2007). Coral reefs under rapid climate change and ocean acidification. *Science, 318,* 1737–1742.

Hutchins, D. A., Mulholland, M. R., & Fu, F. (2009). Nutrient cycles and marine microbes in a CO_2-enriched ocean. *Oceanography, 22,* 128–145.

Jokiel, P. L., Jury, C. P., & Kuffner, I. B. (2016). Coral calcification and ocean acidification. In D. K. Hubbard, et al. (Eds.), *Coral reefs at the crossroads, coral reefs of the world* (pp. 7–45). Berlin: Springer. https://doi.org/10.1007/978-94-017-7567-0_2.

Lohbeck, K. T., Riebesell, U., & Reusch, B. H. (2012). Adaptive evolution of a key phytoplankton species to ocean acidification. *Nature Geoscience, 5,* 346–351.

Millero, F. J., et al. (2009). Effect of ocean acidification on the speciation of metals in seawater. *Oceanography, 22,* 72–85.

Nagelkerken, I., et al. (2015). Ocean acidification alters fish populations indirectly through habitat modification. *Nature Climate Change, 6,* 89–93.

Parson, E. A., & Keith, D. W. (2013). End the deadlock on governance of geoengineering research. *Science, 339,* 1278–1279.

Schlüter, L., et al. (2014). Adaptation of a globally important coccolithophore to ocean warming and acidification. *Nature Climate Change, 4,* 1024–1030.

Stoll, H. M., et al. (2007). Coccolitophore productivity response to greenhouse event of the Paleocene-Eocene thermal maximum. *Earth and Planetary Science Letters, 258,* 192–206.

Zachos, J. C., Dickens, G. R., & Zeebe, R. (2008). An early Cenozoic perspective on greenhouse warming and carbon-cycle dynamics. *Nature, 451,* 279–283.

Chapter 20

Belwood, D. R., et al. (2004). Confronting the coral reef crisis. *Nature, 429,* 827–833.

Bianchi, C. N., et al. (2017). Resilience of the marine animal forest. In S. Rossi, L. Bramanti, A. Gori, & C. Orejas (Eds.), *Marine animal forests* (pp. 1241–1270). Switzerland: Springer International Publishing. https://doi.org/10.1007/978-3-319-17001-5_35-1.

Bozec, Y. M., Alvarez-Filip, L., & Mumby, P. J. (2015). The dynamics of architectural complexity on coral reefs under climate change. *Global Change Biology, 21,* 223–235.

Cerrano, C., et al. (2000). A catastrophic mass-mortality episode of gorgonians and other organisms in the Ligurian Sea (NW Mediterranean), summer 1999. *Ecology Letters, 3,* 284–293.

Coles, S. L., & Brown, B. E. (2003). Coral bleaching—Capacity for acclimatization and adaptation. *Advances in Marine Biology, 46,* 183–223.

Douglas, A. E. (2003). Coral bleaching—How and why? *Marine Pollution Bulletin, 46,* 385–392.

Edmunds, P., et al. (2014). Evaluating the causal basis of ecological success within the scleractinia: An integral projection model approach. *Marine Biology, 161,* 2719–2734.

Garrabou, J., et al. (2009). Mass mortalities in Northwestern Mediterranean rocky benthic communities: Effects of the 2003 heat wave. *Global Change Biology, 15,* 1090–1103.

Glynn, P. W., & Manzello, D. P. (2015). Bioerosion and coral reef growth: A dynamic balance. In C. Birkeland (Ed.), *Coral reefs in the anthropocene* (pp. 67–97). Berlin: Springer. https://doi.org/10.1007/978-94-017-7249-5_4.

Grottoli, A. G., Rodriguez, L. J., & Palardy, J. E. (2006). Heterotrophic plasticity and resilience in bleached corals. *Nature, 440,* 1186–1189.

Hoegh-Guldberg, O. (1999). Climate change, coral bleaching and the future of the world's coral reefs. *Marine and Freshwater Research, 50,* 839–866.

Hoegh-Guldberg, O. (2009). Climate change and coral reefs: Trojan horse or false prophecy? *Coral Reefs, 28,* 569–575.

Hughes, T. P., Ayre, D. J., & Connell, J. H. (1992). The evolutionary ecology of corals. *Trends in Ecology and Evolution, 7,* 292–295.

Hughes, T. P., et al. (2003). Climate change, human impacts, and the resilience of coral reefs. *Science, 301,* 929–933.

IPCC. (2007). Synthesis report. Contribution of working groups I, II and III to the fourth assessment report of the intergovernmental panel on climate change. In C. W. Team, R. K. Pachauri, & A. Reisinger (Eds.), *Intergovernmental panel on climate change, 2008* (p. 104). Geneva, Switzerland.

Knowlton, N. (2001). The future of coral reefs. *PNAS, 98,* 5419–5425.

Knowlton, N., & Jackson, J. B. C. (2008). Shifting baselines, local impacts, and global change on coral reefs. *Plos Biology, 6*(e54), 215–220.

Linares, C., Coma, R., & Zabala, M. (2008). Effects of a mass mortality event on gorgonian reproduction. *Coral Reefs, 27,* 27–34.

Pairaud, I. L., et al. (2014). Impacts of climate change on coastal benthic ecosystems: Assessing the current risk of mortality outbreaks associated with thermal stress in NW Mediterranean coastal areas. *Ocean Dynamics, 64,* 103–115.

Palumbi, S. R., et al. (2014). Mechanisms of reef coral resistance to future climate change. *Science, 344,* 895–898.

Pandolfi, J. M. (2015). Incorporating uncertainty in predicting the future response of coral reefs to climate change. *Annual Review of Ecology, Evolution, and Systematics, 46,* 281–303.

Ponti, M., et al. (2014). Ecological shifts in Mediterranean coralligenous assemblages related to gorgonian forest loss. *PLoS ONE, 9*(7), e102782. https://doi.org/10.1371/journal.pone.0102782.

Rivetti, I., et al. (2014). Global warming and mass mortalities of benthic invertebrates in the Mediterranean Sea. *PLoS ONE, 9*(12), e115655. https://doi.org/10.1371/journal.pone. 0115655.

Rosenberg, E., & Ben-Haim, Y. (2002). Microbial diseases of corals and global warming. *Environmental Microbiology, 4,* 318–326.

Ruzicka, R. R., et al. (2013). Temporal changes in benthic assemblages on Florida Keys reefs 11 years after the 1997/1998 El Niño. *Marine Ecology Progress Series, 489,* 125–141.

Schubert, N., Brown, D., & Rossi, S. (2017). Symbiotic versus non-symbiotic octocorals: Physiological and ecological implications. In S. Rossi, L. Bramanti, A. Gori, & C. Orejas (Eds.), *Marine animal forests* (pp. 887–918). Switzerland: Springer International Publishing.

Veron, J. E. N., et al. (2009). The coral reef crisis: The critical importance of <350 ppm CO_2. *Marine Pollution Bulletin, 58,* 1428–1436.

Vollmer, S. V., & Kline, D. I. (2009). Natural disease resistance in threatened staghorn corals. *Plos One, 3*(e3718), 1–5.

Wakeford, M., Done, T. J., & Johnson, C. R. (2008). Decadal trends in a coral community and evidence of changed disturbance regime. *Coral Reefs, 27,* 1–13.

Weil, E., Rogers, C. S., & Croquer, A. (2017). Octocoral diseases and other stressors in a changing ocean. In S. Rossi, L. Bramanti, A. Gori, & C. Orejas (Eds.), *Marine animal forests*. Switzerland: Springer International Publishing. https://doi.org/10.1007/978-3-319-17001-5_43-1.

Chapter 21

Bennett, E. M., & Balvanera, P. (2007). The future of production systems in a globalized world. *Frontiers in Ecology and the Environment, 5*(4), 191–198.

Borja, A., et al. (2009). Assessing the suitability of a range of benthic indices in the evaluation of environmental impact of fin shellfish aquaculture located in sites across Europe. *Aquaculture, 293,* 231–240.

Buck, B. H., & Buchholtz, C. M. (2004). The offshore-ring: A new system design fir the open ocean aquaculture of macroalgae. *Journal of Applied Phycology, 16,* 355–368.

Bush, S. R., et al. (2013). Certify sustainable aquaculture? *Science, 341,* 1067–1068.

Cao, L., et al. (2007). Environmental impact of aquaculture and countermeasures to aquaculture pollution in China. *Environmental Science and Pollution Research-International, 14,* 452–462.

Cao, L., et al. (2015). China's aquaculture and the world's wild fisheries. *Science, 347,* 133–135.

Chopin, T. (2006). Rationale for developing integrated multi-trophic aquaculture (IMTA): An example from Canada. *Fish Farmer Mag, 29,* 20–21.

Diana, J. M. (2009). Aquaculture production and biodiversity conservation. *BioScience, 59,* 27–38.

Duarte, C. M., Marbá, N., & Holmer, M. (2007). Rapid domestication of marine species. *Science, 316,* 382–383.

Fang, J., et al. (2016). Integrated multi-trophic aquaculture (IMTA) in Sanggou Bay, China. *Aquaculture Environment Interactions, 8,* 201–205.

FAO. (2009). Integrated mariculture: A global review. In D. Soto (Ed.), FAO fisheries and technical paper (p. 194). Technical paper 529.

Ford, J. S., & Myers, R. A. (2008). A global assessment of salmon aquaculture impacts on wild salmonids. *PLoS Biology, 6,* 1–7.

Irisarri, J., et al. (2015). Availability and utilization of waste fish feed by mussels *Mytilus edulis* in a commercial integrated multi-trophic aquaculture (IMTA) system: A multi-indicator assessment approach. *Ecological Indicators, 48,* 673–686.

Iwama, G. K. (1991). Interactions between aquaculture and the environment. *Critical Reviews in Environmental Science and Technology, 21,* 177–216.

Joffre, O. M., et al. (2015). What drives the adoption of integrated shrimp mangrove aquaculture in Vietnam? *Ocean & Coastal Management, 114,* 53–63.

Kalam-Azad, A., Jensen, K. R., & Lin, C. K. (2009). Coastal aquaculture development in Bangladesh: Unsustainable and sustainable practices. *Environmental Management, 44,* 800–809.

Lan, T. D. (2009). Coastal aquaculture and shrimp farming in North Vietnam and environmental cost estimation. *Aquatic Ecosystem Health & Management, 12,* 235–242.

Liao, I. C., & Chao, N. H. (2009). Aquaculture and food crisis: Opportunities and constraints. *Asia Pacific Journal of Clinical Nutrition, 18,* 564–569.

Lüning, K., & Pang, S. (2003). Mass cultivation of seaweeds: Current aspects and approaches. *Journal of Applied Phycology, 15,* 115–119.

Maisashvili, A., et al. (2015). The values of whole algae and lipid extracted algae meal for aquaculture. *Algal Research, 9,* 133–142.

Martínez-Córdova, L. R., et al. (2015). Microbial-based systems for aquaculture of fish and shrimp: An updated review. *Reviews in Aquaculture, 7,* 131–148.

Martínez-Espiñeira, R., et al. (2015). Estimating the biomitigation benefits of Integrated Multi-Trophic Aquaculture: A contingent behavior analysis. *Aquaculture, 437,* 182–194.

Naylor, R., et al. (2000). Effect of aquaculture on world fish supplies. *Nature, 405,* 1017–1024.

Naylor, R. L., et al. (2009). Feeding aquaculture in an era of finite resources. *PNAS, 106,* 15103–15110.

Primavera, J. H. (1993). A critical review of shrimp pond culture in the Philippines. *Reviews in Fisheries Science, 1,* 151–201.

Queiroz, L., et al. (2013). Shrimp aquaculture in the state of Ceará during the period 1970–2012: Trends of the privatization of mangrove forest in Brazil. *Ocean and Coastal Management, 73,* 54–62.

Rivera-Ferre, M. G. (2009). Can export-oriented aquaculture in developing countries be sustainable and promote sustainable development? The shrimp case. *Journal of Agricultural and Environmental Ethics, 22,* 301–321.

Santaella, E. (2008). Acuicultura marina. In Ministerio de Medio Ambiente, Medio Rural y Marino-Secretaría General del Mar-Dirección General de Sostenibilidad de la Costa y del Mar (Eds.), *Actividades humanas en los mares de España.* Madrid, Spain.

Soylent Green. (1973). Film, dir. Richard Fleischer. US: TV-MA.

Stonich, S. C., & Bailey, C. (2000). Resisting the blue revolution: Contending coalitions surrounding industrial shrimp farming. *Human Organization, 59,* 23–36.

Tacon, A. G. J., & Metian, M. (2009). Fishing for aquaculture: Non-food use of small pelagic forage fish-A global perspective. *Reviews in Fisheries Science, 13,* 305–317.

Treeck, P., et al. (2003). Mariculture trials with Mediterranean sponge species: The exploitation of an old natural resource with sustainable and novel methods. *Aquaculture, 218,* 439–455.

Trentacoste, E. M., Martinez, A. M., & Zenk, T. (2015). The place of algae in agriculture: Policies for algal biomass production. *Photosynthesis Research, 123,* 305–315.

Whitmarsh, D., & Palmieri, M. G. (2009). Social acceptability of marine aquaculture: The use of Surrey-based methods for eliciting public and stakeholder preferences. *Marine Policy, 33,* 452–457.

World Bank. (2006). Aquaculture: Changing the face of the waters. Meeting the Promise and Challenge of Sustainable Aquaculture. Report 36622-GLB.

Wu, R. S. S. (1995). The environmental impact of marine fish culture: Towards a sustainable future. *Marine Pollution Bulletin, 31,* 159–166.

Xie, B., et al. (2013). Organic aquaculture in China: A review from a global perspective. *Aquaculture, 414–415,* 243–253.

Chapter 22

Aburto-Oropenza, O., et al. (2008). Mangroves in the Gulf of California increase fishery yields. *PNAS, 105,* 10456–10459.

Agardy, T., et al. (2003). Dangerous targets? Unresolved issues and ideological clashes around marine protected areas. *Aquatic Conservation: Marine and Freshwater Ecosystems, 13,* 353–367.

Airoldi, L., et al. (2005). Impact of recreational harvesting on assemblages in artificial rocky habitats. *Marine Ecology Progress Series, 299,* 55–66.

Bascompte, J., Melián, C. J., & Sala, E. (2005). Interaction strength combinations and the overfishing of marine food web. *PNAS, 102,* 5443–5447.

Baskett, M. L., et al. (2005). Marine reserve design and the evolution of size at maturation in harvested fish. *Ecological Applications, 15,* 882–901.

Bearzi, G. (2007). Marine conservation on paper. *Conservation Biology, 21,* 1–3.

Beaumont, N. J., et al. (2008). Economic valuation for conservation of marine biodiversity. *Marine Pollution Bulletin, 56,* 386–396.

Bramanti, L., et al. (2011). Approaching recreational scuba divers to emblematic species conservation: The case of red coral (*Corallium rubrum*). *Journal for Nature Conservation, 19,* 312–318.

Cárcamo, P. F., et al. (2014). Using stakeholders' perspective of ecosystem services and biodiversity features to plan a marine protected area. *Environmental Science & Policy, 40,* 116–131.

Castilla, J. C. (1999). Coastal marine communities: Trends and perspectives from human-exclusion experiments. *Trends in Ecology & Evolution, 14,* 280–283.

Chuenpadgee, R., et al. (2013). Marine protected areas: Re-thinking their inception. *Marine Policy, 39,* 234–240.

Coleman, F. C., & Williams, S. L. (2002). Overexploiting marine ecosystem engineers: Potential consequences for biodiversity. *Trends in Ecology and Evolution, 17,* 40–44.

Cudney-Bueno, R., et al. (2009). Rapid effects of marine reserves via larval dispersal. *Plos One, 4,* e-140 1–7.

Deguignet, M., et al. (2014). *United Nations list of protected areas.* Cambridge, UK: UNEP-WCMC.

Denny, C. M., & Babcock, R. C. (2004). Do partial marine reserves protect reef fish assemblages? *Biological Conservation, 116,* 119–129.

Dietrich, A. (2007). The impacts of tourism on coral reef conservation awareness and support in coastal communities in Belize. *Coral Reefs, 26,* 985–996.

Edgar, G. J., et al. (2014). Global conservation outcomes depend on marine protected areas with five key features. *Nature, 506,* 216–220.

García-Rubies, A., & Zabala, M. (1990). Effects of total fishing prohibition on the rocky fish assemblages of Medes Islands Marine Reserve (NW Mediterranean). *Scientia Marina, 54,* 317–328.

Harmelin, J.-G., Bachet, F., & García, F. (1995). Mediterranean marine reserves: Fish indices as tests of protection efficiency. *Marine Ecology, 16,* 233–250.

Higgins, R. M., et al. (2008). Priorities for fisheries in marine protected area design and management: Implications for artisanal-type fisheries as found in southern Europe. *Journal for Nature Conservation, 16,* 222–233.

Hyrenbach, K. D., Forney, K. A., & Dayton, P. K. (2000). Marine protected areas and ocean basin management. *Aquatic Conservation: Marine and Freshwater Ecosystems, 10,* 437–458.

Leenhardt, P., et al. (2013). The rise of large-scale marine protected areas: Conservation or geopolitics? *Ocean & Coastal Management, 85,* 112–118.

Liddle, M. J. (1991). Recreational ecology: Effects of trampling on plants and corals. *TREE, 6,* 13–17.

Mumby, P. J., et al. (2004). Mangroves enhance the biomass of coral reef fish communities in the Caribbean. *Nature, 427,* 533–536.

Potts, T., et al. (2014). Do marine protected areas deliver flows of ecosystem services to support human welfare? *Marine Policy, 44,* 139–148.

Pendoley, K. L., et al. (2014). Protected species use of a coastal marine migratory corredor connecting marine protected areas. *Marine Biology, 161,* 1455–1466.

Rius, M., & Zabala, M. (2008). Are marine protected areas useful for the recovery of the Mediterranean mussel populations? *Aquatic Conservation: Marine and Freshwater Ecosystems, 18,* 527–540.

Sala, E., Boudouresque, C. F., & Harmelin-Vivien, M. (1998). Fishing, trophic cascades, and the structure of algal assemblages: Evaluation of an old but untested paradigm. *Oikos, 82,* 425–439.

Santangelo, G., et al. (2012). Patterns of variation in recruitment and post-recruitment processes of the Mediterranean precious gorgonian coral *Corallium rubrum. Journal of Experimental Marine Biology and Ecology, 411,* 7–13.

Watson, J. E., et al. (2014). The performance and potential of protected areas. *Nature, 515,* 67–71.

Zorrilla-Pujana, J., & Rossi, S. (2016). Environmental education indicators system for protected areas management. *Ecological Indicators, 67,* 146–155.

Chapter 23

Abeloson, A., et al. (2016). Upgrading marine ecosystem restoration using ecological–social concepts. *BioScience, 66,* 156–163.

Airoldi, L., Connell, S. D., & Beck, M. W. (2007). The loss of natural habitats and the addition of artificial substrata. *Ecological Studies, 206,* 269–280 (M. Wahl (Ed.), *Marine hard bottom communities*).

Alagna, A., et al. (2015). Assessing *Posidonia oceanica* seedling substrate preference: An experimental determination of seedling anchorage success in rocky vs. sandy substrates. *PLoS ONE, 10*(4), e0125321. https://doi.org/10.1371/journal.pone.0125321.

Balestri, E., Piazzi, L., & Cinelli, F. (1998). Survival and growth of transplanted and natural seedlings of *Posidonia oceanica* (L.) Delile in a damaged coastal area. *Journal of Experimental Marine Biology and Ecology, 228,* 209–225.

Beier, P., & Noss, R. F. (1998). Do habitat corridors provide connectivity? *Conservation Biology, 12,* 1241–1252.

Bulleri, F., & Airoldi, L. (2005). Artificial marine structures facilitate the spread of a non-indigenous green alga, *Codium fragile* ssp. *Tomentosoides,* in the north Adriatic Sea. *Journal of Applied Ecology, 42,* 1063–1072.

Burt, J., et al. (2009). Are artificial reefs surrogates of natural habitats for corals and fish in Dubai, United Arab Emirates? *Coral Reefs, 28,* 663–675.

Claudet, J., & Pelletier, D. (2004). Marine protected areas and artificial reefs: A review of the interactions between management and scientific studies. *Aquatic Living Resources, 17,* 129–138.

De Groot, R. S., et al. (2013). Benefits of investing in ecosystem restoration. *Conservation Biology, 27,* 1286–1293.

Duarte, C. M., Sintes, T., & Marbà, N. (2013). Assessing the CO_2 capture potential of seagrass restoration projects. *Journal of Applied Ecology, 50,* 1341–1349.

Edwards, A. J., & Clark, S. (1998). Coral transplantation: A useful management tool or misguided meddling? *Marine Pollution Bulletin, 37,* 474–487.

Falcy, M. R., & Estades, A. F. (2007). Effectiveness of corridors relative to enlargement of habitat patches. *Conservation Biology, 21,* 1341–1346.

FAO. (2016). Practical guidelines for the use of artificial reefs in the Mediterranean and the Black Sea. In G. Fabi, et al. (Eds.), *Studies and reviews*. General Fisheries Commission for the Mediterranean. No. 96. Rome, Italy.

Horozowski-Fridman, Y. B., & Rinkevich, B. (2017). Restoration of the animal forests: Harnessing silviculture biodiversity concepts for coral transplantation. In S. Rossi, L. Bramanti, A. Gori, & C. Orejas (Eds.), *Marine animal forests* (pp. 1313–1336). Switzerland: Springer International Publishing. https://doi.org/10.1007/978-3-319-17001-5_36-1.

Inger, R. (2009). Marine renewable energy: Potential benefits to biodiversity? An urgent call for research. *Journal of Applied Ecology, 46,* 1145–1153.

Jorgensen, D. (2009). An oasis in a water desert? Discourses on an industrial ecosystem in the Gulf of Mexico Rings-to-Reefs program. *History and Technology, 25,* 343–364.

Leitâo, F., et al. (2009). *Diplodus* spp. Assemblages on artificial reefs: Importance for near shore fisheries. *Fisheries Management and Ecology, 16,* 88–99.

Okubo, N., & Onuma, A. (2015). An economic and ecological consideration of commercial coral transplantation to restore the marine ecosystem in Okinawa, Japan. *Ecosystem Services, 11,* 39–44.

Rinkevich, B. (2005). Conservation of coral reefs through active restoration measures: Recent approaches and last decade progress. *Environmental Science & Technology, 39,* 4333–4342.

Rinkevich, B. (2008). Management of coral reefs: We have gone wrong when neglecting active reef restoration. *Marine Pollution Bulletin, 56,* 1821–1824.

Sánchez-Lizaso, J. L., Fernández-Torquemada, Y., & González-Correa, J. M. (2009). Evaluation of the viability of *Posidonia oceanica* transplants associated with a marina expansion. *Botanica Marine, 52,* 471–476.

van Katwijk, M. M., et al. (2009). Guidelines for seagrass restoration: Importance of habitat selection and donor population, spreading of risks, and ecosystem engineering effects. *Marine Pollution Bulletin, 58,* 179–188.

Vogt, P., et al. (2007). Mapping landscape corridors. *Ecological Indicators, 7,* 481–488.

Chapter 24

Alam, F., et al. (2012). Biofuel from algae: Is it a viable alternative? *Procedia Engineering, 49,* 221–227.

Boschen, R. E., et al. (2015). Megabenthic assemblage structure on three New Zealand seamounts: Implications for seafloor massive sulfide mining. *Marine Ecology Progress Series, 523,* 1–14.

Buck, B. H., et al. (2008). Meeting the quest for spatial efficiency: Progress and prospects of extensive aquaculture within offshore wind farms. *Helgoland Marine Research, 62,* 269–281.

Camuffo, D., et al. (2000). Sea storms in the Adriatic sea and the Western Mediterranean during the last millennium. *Climatic Change, 46,* 209–223.

Castelletto, N., et al. (2008). Can Venice be raised by pumping water underground? A pilot project to help decide. *Water Resources Research, 44,* 1–11.

Chisti, Y. (2007). Biodiesel from microalgae beats bioethanol. *Trends in Biotechnology, 26,* 126–131.

Clark, M. R., et al. (2010). The ecology of seamounts: Structure, function, and human impacts. *Annual Review of Marine Science, 2,* 253–278.

Finkl, C. W., & Charlier, R. (2009). Electrical power generation from ocean currents in the Straits of Florida: Some environmental considerations. *Renewable and Sustainable Energy Reviews, 13,* 2597–2604.

Fuentes-Grünewald, C., et al. (2012). Biomass and lipid production of dinoflagellates and raphidophytes in indoor and outdoor photobioreactors. *Marine Biotechnology, 15,* 37–47.

Griffin, R., Buck, B., & Krause, G. (2015). Private incentives for the emergence of co-production of offshore wind energy and mussel aquaculture. *Aquaculture, 436,* 80–89.

Lee, M. Q., Lu, C. N., & Huang, H. S. (2009). Reliability and cost analysis of electricity collection systems of marine current farm-a Taiwanese case study. *Renewable and Sustainable Energy Reviews, 13,* 2012–2021.

Levin, L. A., et al. (2016). Defining "serious harm" to the marine environment in the context of deep-sea bed mining. *Marine Policy, 74,* 245–259.

Li, Q., Du, W., & Liu, D. (2009). Perspectives of microbial oils for biodiesel production. *Applied Microbiology and Biotechnology, 80,* 749–756.

López, I., et al. (2013). Review of wave energy technologies and the necessary power-equipment. *Renewable and Sustainable Energy Reviews, 27,* 413–434.

Maity, J. P., et al. (2014). Microalgae for third generation biofuel production, mitigation of greenhouse gas emissions and wastewater treatment: Present and future perspectives: A mini review. *Energy, 78,* 104–113.

Michler-Cieluch, T., & Kodeih, S. (2008). Mussel and seaweed cultivation in offshore wind farms: An opinion survey. *Coastal Management, 36,* 392–411.

Michler-Cieluch, T., & Krause, G. (2008). Perceived concerns and possible management strategies for governing 'wind farm-mariculture integration'. *Marine Policy, 32,* 1013–1022.

Murton, B. (2013). Seafloor mining: The future or just another pipe dream? *International Journal of the Society for Underwater Technology, 31,* 53–54.

Nunneri, C., et al. (2008). Ecological risk as a tool for evaluating the effects of offshore wind farm construction in the North Sea. *Regional Environmental Change, 8,* 31–43.

Pérez-Collazo, C., Greaves, D., & Iglesias, P. (2015). A review of combined wave and offshore wind energy. *Renewable and Sustainable Energy Reviews, 42,* 141–153.

Plieninger, T., & Bens, O. (2008). How the emergence of biofuels challenges environmental conservation. *Environmental Conservation, 34,* 273–275.

Santos-Ballardo, D., et al. (2016). Microalgae potential as a biogas source: Current status, restraints and future trends. *Reviews of Environmental Science and Biotechnology, 15,* 243–264.

Schalermann, J. P. W., & Laurance, W. F. (2008). How green are biofuels? *Science, 319,* 43–44.

Sevigné-Itoiz, E., et al. (2012). Energy balance and environmental impact analysis of marine microalgal biomass production for biodiesel generation in a photobioreactor pilot plant. *Biomass and Bioenergy, 39,* 324–335.

Snyder, B., & Kaiser, M. J. (2009). Ecological and economic cost-benefit analysis of offshore wind energy. *Renewable Energy, 34,* 1567–1578.

Thistle, D. (2003). The deep-sea floor: An overview. In P. A. Tyler (Ed.), *Ecosystems of the World* (pp. 1–37). New York: Elsevier.

Vasudevan, P. T., & Briggs, M. (2008). Biodiesel production-current state of the art and challenges. *Journal of Industrial Microbiology & Biotechnology, 35,* 421.

Warren, C. R., & Birnie, R. V. (2009). Re-powering Scotland: Wind farms and the 'energy or environment' debate. *Scottish Geographical Journal, 125,* 97–126.

Chapter 25

Boero, F. (2009). Recent innovations in marine biology. *Marine Ecology, 30*(suppl 1), 1–12.

Dayton, P. K. (2003). The importance of natural sciences to conservation. *American Naturalist, 162,* 1–13.

Dayton, P. K., & Sala, E. (2001). Natural history: The sense of wonder, creativity and progress in ecology. *Scientia Marina, 65,* 199–206.

Helvarg, D. (2006). *50 ways to save the ocean*. Maui, US: Inner Ocean Publishing Inc.

Jacques, P. J., Dunlap, R. E., & Freeman, M. (2008). The organization of denial: Conservative think tanks and environmental scepticism. *Environmental Politics, 17,* 349–385.

May, R. M. (1988). How many species are there on Earth? *Science, 241,* 1441–1449.

Pope, C., & Rauber, P. (2004). *Strategic ignorance: Why the Bush administration is recklessly destroying a century of environmental progress.* Washington, DC: Sierra Club Books.

Safina, C., et al. (2005). U.S. Ocean fish recovery: Staying the course. *Science, 309,* 707–708.

Schiermeier, Q. (2010). The real holes in climate nature. *Nature, 463,* 284–287.

Shelton, P. A. (2009). Eco-certification of sustainably managed fisheries-redundancy or synergy? *Fisheries Research, 100,* 185–190.

Steneck, R. S. (1998). Human influences on coastal ecosystems: Does overfishing create trophic cascades? *Trends in Ecology & Evolution, 13,* 429–430.

Uhl, C. (1998). Conservation biology in your own yard. *Conservation Biology, 12,* 1175–1177.

Williams, S. E., et al. (2009). Towards an integrated framework for assessing the vulnerability of species to climate change. *Plos Biology, 6*(e325), 2621–2626.

Zanetell, B. A., & Rassam, G. (2003). Taxonomists: The unsung heroes of our quest to save biodiversity. *Fisheries, 28,* 29.

Epilogue

Bearzi, G. (2009). When swordfish conservation biologist eat swordfish. *Conservation Biology, 23,* 1–2.

Costello, M. J., et al. (2010). A census of marine biodiversity knowledge, resources, and future challenges. *PLos One, 5,* e12110.

Fricke, H. (1997). Living coelacanths: Values, eco-ethics and human responsibility. *Marine Ecology Progress Series, 161,* 1–15.

Jackson, J. B. C. (2008). Ecological extinction and evolution in the brave new ocean. *PNAS, 105,* 11458–11465.

Jackson, J. B. C., & Sala, E. (2001). Unnatural oceans. *Scientia Marina, 65*(Suppl. 2), 273–281 (J. M. Gili, J. L. Pretus, & T. T. Packard (Eds.), *A marine science odyssey into the 21st century*).

Kennedy, R. (2004). *Crimes against nature.* New York: Harper Collins.

McCright, A., & Dunlap, R. E. (2003). Defeating Kyoto: The conservative movement's impact on US climate change policy. *Social Problems, 50,* 348–373.

Pitcher, T. J. (2008). The sea ahead: Challenges to marine biology from seafood sustainability. *Hydrobiologia, 606,* 161–185.

Pauly, D. (2009). Beyond duplicity and ignorance in global fisheries. *Scientia Marina, 73,* 215–224.

Prins, G., & Rayner, S. (2007). Time to ditch Kyoto. *Nature, 449,* 973–975.

Rind, D. (1999). Complexity and climate. *Science, 284*(5411), 105–107.

Stern, N. (2007). *The economics of climate change: The Stern Review.* Cambridge: Cambridge University Press.

Thrush, S. M., & Dayton, P. K. (2010). What can ecology contribute to ecosystem-based management? *Annual Review of Marine Science, 2,* 419–441.

Van den Berg, J. C. J. M. (2009). Safe climate policy is affordable—12 reasons. *Climatic Change, 1,* 339–385.

Weaver, R. K. (1988). The changing world of think tanks. *Political Science and Politics, 22,* 563–578.

Printed in the United States
By Bookmasters